Airplane Design

Part III: Layout Design of Cockpit, Fuselage, Wing and Empennage: Cutaways and Inboard Profiles

Dr. Jan Roskam
Ackers Distinguished Professor Emeritus of Aerospace Engineering
The University of Kansas, Lawrence

2018

1440 Wakarusa Drive, Suite 500 • Lawrence, Kansas 66049, U.S.A.

PUBLISHED BY
Design, Analysis and Research Corporation (*DARcorporation*)
1440 Wakarusa Drive, Suite 500
Lawrence, Kansas 66049
U.S.A.
Phone: (785) 832-0434
info@darcorp.com
www.darcorp.com

Library of Congress Catalog Card Number: 97068580

ISBN: 978-1-884885-56-3
ISBN: 978-1-884885-67-9 (E-book)

In all countries, sold and distributed by
Design, Analysis and Research Corporation
1440 Wakarusa Drive, Suite 500
Lawrence, Kansas 66049
U.S.A.

The information presented in this book has been included for their instructional value. They are not guaranteed for any particular purpose. The publisher does not offer any warranties or representations, nor does it accept any liabilities with respect to the information. It is sold with the understanding that the publisher is not engaged in rendering engineering or other professional services. If such services are required, the assistance of an appropriate professional should be sought.

Copyright © 2018 by Design, Analysis and Research Corporation.
All rights reserved
Printed in the United States of America
First Printing, 1986
Second Printing, 1989
Third Printing, 2002
Fourth Printing, 2011
E-book, 2017
Fifth Printing, 2017
Sixth Printing, 2018

Information in this document is subject to change without notice and does not represent a commitment on the part of *DARcorporation*. No part of this book may be reproduced, stored in a retrieval system or transmitted in any form or by any means, electronic or mechanical, photocopying, recording, or otherwise, without written permission of the publisher, *DARcorporation*.

TABLE OF CONTENTS

	TABLE OF SYMBOLS	v
	ACKNOWLEDGEMENT	vi
1.	INTRODUCTION	1
2.	COCKPIT (OR FLIGHT DECK) LAYOUT DESIGN	3
	2.1 DIMENSIONS AND WEIGHTS FOR CREW MEMBERS	4
	2.2 LAYOUT OF COCKPIT SEATING AND COCKPIT CONTROLS	12
	2.2.1 Civil Cockpit Layouts	12
	2.2.2 Military Cockpit Layouts	13
	2.3 DETERMINATION OF VISIBILITY FROM THE COCKPIT	23
	2.4 EXAMPLES OF COCKPIT LAYOUTS	29
3.	FUSELAGE LAYOUT DESIGN	35
	3.1 AERODYNAMIC DESIGN CONSIDERATIONS	36
	3.1.1 Friction Drag	36
	3.1.2 Profile and Base Drag	38
	3.1.3 Compressibility Drag	39
	3.1.4 Induced Drag	40
	3.2 GUIDELINES FOR FLYING BOAT HULL AND FLOAT DESIGN	42
	3.3 INTERIOR LAYOUT DESIGN OF THE FUSELAGE	45
	3.3.1 Layout of the Cross Section	45
	3.3.1.1 Passenger cabin	46
	3.3.1.2 Cargo hold	53
	3.3.1.3 Military	53
	3.3.1.4 Supersonic airplanes	53
	3.3.2 Seating Layouts, Seats and Restraint Systems	57
	3.3.2.1 Seating arrangements and seats for general aviation airplanes	57
	3.3.2.2 Seating arrangements and seats for transports	57
	3.3.2.3 Restraint systems	67
	3.3.3 Layout of Doors, Emergency Exits and Windows	68
	3.3.3.1 General aviation airplanes	68
	3.3.3.2 Transport airplanes	68
	3.3.3.3 Military airplanes	71
	3.3.4 Galley, Lavatory and Wardrobe Layouts	73
	3.3.5 Layout of Cargo/Baggage Holds Including Data on Cargo Containers and Pallets	76
	3.3.5.1 Cargo and baggage volume requirements	76
	3.3.5.2 Data on standard containers and pallets	77

                3.3.5.3 Typical loading/unloading
                        configurations   77
        3.3.6 Inspection, Maintenance and Servicing
              Considerations   82
3.4 DESIGN DATA FOR FUSELAGE CROSS SECTIONS, CABIN
    AND CARGO HOLD LAYOUTS, WINDOW AND DOOR
    LAYOUTS   85
3.5 STRUCTURAL DESIGN CONSIDERATIONS AND EXAMPLES
    OF STRUCTURAL LAYOUT DESIGN OF FUSELAGES   123
        3.5.1 Typical Frame Depths, Frame Spacings
              and Longeron Spacings   124
        3.5.2 Examples of Fuselage Structural
              Arrangements   126
        3.5.3 Examples of Fuselage Shell Layout   132
        3.5.4 Examples of Door and Stair Design   137
        3.5.5 Examples of Cockpit and Cabin Window
              Design   143
        3.5.6 Examples of Floor Design   148
3.6 EXAMPLES OF INBOARD PROFILES   153

4. WING LAYOUT DESIGN   163
4.1 WING CONFIGURATION: AERODYNAMIC AND
    OPERATIONAL DESIGN CONSIDERATIONS   164
        4.1.1 Wing Size: Large or Small? Or, Wing
               Loading: Low or High?   165
        4.1.2 High, Mid or Low Wing?   170
        4.1.3 Forward Sweep, No Sweep or Aft Sweep?   175
        4.1.4 Variable Sweep: One Pivot or Two?   178
        4.1.5 Bi-plane, Braced Wing or Joined Wing?   184
        4.1.6 Wing Aspect Ratio: High, Low and/or
               Winglets?   185
        4.1.7 Wing Thickness Ratio: Large or Small?   187
        4.1.8 Wing Taper Ratio: Large or Small?   189
        4.1.9 Straight Taper or Variable Taper?   191
        4.1.10 Twist: How Much?   193
        4.1.11 Wing Dihedral: How Much?   194
        4.1.12 Wing Incidence on the Fuselage:
               How Much?   195
        4.1.13 Variable Camber (MAW = Mission
               Adaptive Wing)?   199
        4.1.14 Leading Edge Strakes (Lexes)   199
        4.1.15 Planform Tailoring: Why and How?   201
        4.1.16 Area Ruling: When is it Required?   204
        4.1.17 Wing Span: When is it Too Large?   204
        4.1.18 Aerodynamic Coupling   206
        4.1.19 Flaps: What Size and Which Type?   206
        4.1.20 Lateral Controls: Type, Size and
               Location?   208
        4.1.21 Review of Wing Drag Contributions   214

- 4.2 STRUCTURAL DESIGN CONSIDERATIONS AND EXAMPLES OF STRUCTURAL LAYOUT DESIGN — 218
  - 4.2.1 Typical Spar, Rib and Stiffener Spacings — 218
  - 4.2.2 Examples of Wing Structural Arrangements — 220
  - 4.2.3 Examples of Wing/Fuselage Integration — 226
  - 4.2.4 Examples of Wing Cross Section Design — 226
  - 4.2.5 Examples of Lateral Control Mechanizations — 232
  - 4.2.6 Examples of High Lift Device Mechanizations — 232
  - 4.2.7 Examples of Wing Skin Gages — 232
  - 4.2.8 Maintenance and Access Requirements — 232
- 4.3 MILITARY DESIGN CONSIDERATIONS — 239
- 4.4 DETAILED OVERALL STRUCTURAL ARRANGEMENTS — 239

5. EMPENNAGE LAYOUT DESIGN — 249
  - 5.1 EMPENNAGE CONFIGURATION: AERODYNAMIC AND OPERATIONAL DESIGN CONSIDERATIONS — 249
    - 5.1.1 Conventional (Tails Aft), Canard or Three-surface? — 250
    - 5.1.2 Additional Empennage Configuration Choices — 254
    - 5.1.3 Empennage Size: Stability, Control and Handling Considerations — 259
      - 5.1.3.1 Longitudinal considerations — 260
      - 5.1.3.2 Lateral-Directional considerations — 261
    - 5.1.4 Stall and Spin Characteristics — 263
    - 5.1.5 Empennage Planform Design — 272
    - 5.1.6 Empennage Airfoil Design or Selection — 272
    - 5.1.7 Review of Empennage Drag Contributions — 273
  - 5.2 STRUCTURAL AND INTEGRATION DESIGN CONSIDERATIONS FOR THE EMPENNAGE — 275
    - 5.2.1 Typical Spar, Rib and Stiffener Spacings — 275
    - 5.2.2 Examples of Empennage Structural Arrangements — 277
    - 5.2.3 Examples of Fuselage/Empennage Integration and/or Vertical/Horizontal Tail Integration — 278
    - 5.2.4 Examples of Empennage Cross Section Design — 287
    - 5.2.5 Examples of Longitudinal Control Mechanizations — 287
    - 5.2.6 Examples of Directional Control Mechanizations — 287
    - 5.2.7 Examples of Empennage Skin Gages — 287
    - 5.2.8 Maintenance and Access Requirements — 288

6. INTEGRATION OF THE PROPULSION SYSTEM — 291
   6.1 PRESENTATION OF ENGINE AND PROPELLER DATA — 291
      6.1.1 Propellers — 292
      6.1.2 Piston Engines — 300
      6.1.3 Turbopropeller Engines — 301
      6.1.4 Turbojet and Turbofan Engines — 301
      6.1.5 Propfan Engines — 302
   6.2 RELATION BETWEEN FLIGHT ENVELOPE AND ENGINE TYPE — 328
   6.3 INSTALLED THRUST, POWER AND EFFICIENCY CONSIDERATIONS — 330
      6.3.1 Power Extraction — 330
      6.3.2 Propeller Installations — 331
      6.3.3 Piston-Engine Installations — 331
      6.3.4 Subsonic and Supersonic Turbojet and Turbofan Installations — 331
   6.4 STABILITY AND CONTROL CONSIDERATIONS — 333
      6.4.1 Effect of One or More Engines Inoperative and Effects of Power Transients — 333
      6.4.2 Tractor Versus Pusher — 333
      6.4.3 Effect of Engine/Propeller Thrust Line Location and Inclination — 334
   6.5 STRUCTURAL CONSIDERATIONS — 335
      6.5.1 Transmission of Thrust into the Airframe — 335
      6.5.2 Lateral Disposition of Engines over the Wing — 337
      6.5.3 Extension Shafts and Propeller Blade Excitation — 337
      6.5.4 Flutter — 339
   6.6 MAINTENANCE AND ACCESSIBILITY CONSIDERATIONS — 340
   6.7 SAFETY CONSIDERATIONS — 344
      6.7.1 Installation Safety — 344
      6.7.2 Safety During Ground Operation — 346
      6.7.3 Foreign Object Damage (FOD) — 346
      6.7.4 Engine Reliability and Shutdown Rates — 349
   6.8 NOISE CONSIDERATIONS — 350
      6.8.1 Interior Noise Design Considerations — 350
      6.8.2 Exterior Noise Design Considerations — 353
   6.9 EXAMPLE ENGINE INSTALLATIONS — 356
      6.9.1 Piston-Propeller Installations — 356
      6.9.2 Turbo-Propeller Installations — 363
      6.9.3 Turbojet and Turbofan Installations — 363
      6.9.4 Propfan and Ultra-Bypass Installations — 369
      6.9.5 Nozzles and Thrust Reversers — 376

7. PRELIMINARY STRUCTURAL ARRANGEMENT, MATERIAL SELECTION AND MANUFACTURING BREAKDOWN — 381
   7.1 PREPARING A PRELIMINARY STRUCTURAL ARRANGEMENT — 381
   7.2 PRELIMINARY SELECTION OF STRUCTURAL MATERIALS — 386

7.3	PRELIMINARY SELECTION OF MANUFACTURING BREAKDOWN	393
8.	COLLECTION OF CUTAWAY DRAWINGS	399
9.	REFERENCES	445
10.	INDEX	451

TABLE OF SYMBOLS
================

Symbol   Definition   Dimension

The symbols for flying boat hull geometry are defined in Figure 3.9.

All other symbols used in this part are identical to the symbols used in Parts I, II, V, VI and VII.

Symbols used in Chapter 2 are defined in the text.

Symbols used in Chapter 3 are defined in Part II.

Symbols used in Chapter 4 are defined in Part I.

Symbols used in Chapter 5 are defined in Part II, in Part VI and in Part VII.

Symbols used in Chapter 6 are defined in Part II.

ACKNOWLEDGEMENT

Writing a book on airplane design is impossible without the supply of a large amount of data. The author is grateful to the following companies for supplying the raw data, manuals, sketches and drawings which made the book what it is:

Aerospatiale
Beech Aircraft Corp.
The Boeing Company
British Aerospace Corp.
Cessna Aircraft Company
Fairchild Republic Co.
Gates Learjet Corporation
General Electric Corp.
Pratt and Whitney
Avco Lycoming
Detroit Diesel Allison
The Garrett Corporation

Fairchild Republic
Gates Learjet Corporation
Grumman Aerospace Corp.
Gulfstream Aerospace Corp.
Lockheed Aircraft Corp.
McDonnell Douglas Corp.
Royal Netherlands Aircraft Factory: Fokker
SIAI Marchetti S.p.A.
Teledyne Continental
Hamilton Standard
NASA

A significant amount of airplane design information has been accumulated by the author over many years from the following magazines:

Interavia (Swiss, monthly)
Flight International (British, weekly)
Business and Commercial Aviation (USA, monthly)
Aviation Week and Space Technology (USA, weekly)
Journal of Aircraft (USA, AIAA, monthly)

The author wishes to acknowledge the important role played by these magazines in his own development as an aeronautical engineer. Aeronautical engineering students and graduates should read these magazines regularly.

Nearly all cockpit and fuselage design data as well as most inboard profiles in this book were drawn by Mr. Govert Tukker of Molenaarsgraaf, The Netherlands. The author is grateful to Mr. Tukker for his skill and patience in carrying out this most difficult assignment.

A number of drawings in this book are credited to companies which merged into other companies or ceased operation.

2. COCKPIT (OR FLIGHT DECK) LAYOUT DESIGN

The following considerations play an important role in the layout of a cockpit or a flight deck:

1. The pilot(s) and other crew members must be positioned so that they can reach all controls comfortably, from some reference position.

2. The pilot(s) and other crew members must be able to see all 'flight essential' instruments without undue effort.

3. Communication by voice or by touch must be possible without undue effort.

4. Visibility from the cockpit must adhere to certain minimum standards.

This chapter provides the information, necessary to make realistic layouts of cockpits and/or flightdecks for civil and for military airplanes.

The word 'cockpit' is usually associated with small to medium sized airplanes. The word 'flightdeck' is usually associated with large airplanes. In this textbook these words are used interchangeably.

Section 2.1 contains baseline data on the dimensions and weights of crew members. In laying out cockpit arrangements it is essential that these data are accounted for.

Section 2.2 contains data needed to prepare realistic layouts for cockpit seating and for cockpit controls. Civilian as well as military arrangements are covered.

Section 2.3 presents methods for assuring that minimum standards of visibility from the cockpit are met.

Section 2.4 shows example cockpit layouts for several airplane types.

2.1 DIMENSIONS AND WEIGHTS FOR CREW MEMBERS

Figure 2.1 and Table 2.1 provide baseline data for weights and for dimensions of 'standing' (male) crew members. Notice that the center of gravity of a 'standing' crew member is roughly at the hip joint.

Figure 2.2 and Table 2.2 provide baseline data for dimensions of 'sitting' (male) crew members. Observe, that the center of gravity of a 'sitting' crew member is roughly at the forward intersection of the lower torso and the upper legs.

Note: For female crew members it is suggested to multiply all weight and dimension data by 0.85.

With this information the designer can ensure that all crew members 'fit' into their assigned positions.

Particularly when developing new cockpit or flight deck arrangements it is essential (before going into the mock-up stage) to validate the proposed arrangement. This is done by constructing a 'puppet'. Figure 2.1 can serve as the model for such a puppet. The puppet must be made to the same scale as the drawings of the proposed cockpit. The puppet should be made with rotating joints, using the joint rotation points shown in Figure 2.1. This way it is possible to position the puppet on the drawing board or on the CAD screen in relationship to the proposed interior cockpit contours. Checks for conflicts can then be easily made.

Once the puppet has been positioned in its reference position, a check should be made to ensure that arm and leg motions needed to carry out control manipulation of throttles, stick or wheel, side-arm controller and rudder pedals are indeed feasible. Data for such required control manipulations are given in Section 2.2. Figure 2.3 shows which areas are easy and which areas are difficult to reach for the 'average' crew member.

Figure 2.4 shows scaled views for 'standing' military crew members in typical military gear. Figures 2.5 and 2.6 present scaled views for 'sitting' military crew members. The sizes reflected in Figures 2.4 through 2.6 are for the '90-percentile, male' crew member. For females, the factor 0.85 previously suggested may be used.

References 8 and 9 provide more detailed information about the human body.

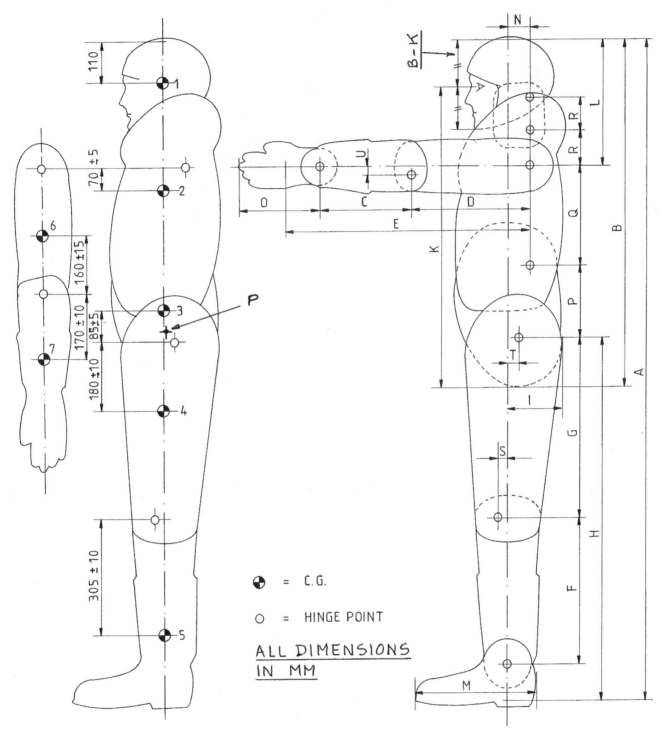

Figure 2.1 Dimensions of Standing, Male Crew Member, Winter Clothing and Light Helmet Included

Table 2.1 Dimensions and Weights for Male Crew Members as Shown in Figure 2.1
===

A	B	C	D	E	F	G	H	I	K	L
1,600	870	230	300	620	350	435	850	140	760	300
1,750	920	255	335	685	390	475	950	150	805	330
1,900	990	280	370	750	430	515	1,050	160	875	360

A	M	N	O	P	Q	R	S	T	U
1,600	300	50	200	190	260	80	25	20	20
1,750	325	60	220	200	270	90	30	30	20
1,900	350	70	240	210	280	100	30	30	20

Body width across shoulders: 533 mm, across elbows: 561 mm and across hips: 457 mm.

Body component weights are for a male pilot with a weight of 179.3 lbs.

Body Component	Number in Figure 2.1	Weight in lbs.
Head and neck	1	15.0
Upper torso	2	49.0
Lower torso	3	28.0
Upper legs	4	39.9
Lower legs and feet	5	29.8
Upper arms	6	9.9
Lower arms and hands	7	7.7
Total		179.3

Notes:
1. All dimensions in mm. (1 in.=25.4 mm.)
2. The c.g. of the 'upright' pilot of Figure 2.1 is at point P.
3. For pilot positions differing from the upright, the new c.g. can be computed with the help of the table to the left.
4. All weights include helmet and flight clothing.
5. For a female pilot multiply all weight data by 0.81.
6. Data source: Design Requirements for the RAF and RN (England).

Part III   Chapter 2   Page 6

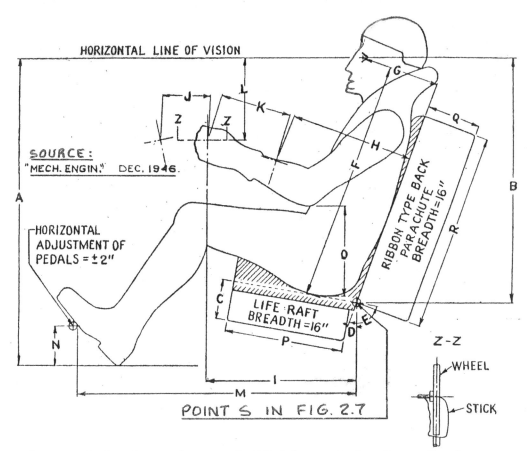

Figure 2.2 Dimensions of Sitting, Male Crew Member, Winter Clothing and Light Helmet Included

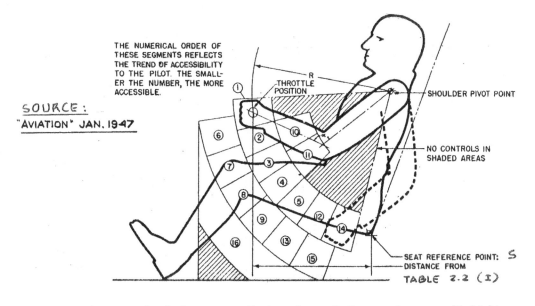

Figure 2.3 Areas of Good and Poor Accessibility

Table 2.2 Dimensions and Weights for Male Crew Members as Shown in Figure 2.2
==

For Wheel Type Controllers:

A	B	C	D deg.	E deg.	F	G	H	I	J	K
37	30.25	5	21	101	29.75	10.00	16.63	19	6	9
39	30.75	5	19	101	30.25	9.75	15.75	19	6	9
41	31.50	5	16	101	31.00	9.75	15.13	19	6	9
43	31.75	5	16	101	31.25	10.00	15.13	19	6	9

A	L	M	N	O	P	Q	R
37	10.00	36.0	5	9.25	15	7	25
39	10.50	35.0	5	9.25	15	7	25
41	10.75	34.5	5	9.25	15	7	25
43	11.00	34.5	5	9.25	15	7	25

For Stick Type Controllers:

A	B	C	D deg.	E deg.	F	G	H	I	J	K
37	30.25	5	21	101	29.75	10.00	14.50	19	6	9
39	30.75	5	19	101	30.25	9.75	13.75	19	6	9
41	31.50	5	16	101	31.00	9.75	13.50	19	6	9
43	31.75	5	16	101	31.25	10.00	13.00	19	6	9

A	L	M	N	O	P	Q	R
37	11.50	36.0	5	9.25	15	7	25
39	13.75	35.0	5	9.25	15	7	25
41	15.50	34.5	5	9.25	15	7	25
43	17.50	34.5	5	9.25	15	7	25

Seat adjustment: Horizontal: +/- 1.5 in. and Vertical: +/- 3.5 in.

DATA SOURCE:
BOEING WICHITA
1/20

BASED ON
6.0 FT MALE
90 PERCENTILE

1/40

SCALE 1/10

Figure 2.4 Scaled Views of Standing, Male Crew Member in Military Gear

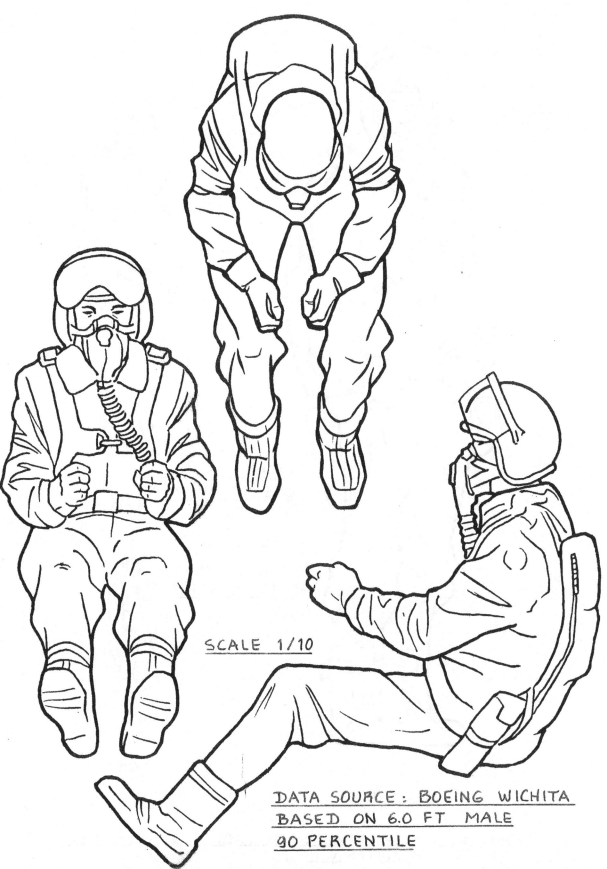

Figure 2.5 Scaled Views of Sitting, Male Crew Member in Military Gear

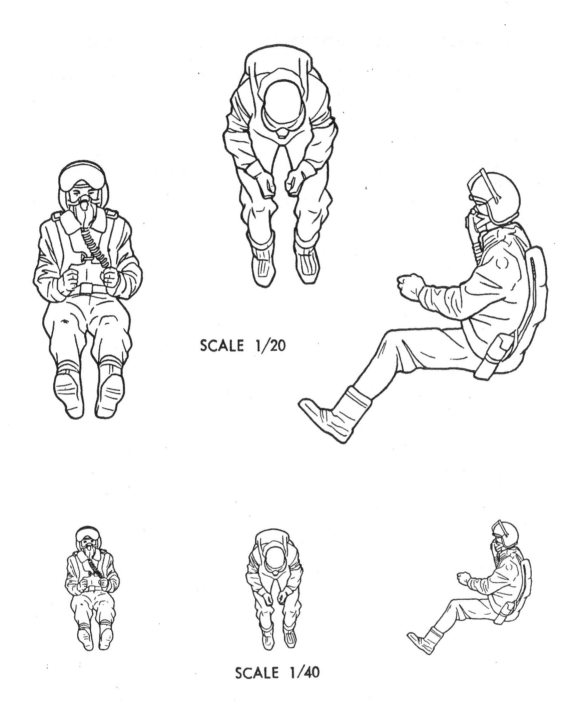

SCALE 1/20

SCALE 1/40

DATA SOURCE: BOEING WICHITA
BASED ON 6.0 FT MALE, 90 PERCENTILE

Figure 2.6 Scaled Views of Sitting, Male Crew Member in Military Gear

2.2 LAYOUT OF COCKPIT SEATING AND COCKPIT CONTROLS

Two types of cockpit (or flight deck) layout will be considered:

2.2.1 Civil Cockpit Layouts
2.2.2 Military Cockpit Layouts

All cockpit layouts must account for dimensional limitations of the human body. Since humans come in widely differing sizes, the design of cockpits must, to some extent, allow for these variations. This is accomplished by arranging for seat position adjustment and, where needed also for rudder pedal adjustment.

2.2.1 Civil Cockpit Layouts

Figure 2.7 shows a typical arrangement of pilot seat and pilot controls for civil airplanes. The geometric quantities in Figure 2.7 are defined in Table 2.3.

It is not practical to employ a fixed relationship between pilot seating and pilot controls. The reason is that human bodies vary greatly in geometrical dimensions. Typical of the variations that have been measured in adults are:

variation in armlength (C+D+O in Fig.2.1): +/- 15 cm
variation in leglength (H in Fig.2.1): +/- 20 cm
variation in seat-eye distance (c in Fig.2.7): +/- 12 cm

It is of interest to note that no systematic relationship between (C+D+O), H and c has been observed by human factors researchers. This implies that a considerable amount of adjustment must be designed into cockpits. Table 2.3 defines the most important adjustment requirements.

Figure 2.7 applies to wheel controlled and to center-stick controlled airplanes. Adjustment requirements are listed in Table 2.3. Note that a wheel rotation of more than 85 deg. is not acceptable!

For homebuilt, center-stick controlled airplanes the layout of Figure 2.8 can be used. The adjustment requirements previously suggested, apply here also.

In several airplanes, side-stick controllers are now being used. Figure 2.9 shows a typical side-stick controller. In a transport cockpit (such as the Airbus 320) the side-stick controllers are arranged to the left for the captain (port side) and to the right for the co-pilot (starboard side).

In a side-stick controller layout the pilot's arm must rest on a flat surface. It may be assumed that this 'arm-rest' surface is a distance {L-(B-K)+D} below the pilot's eye position (point C in Figure 2.7).

2.2.2 Military Cockpit Layouts

Guidelines for military cockpit layouts are given in Ref.10. Figure 2.10 shows the recommended layout for center-stick controllers.

Figure 2.11 presents the recommended layout for wheel controllers. Although Figure 2.11 shows that wheel rotation should not exceed 90 deg., MIL-STD-1472B indicates that a limit of 120 deg. is acceptable.

Figure 2.10 applies primarily to fighter and trainer type airplanes, while Figure 2.11 applies primarily to cargo (including transport) and bomber (including patrol) type airplanes.

Many military airplanes are equipped with ejection seats. In that case it is essential that certain minimum clearances are provided, to facilitate ejection. Fig.2.12 shows what these minimum clearance requirements are. Observe that Figure 2.12 also defines the clearance requirements for tandem seat arrangements.

Figure 2.13 gives typical dimensions for an ejection seat.

CAUTIONARY NOTES:

1) Flight essential crew members and their primary cockpit controls should not be located within the 5 degree arcs shown in Figure 2.14.

2) In civil airplanes (FAR 23.771 and FAR 25.771, Ref.11) this 'arc' requirement must be met for propeller driven airplanes only.

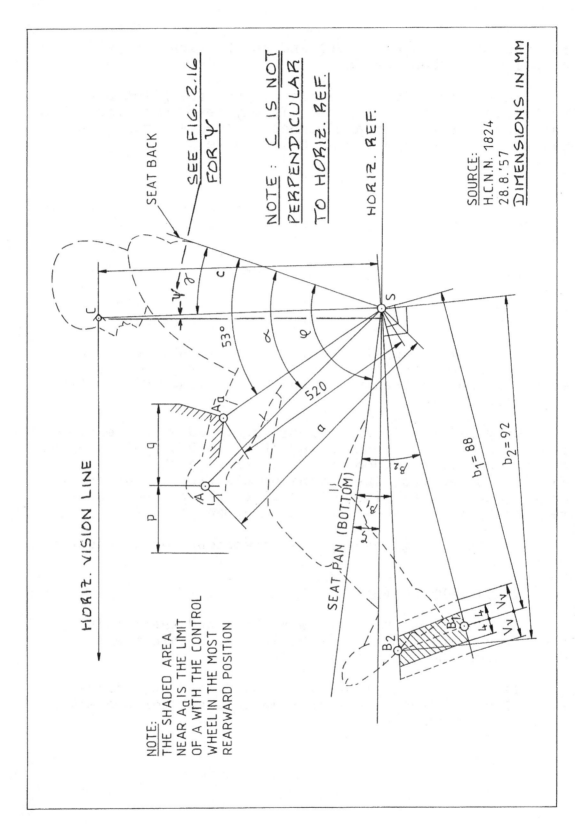

Figure 2.7 Recommended Seat Arrangement for Civil, Wheel and Center-stick Controlled Airplanes

Table 2.3 Dimensions for Civil Cockpit Controls and for Seat Adjustments

Notes:
1) See Figure 2.7 for explanation of symbols.
2) All linear dimensions are in cm.
3) All angular dimensions are in deg.

Symbol		Wheel Control	Stick Control
a		67 (+/- 4)	63 (+/- 4)
ξ		$7°$ (+/- $2°$)	$7°$ (+/- $2°$)
p	= Forward motion of point A:	18 (+/- 2)	16 (+/- 2)
q	= Rearward motion of point A:	22 (+/- 2)	20 (+/- 2)
r	= Sidewise motion of point A from center*:	-----	15 (+/- 2)
d	= Distance between handgrips of wheel*:	38 (+/- 5)	-----
ε	= Wheel rotation from center*:	$85°$ (max.)	-----
v	= Distance between rudder pedal center lines*:	38 (+/- 12)	45 (+/- 5)
α		$64°$ (+/- $3°$)	$70°$ (+/- $3°$)
β_1		$10°$	same
β_2		$22°$	same
c		77 (+/- 2)	same
γ		$21°$ (+/- $1°$)	same
φ		$102°$ (+/- $2°$)	same
V_v	= Adjustment range of pedals from center position B:	7 (+/- 2)	same
U_v	= Forward and aft pedal motion from center position B*:	10 (+/- 2)	same
S_h	= Horizontal adjustment range of S from center position*:	< 10	same
S_v	= Vertical adjustment range of S from center position*:	8 (+/- 1)	same

* Not shown in Figure 2.7.

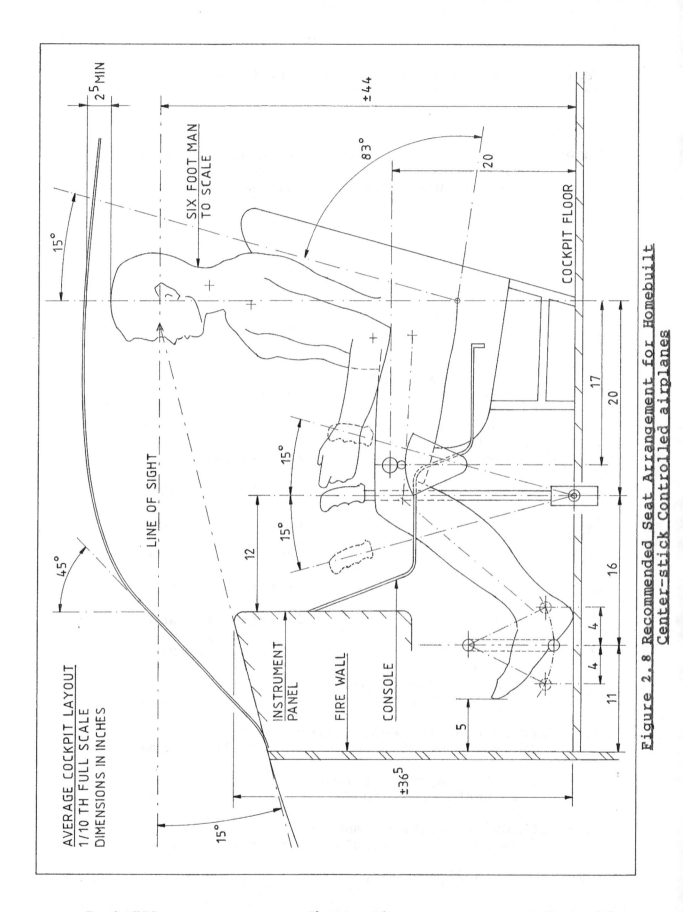

Figure 2.8 Recommended Seat Arrangement for Homebuilt Center-stick Controlled airplanes

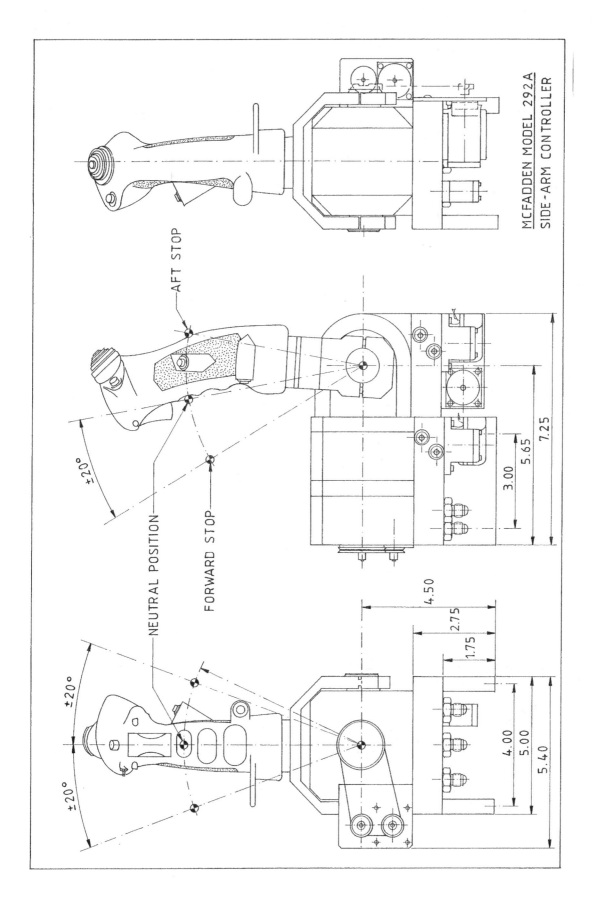

Figure 2.9 Typical Side-stick Dimensions

Part III　　　Chapter 2　　　Page 17

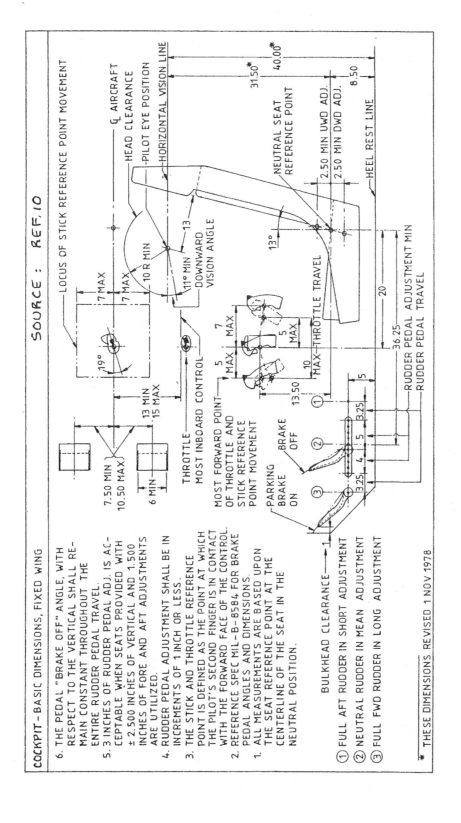

Figure 2.10 Recommended Seat Arrangement for Military, Center-stick Controlled Airplanes

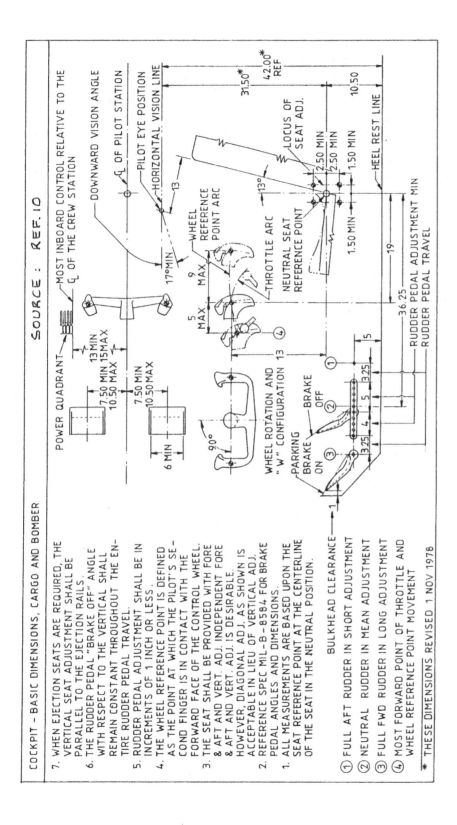

Figure 2.11 Recommended Seat Arrangement for Military Wheel Controlled Airplanes

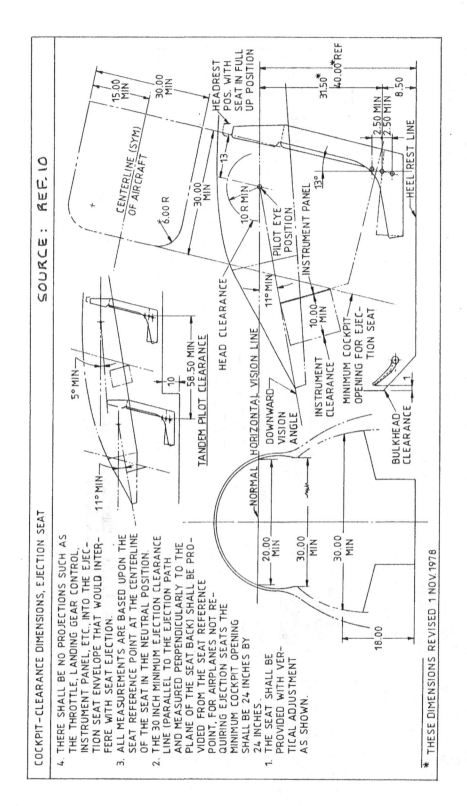

Figure 2.12 Recommended Clearances for Ejection Seats

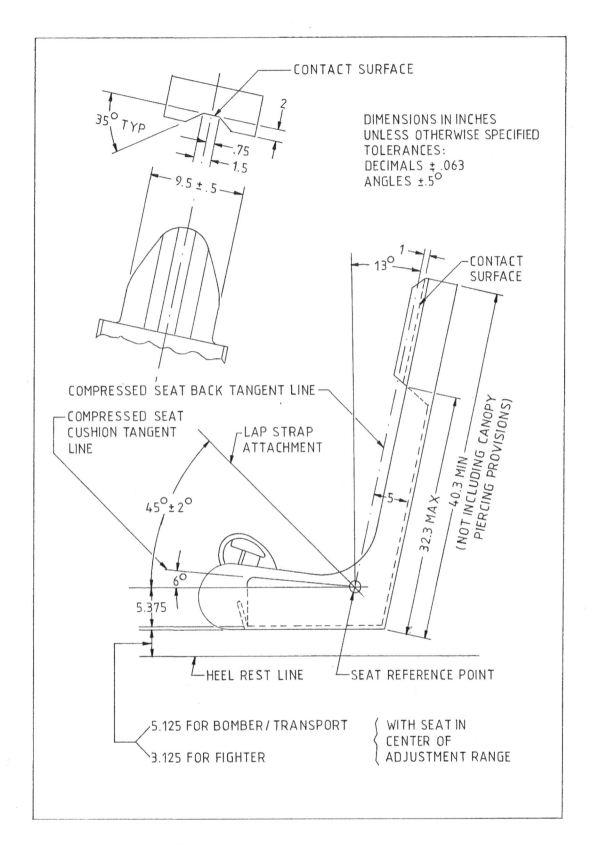

Figure 2.13 Typical Ejection Seat Dimensions

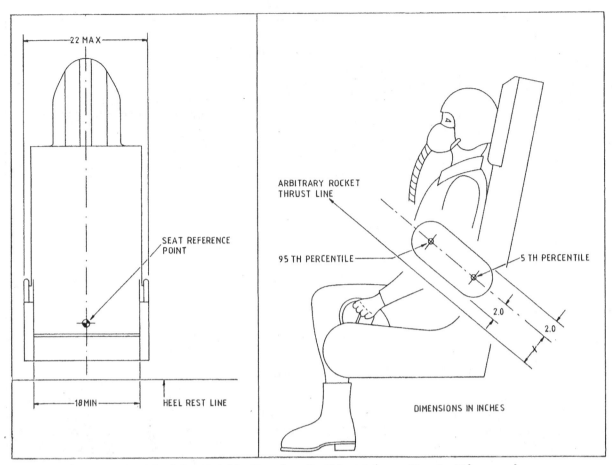

Figure 2.13 (Cont'd) Typical Ejection Seat Dimensions

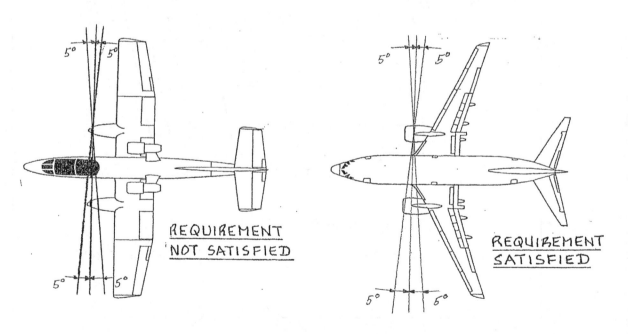

Figure 2.14 Areas Where Pilots and Their Primary Flight Controls Should NOT be Located

2.3 DETERMINATION OF VISIBILITY FROM THE COCKPIT

Good visibility from the cockpit is essential for a number of reasons:

1. During take-off and landing operations a pilot must have a good view of the immediate surroundings.

2. During en-route operations the pilot must be able to observe conflicting traffic.

3. In fighters, success in combat may depend on good visibility. Formation flying is impossible without it.

Minimum cockpit visibility rules are in force for civil as well as for military airplanes. The following definition and visibility design procedure applies to civil and to military transport airplanes. For fighters, trainers and aerobatic airplanes much more stringent visibility requirements are used. The customer defines the minimum required visibility in each specific case.

Visibility from the cockpit is defined as the angular area obtained by intersecting the airplane cockpit with radial vectors emanating from the eyes of the pilot. These radial vectors are assumed to be centered on the pilot's head. Fig.2.15 illustrates a typical situation.

Reference 11 contains rules for minimum visibility of civil airplanes. In implementing these rules, use is made of a horizontal reference plane H, defined in Fig.2.15a. The azimuth and vertical inclination of unobstructed eye vectors are determined as shown in Figure 2.15b. Note in Figure 2.15a that the pilot's eye center is assumed to move on a circle with a radius of 84 mm about the axis of rotation of the pilot's head.

Although pilots generally see through both eyes, it is customary to construct the visibility pattern by assuming that point C (Figure 2.15) is the center of vision.

In laying out a cockpit for acceptable visibility it is important to locate Point C as shown in Figure 2.15b. Point C can then be used to locate the pilot seat. The seat itself must be located relative to the floor and relative to the cockpit controls using the dimensions discussed in Section 2.2. The process is shown in Fig.2.16. It can be broken down into the following steps:

<u>Step 1:</u> Locate point C on the horizontal vision axis as shown in Figure 2.16.

Step 2: Make sure that the distance labelled L_c in Figure 2.15b is within the indicated range.

Step 3: Draw the angle $\psi = 8.75$ degrees.

Step 4: Locate point S with the help of the distance 'c' as defined in Fig. 2.7 and in Table 2.3. The maximum allowable value for c is 80 cm.

Step 5: Orient the pilot seat in accordance with the dimensions of Figure 2.7.

Step 6: Draw in the areas required for cockpit control and for seat motions and adjustments.

Step 7: Check the minimum required visibility with the visibility rules of Figs 2.15 and 2.16.

1. In transports with side-by-side pilot seating there shall be no obstructing window frames in the area from 30^0 starboard to 20^0 port: See Figure 2.17.

2. In the area from 20^0 port to 60^0 port, window frames may not be wider than 2.5 inches.

3. Figure 2.17 illustrates the combined azimuth and vertical inclination areas where cockpit visibility cannot be obstructed.

Figure 2.18 shows the 'ideal' visibility pattern.

In reality it is very difficult to achieve the 'ideal' visibility pattern. Large windows require very stiff frames. Both windows and frames must meet the 'birdstrike' requirement (Ref.11, FAR 25.775). The larger the windows and the stiffer the frames, the more weight this means.

Another problem is drag. Large flat windows result in large drag increments. Curved windows on the other hand offer low drag but may lead to image distortions.

In reality it is therefore necessary to strike a compromise. Figure 2.19 illustrates typical visibility patterns in certified transports. These patterns illustrate that compromises are indeed accepted.

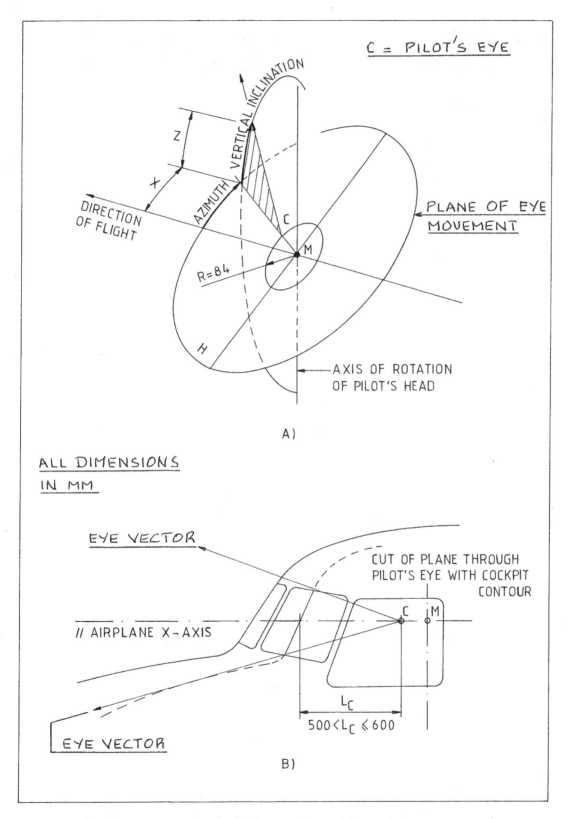

Figure 2.15 Definition of Radial Eye Vectors

Part III Chapter 2 Page 25

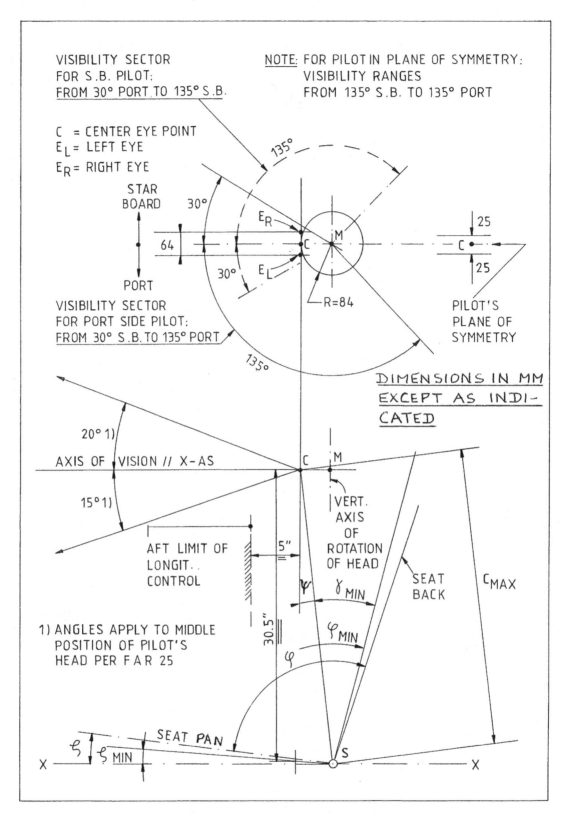

Figure 2.16 Visibility Requirements for the Port and for the Starboard Side and the Connection with Acceptable Seat Arrangements

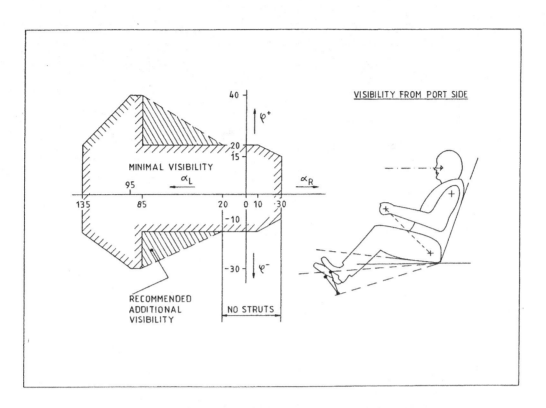

Figure 2.17 Minimum Recommended Visibility Pattern for for the Port Side

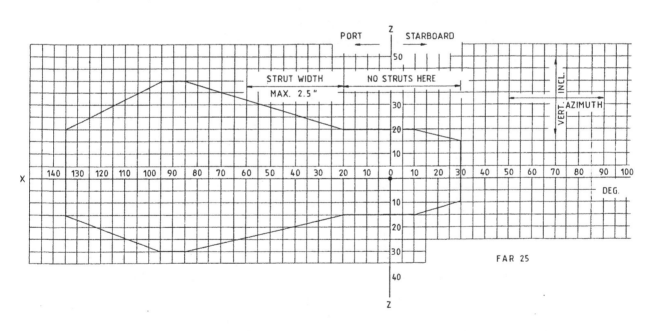

Figure 2.18 Ideal Minimum Visibility Pattern for Transport Airplanes

Part III   Chapter 2   Page 27

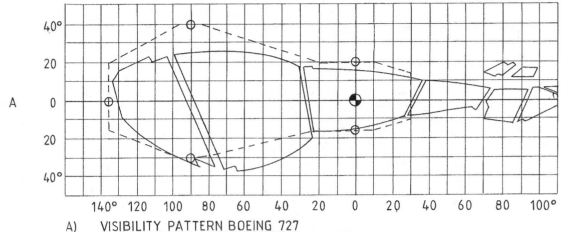

A) VISIBILITY PATTERN BOEING 727
POINT C IS 43" ABOVE HEEL REST AND 5" BEHIND THE MOST AFT
POSITION OF THE LONGITUDINAL CONTROLS

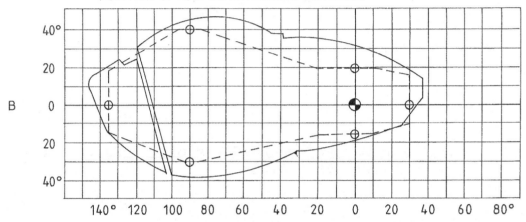

B) VISIBILITY PATTERN BOEING 727
POINT C IS 43" ABOVE HEEL REST AND IMMEDIATELY ABOVE THE
MOST AFT POSITION OF THE LONGITUDINAL CONTROLS

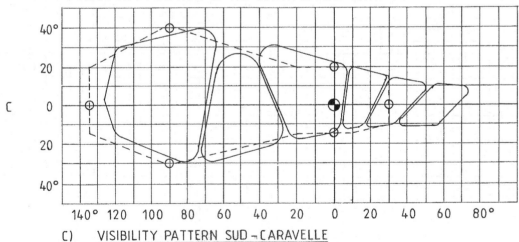

C) VISIBILITY PATTERN SUD-CARAVELLE

SOURCES: A), B) FROM BOEING
C) FROM TECHNIQUE ET SCIENCE "AERONAUTIQUES
ET SPATIALES" MAY-JUNE 1962

- - - - - REQUIRED BY FAR 25
———— MEASURED

Figure 2.19 Typical Visibility Patterns for Three Transport Airplanes

2.4 EXAMPLES OF COCKPIT LAYOUTS

Figures 2.20 through 2.27 present examples of cockpit layouts for FAR 25 certified transports and for fighters. For typical cockpit layouts of FAR 23 certified airplanes the reader should refer to the examples of cabin arrangements for these airplanes as presented in Chapter 3.

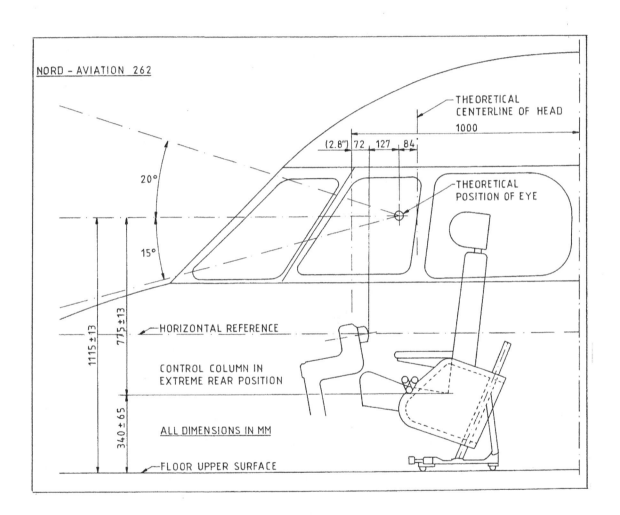

Figure 2.20 Cockpit Layout: Nord 262

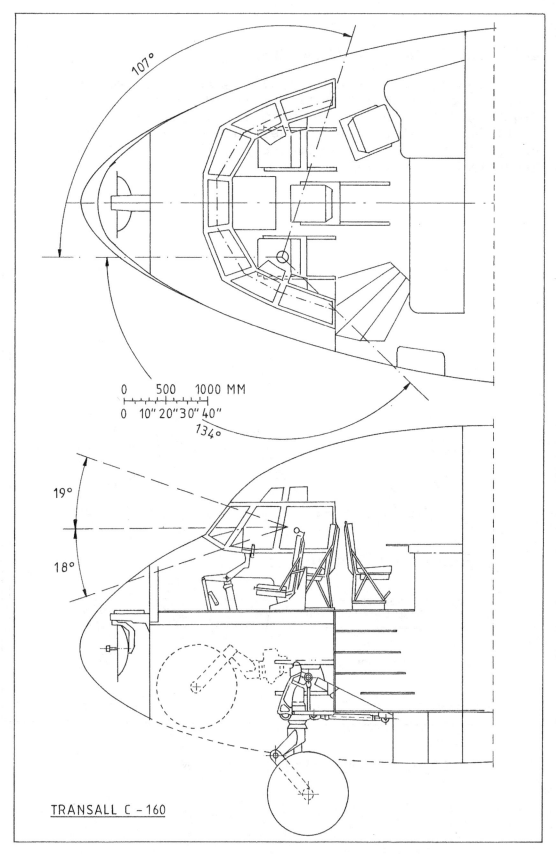

Figure 2.21 Cockpit Layout: Transall C-160

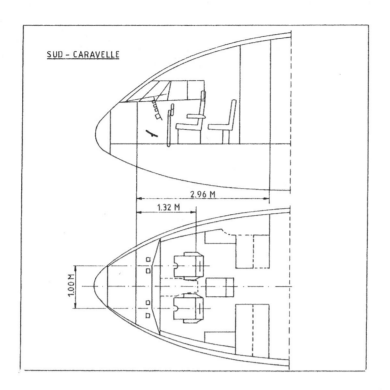

Figure 2.22 Cockpit Layout: Sud Caravelle

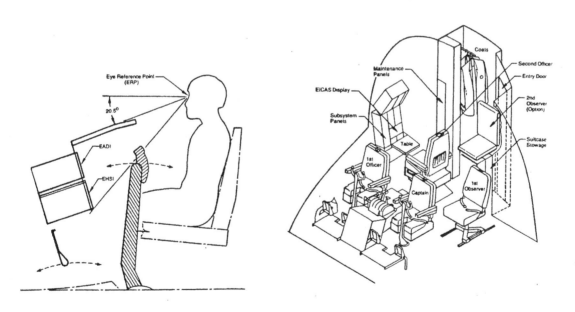

Figure 2.23 Cockpit Layout: Boeing 767 (Three Crew Alt.)

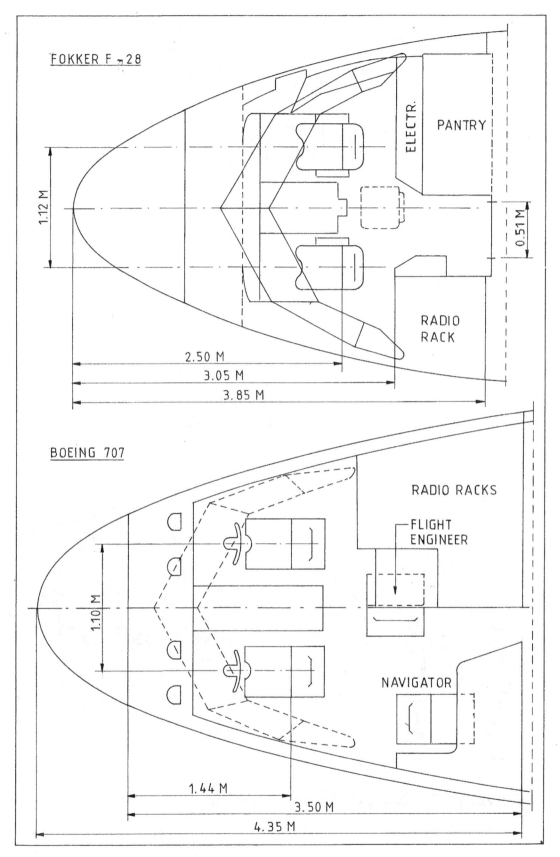

Figure 2.24 Cockpit Layout: Fokker F28 and Boeing 707

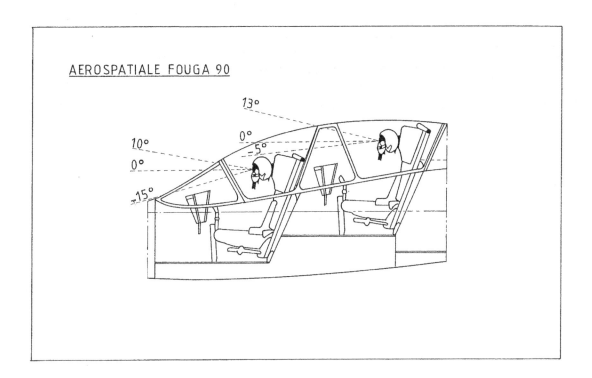

Figure 2.24 Cockpit Visibility and Layout: Fouga 90

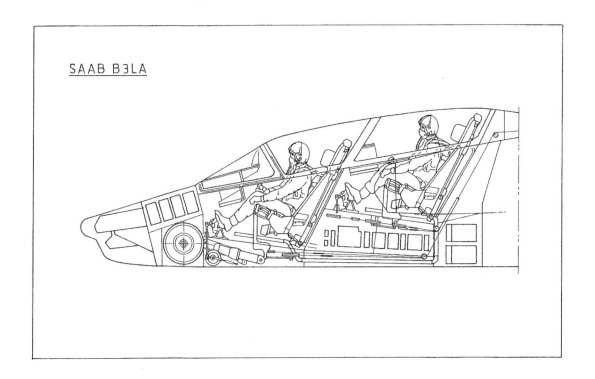

Figure 2.25 Cockpit Layout and Inboard Profile: SAAB B3LA

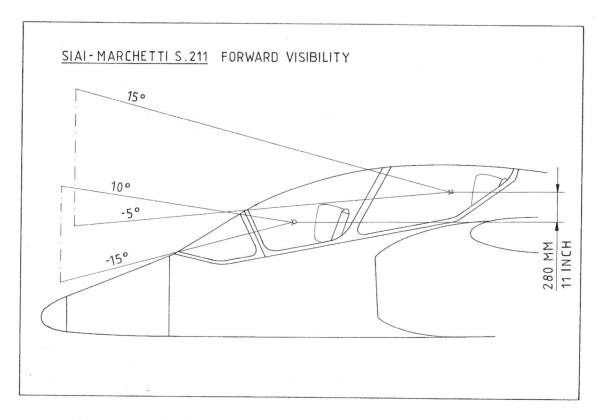

Figure 2.26 Visibility Pattern: SIAI-Marchetti S211

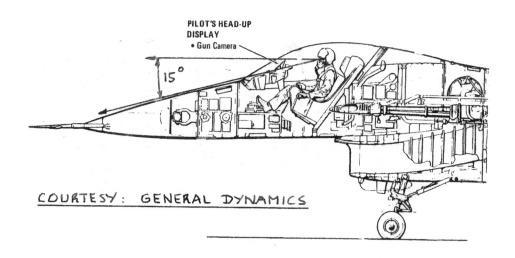

Figure 2.27 Cockpit Layout and Inboard Profile: GD F-16

3. FUSELAGE LAYOUT DESIGN

The purpose of this chapter is to provide the rationale behind the selection of fuselage size, shape, interior and structural arrangement of airplane fuselages.

A step-by-step procedure for arriving at a satisfactory fuselage layout is presented in Chapter 4 of Part II. It is recommended that the reader use that procedure together with the broad range of data on fuselage design given in this chapter.

The reader should also review the configurations presented in Chapter 3 of Part II. It is always useful to find out what has been done by various manufacturers!

Section 3.1 contains a discussion of aerodynamic design considerations.

Section 3.2 contains some guidelines for the exterior layout of flying boat hulls and floats.

Section 3.3 considers the problem of interior fuselage layout design. Data on seat dimensions, exit and door dimensions, galleys, lavatories, wardrobes, cargo containers, maintenance and servicing provisions are provided.

Section 3.4 contains design data on fuselage cross sections, cabin and cargo hold layouts, door and window layouts, as used in a wide range of airplanes.

Section 3.5 presents a number of structural design considerations. Examples of structural layouts of important fuselage details such as: frame and longeron layout, cross section, skin splices, doors, stairs, windows, pressure bulkheads and floors are included.

The fuselage contains not only the payload in most airplanes but also the crew and an assortment of systems needed to operate the airplane. It is not difficult to envision that conflicts for available space will arise. To help visualize the relative arrangement of crew, payload, systems and structure an inboard profile is usually prepared. Section 3.6 presents several example inboard profiles.

For additional information and study references 12 through 15 are highly recommended.

3.1 AERODYNAMIC DESIGN CONSIDERATIONS

The fuselage is responsible for a large percentage of the overall drag of most airplanes: 25 - 50 percent. Since it is desirable to have as little drag as possible, the fuselage should be sized and shaped accordingly.

Fuselages generate the following types of drag:

1. Friction drag
2. Profile drag
3. Base drag
4. Compressibility drag
5. Induced drag

For more detailed data on the relationship between fuselage drag and fuselage shape the reader should consult References 13 and 15. Methods for estimating fuselage drag are presented in Part VI.

3.1.1 Friction Drag

Friction drag is directly proportional to wetted area. Wetted area itself is directly related to fuselage length and to the perimeters of fuselage cross sections. To reduce friction drag, two options are available:

1. Shape the fuselage so that laminar flow is possible.

2. Reduce the length and perimeter as much as possible.

A significant amount of research is being conducted by NASA to determine the conditions for and the extent of laminar flow which can be achieved on a fuselage. Exterior roughness and nose shape are the primary factors which determine the extent of laminar flow which can be achieved at any given combination of Mach Number and Reynold's Number. Without the use of direct boundary layer control it appears that no more than 20 to 30 percent laminar flow (based on fuselage length) can be achieved. A smooth, cambered nose shape (See Figure 3.47, Part II) is required to achieve this.

Most fuselages today have a turbulent boundary layer with correspondingly high friction coefficients. It is shown in Figure 3.1 that the fuselage fineness ratio parameter, l_f/d_f plays an important role in determining

fuselage friction drag. A 'weak' minimum occurs for a fineness ratio of around 6.0. Table 4.1 in Part II shows ranges of fineness ratio employed in fuselages of twelve types of airplanes. Note the trend toward very high fineness ratios as the cruise speed increases.

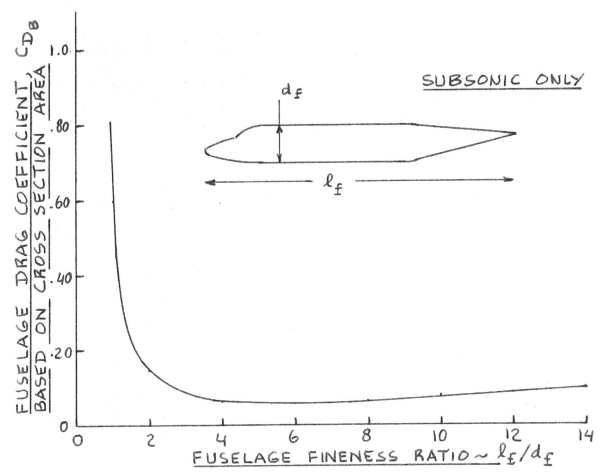
Figure 3.1 Effect of Fineness Ratio on Fuselage Drag

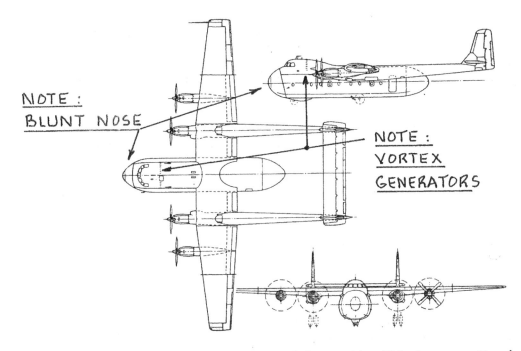
Figure 3.2 Armstrong Whitworth 650 Argosy Freighter

Reference 16 shows that if fuselage length is increased for the same level of static stability, the tail sizes can be decreased thereby reducing overall friction drag. If this factor is accounted for, the optimum fuselage finess ratio from a viewpoint of friction drag (at subsonic speeds) is around 8.0.

In many airplanes design constraints of non-aerodynamic nature will prevent the application of high fineness ratios. Examples are cargo airplanes with rear loading capability, twin boom configurations and carrier based airplanes which must fit on the carrier elevators.

3.1.2 Profile and Base Drag

Profile and base drag are a strong function of front and aft body shape. Blunt fore-bodies and blunt aft-bodies promote flow separations which lead to high profile and base drag.

Fore-body bluntness can be caused by:

1. Poor cockpit window or canopy shaping

2. Requirement for front end loading

The ideal 'streamline' nose shape can be achieved only if the windshields are integrated smoothly into the surface of the fuselage. The DH Comet and the Sud Caravelle (which used the Comet nose and cockpit) are early examples of such a configuration. Recently, the GP180 (Fig.3.47, Part II) uses this concept. Although drag can be considerably reduced by these types of windshield fairing, image distortions may be introduced if the 'fairing angle' becomes too acute.

Figure 3.2 (AW 650 Argosy) illustrates what can happen to the shape of a fuselage nose if front loading is a requirement. It must be remembered that all design decisions are based on a compromise. It would have been possible to streamline the Argosy nose. That would have made it longer, increasing wetted area and thereby friction drag and structural weight. In the case of the Lockheed Galaxy (Fig.3.30, Part II) and the Antonov AN124 (See Jane's, 1985 edition) the longer streamlined nose option was exercised. Reference 15 contains detailed data on the effect of fore-body shape on profile drag.

In the case of fighters and trainers the requirement for good visibility from the cockpit becomes a dominant design criterion. This calls for a large canopy. Canopy drag then becomes an important factor in the design of the fuselage. Part VI contains data with which canopy drag can be related to canopy size.

Figure 3.3 illustrates the effect of aft body bluntness on drag.

Even with a large fuselage fineness ratio, it is possible to increase fuselage drag by using too much 'upsweep' in the aft-body. Figure 3.4 shows what is meant by 'upsweep'. Figure 3.4 also shows how 'upsweep' affects drag.

Upsweep of the aft-body can lead to vortex induced separations. Figure 3.5 illustrates such a vortex flow situation. These separated vortices not only increase drag but can cause lateral oscillation problems. The vortex flow can be stabilized by the use of 'sharp' corners. These 'sharp' corners have been shown to reduce the lateral oscillation problem and also to reduce drag. Figure 3.6 shows an example.

Upsweep is applied to airplanes for the following reasons:

1. to facilitate take-off rotation

2. to facilitate rear cargo loading

Sometimes it becomes necessary to add a 'bulge' to the upper rear fuselage if a large upsweep angle was dictated by rear loading considerations. This bulge is needed to create sufficient structural depth in the fuselage to resist tailloads. Examples of such 'bulged' fuselages are seen in Figures 3.29a and 3.30a in Part II.

3.1.3 Compressibility Drag

A fuselage alone does not experience compressibility drag effects until very high subsonic Mach numbers. Compressibility drag arises from the existence of shocks on the fuselage. The appearance of shocks is strongly coupled to the sweep and thickness of the wing in the area of wing/fuselage juncture. The area rule concept must be used to minimize compressibility drag. The area rule is discussed in Ref.13 and in Part VI. Figure 3.7 shows an area ruled fuselage for a high subsonic transport. Figures 3.34 through 3.36 (Part II) show area ruled fuselages for supersonic applications.

3.1.4 Induced Drag

A fuselage contributes to induced drag primarily because of its adverse effect on wing spanload distribution. Figure 3.8 illustrates a typical effect.

When a fuselage is equipped with leading edge strakes (lexes) and/or the fuselage is sharply blended into the wing (such as is the trend in modern fighter design), there will be significant effects of the fuselage on induced drag.

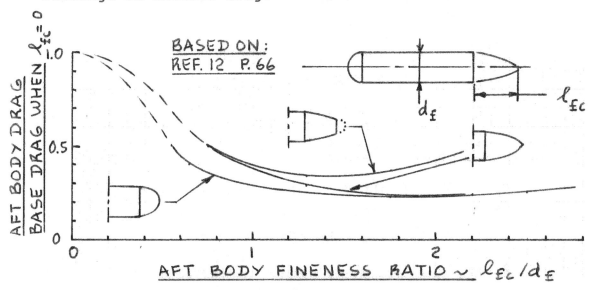

Figure 3.3 Effect of Aft Body Bluntness on Drag

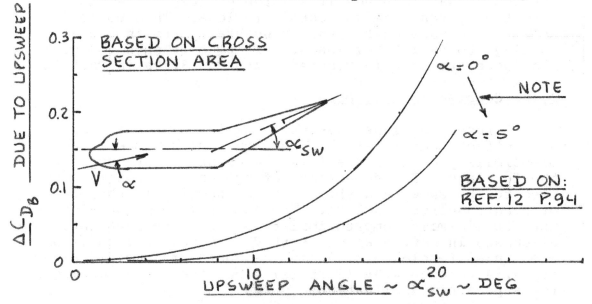

Figure 3.4 Definition of Upsweep and Its Effect on Drag

Part III   Chapter 3   Page 40

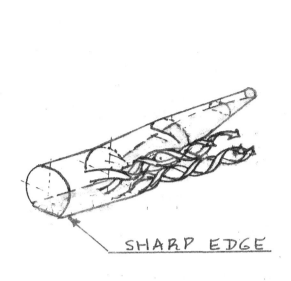

Figure 3.5 Vortex Separation from a Fuselage

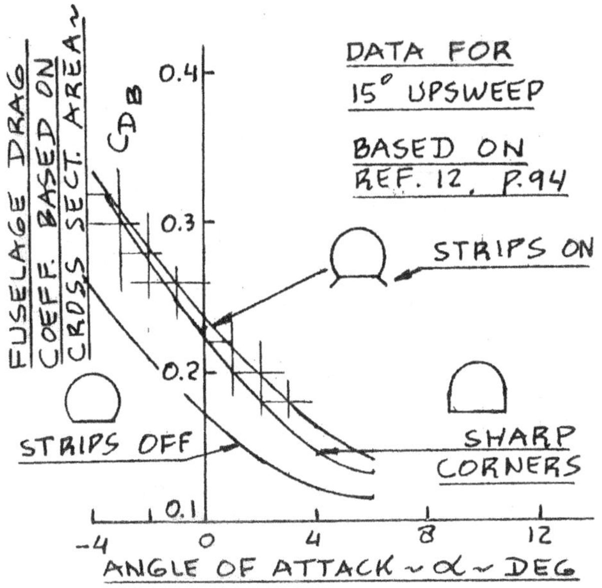

Figure 3.6 Effect of Cross Section Shape on Drag

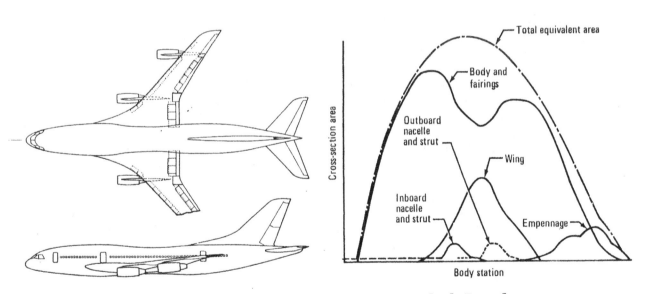

Figure 3.7 Example of an Area Ruled Fuselage

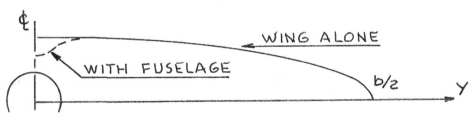

Figure 3.8 Effect of Fuselage on Wing Span Loading

3.2 GUIDELINES FOR FLYING BOAT HULL AND FLOAT DESIGN

Figure 3.9 shows the typical hull geometry employed in flying boats as well as in floats. The following design considerations are important:

1. Buoyancy.   2. Hydrodynamic drag and aerodynamic drag.

3. Effect of hull shape on directional stability.

4. Effect of hull shape on landing and take-off characteristics (air and water).

5. Effect of hull shape on water spray and on where the spray goes: salt water sprays into engines are bad.

6. Effect of hull shape and hull size on ability to operate in certain sea states.

7. It is essential that the hull bottom be designed with enough watertight compartments so that the flooding of one does not result in the sinking of the airplane.

8. Special attention needs to be given to the selection of hull materials: particularly sea water is a very corrosive environment!

The reader should consult references 14, 15 and 17 for information on aerodynamic and hydrodynamic hull and float design. A wealth of design and analysis information on flying boats and on floats may be found in NACA Technical Reports issued as 'annual bound volumes' in the 1920-1958 era. Ref.18 is an example of such data.

Flying boats and float equipped airplanes all have 'built-in' drag and weigt penalties associated with the need for buoyancy and acceptable handling characteristics in the water. Part VI shows that the drag penalties due to wetted area compared to land based airplanes are rather large.

Reference 17 contains data showing the effect of stall speed on hull loading for water landings. This effect needs to be accounted for in the early weight estimates of any new water based airplane.

Figures 3.10 through 3.12 contain design guidelines for the exterior shape of the hull. The nomenclature is explained in Figure 3.9.

Figure 3.13 shows that the ability of a flying boat

to operate in given sea state conditions is directly related to the displacement (i.e. take-off weight, i.e. airplane size).

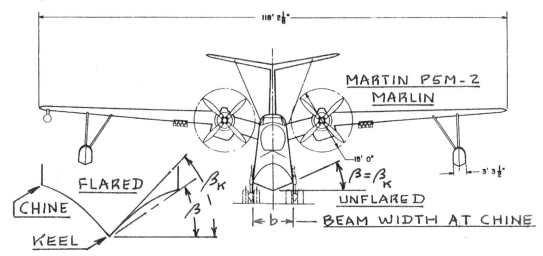

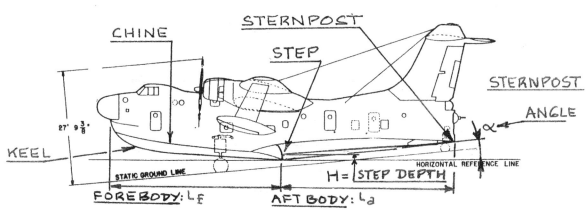

Figure 3.9 Typical Flying Boat and Float Hull Geometry

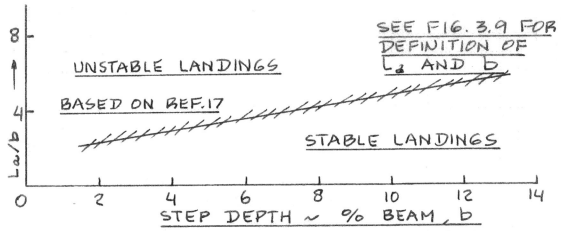

Figure 3.10 Effect of Hull Geometry on Stability of Water Landings: I

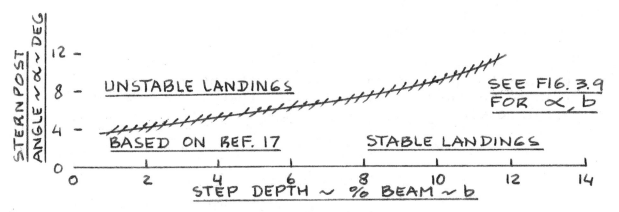

Figure 3.11 Effect of Hull Geometry on Stability of Water Landings: II

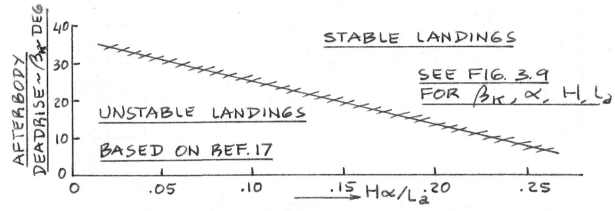

Figure 3.12 Effect of Hull Geometry on Stability of Water Landings: III

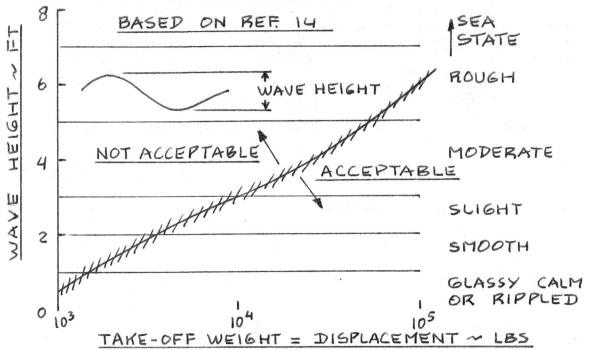

Figure 3.13 Effect of Hull Displacement on Ability to Operate in Sea States

3.3 INTERIOR LAYOUT DESIGN OF THE FUSELAGE

The fuselage in most airplanes carries the crew, the payload (passengers and/or cargo and weapons) and many of the systems needed for the operation of an airplane.

In commercial passenger operations the interior design reflects a compromise between level of creature comforts and the weights and sizes required to create the creature comforts.

In cargo operations the ability to efficiently load and unload cargo plays an important role.

In fighter design a major problem is that of 'packaging' of all required systems so that they operate satisfactorily, don't interfere with another (particularly important with avionics) and can be easily accessed.

In commercial as well as military operations the problems associated with servicing and maintenance dictate where access must be designed into the fuselage. Design for good access, maintenance and inspectability usually conflicts directly with design for low structural weight, low complexity and low drag.

The fuselage normally also houses the cockpit (or flight deck). Design requirements for satisfactory cockpit layouts are provided in Chapter 2. This chapter contains design information for the following aspects of fuselage interior layout design:

    3.3.1 Layout of the cross section
    3.3.2 Seating layouts, seats and restraint systems
    3.3.3 Layout of doors and emergency exits
    3.3.4 Galley, lavatory and wardrobe layouts
    3.3.5 Layout of cargo/baggage holds, including data on cargo containers
    3.3.6 Maintenance and servicing considerations

3.3.1 Layout of the Cross Section

Fuselage cross sections, for commercial airplanes are the result of compromises between weight, drag, systems and creature comfort considerations. In military applications, additional considerations may be those of radar observability and weapons system integration.

For pressurized airplanes the most efficient cross section from a structural viewpoint is the cirle. However, for small airplanes a circular cross section is wasteful in terms of volume. To verify this, draw a circle around the human body in a sitting position.

From a manufacturing viewpoint a flat sided fuselage is the cheapest to build. The Shorts 330 of Fig.3.18d, Part II is an example of such an approach.

3.3.1.1 Passenger cabin

The dimensions of the human body dictate the minimum cabin size that will 'fit around' the occupant(s) after a decision has been made whether the cabin cross section allows for 'stand-up' room or for 'crawl-to-your-seat' room.

Figures 3.14 through 3.18 provide scaled drawings of males and females in a variety of postures.

In small civil airplanes (such as homebuilts, single engine airplanes and most twin engine airplanes) it is usually not practical to design for 'stand-up' room. The added weight, drag and cost are judged not to be acceptable. Figures 3.45 and 3.48-3.52 in Section 3.4 provide dimensioned cross sections for 'small' civil airplanes.

Sailplanes and the BD-5J represent extremes of cabin comfort at the 'low' end of the scale. The inboard profile of Figure 3.87 in Section 3.6 shows the tight fit of the BD-5J around the human body.

For transport airplanes, Figure 3.19 shows a statistical relationship between fuselage width and the number of seats abreast. The minimum allowable width of aisles between seats is dictated by emergency evacuation considerations. Figure 3.20 summarizes the allowable dimensions based on FAR 25.815.

FAR 25.817 states that on each side of an aisle, no more than three seats may be placed abreast.

In passenger transports a critical choice which affects the design of the cross section is the number of seats abreast. The fewer seats abreast, the longer the fuselage and the more difficult 'growing' the airplane becomes. The more seats abreast, the shorter the fuselage and the easier it becomes to 'grow' the airplane.

<u>Important note:</u> In passenger transports it is undesirable to interrupt the fuselage cross section locally by a wing torque box. This can be a real problem in the case of high wing transports.

Figure 3.14 Scaled Views of Standing, Male Passengers

Part III Chapter 3 Page 47

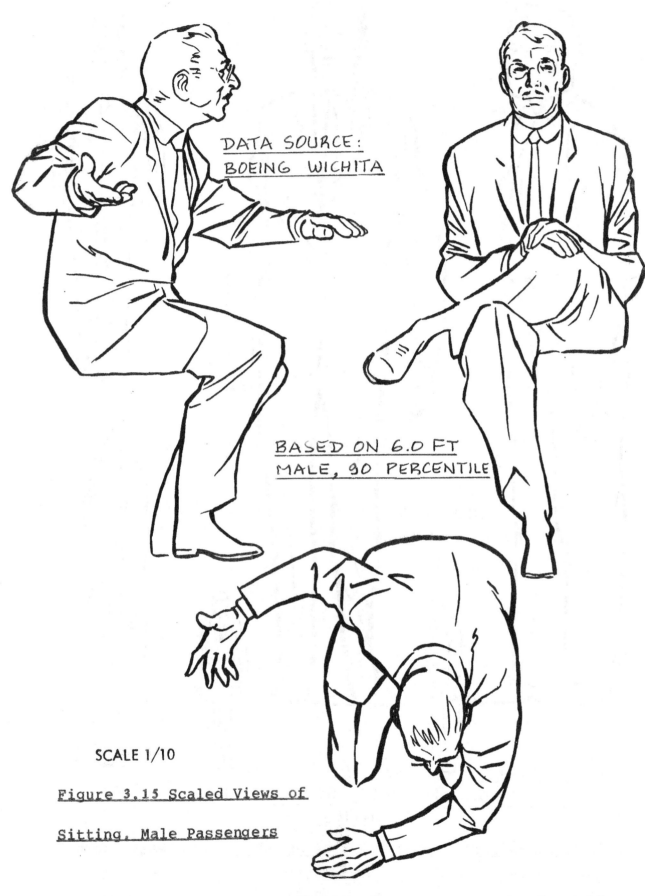

Figure 3.15 Scaled Views of Sitting, Male Passengers

Figure 3.16 Scaled Views of Sitting, Male Passengers

1/20

DATA SOURCE:
BOEING WICHITA

1/40

SCALE 1/10

Figure 3.17 Scaled Views of Standing, Female Passengers

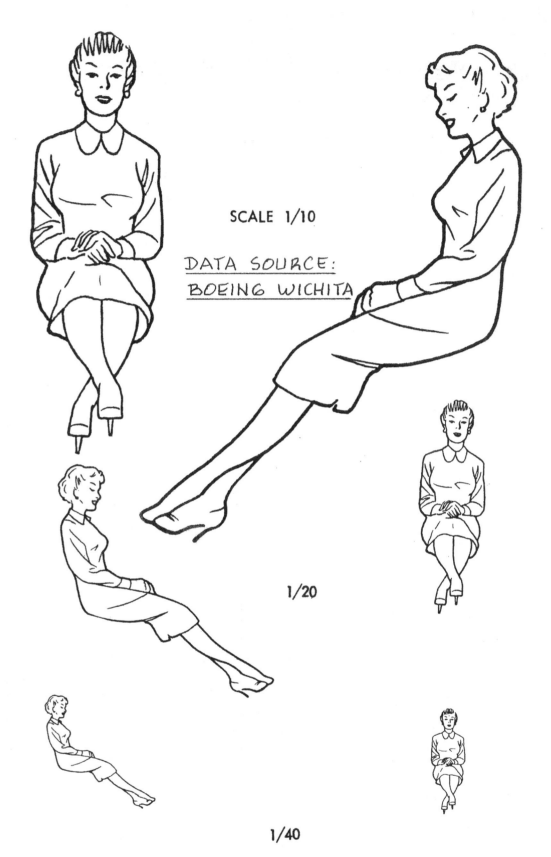

Figure 3.18 Scaled Views of Sitting, Female Passengers

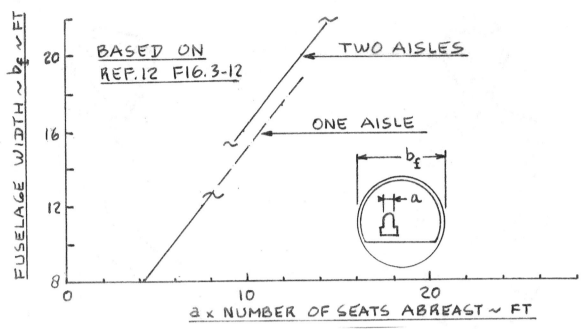

Figure 3.19 Statistical Relationship Between Fuselage Width and Total Seat Width

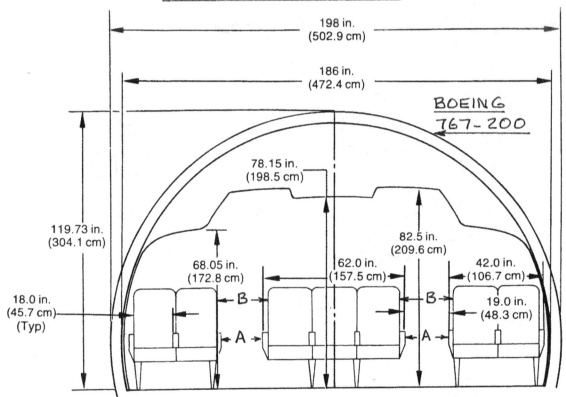

Number of Seats	Minimum Value of A	Minimum Value of B
10 or less	12 inches	15 inches
11 - 19	12 inches	20 inches
20 or more	15 inches	20 inches

Figure 3.20 Minimum Aisle Width Requirements

Part III　　　　　　　　　　Chapter 3　　　　　　　　　　Page 52

3.3.1.2 Cargo hold

In most passenger airplanes baggage and cargo is carried in 'standard' containers. From a competitive viewpoint it is important to be able to carry as many different types of containers as possible. This represents a very difficult design problem. The problem needs to be solved as part of the overall cross section and fuselage layout process. To accomplish this, baseline data on cargo/baggage containers are needed. Sub-section 3.3.5 provides these data.

The location of the wing on the fuselage as well as the amount of space needed for landing gear retraction can help dictate the amount of cargo containers that can be carried in a given volume. If from a competitive viewpoint it becomes desirable to carry one or two more containers, the fuselage length, the cross section or the wing location on the fuselage may have to be reexamined.

3.3.1.3 Military

In laying out the cross section of troop transports it is necessary to account for the dimensions of 'combat-ready' troops. Figures 3.21 - 3.23 provide scaled drawings of combat ready troops in a variety of postures.

In military airplanes the additional problems of pilot visibility (as in fighters) and radar observability influence the design of the cross section. Design rules for achieving a low radar cross section are given in Chapter 7.

Dimensions of a number of military vehicles which may have to be carried in military transports are given in Chapter 7.

Weapons integration needs may further complicate the choice of cross section design. Chapter 7 also deals with weapons integration problems.

3.3.1.4 Supersonic airplanes

In the case of supersonic airplanes the need for area ruling places further constraints on cross section design. Part VI contains a discussion of the area ruling concept. An example of a fuselage with subsonic area ruling can be found in Figure 3.7.

Figure 3.21 Scaled Views of Standing, Combat Ready Troops

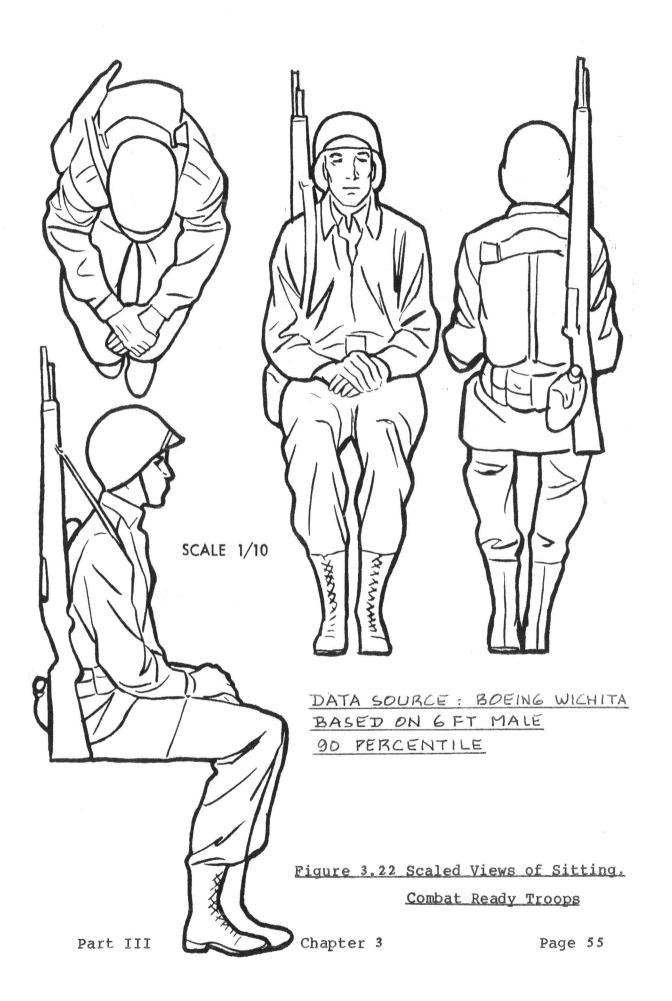

SCALE 1/10

DATA SOURCE: BOEING WICHITA
BASED ON 6 FT MALE
90 PERCENTILE

Figure 3.22 Scaled Views of Sitting, Combat Ready Troops

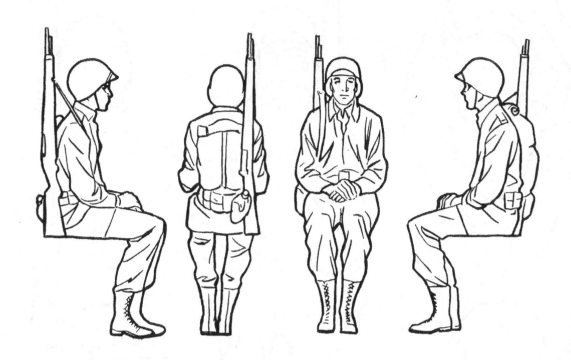

SCALE 1/20

DATA SOURCE: BOEING WICHITA
BASED ON 6 FT MALE, 90 PERCENTILE

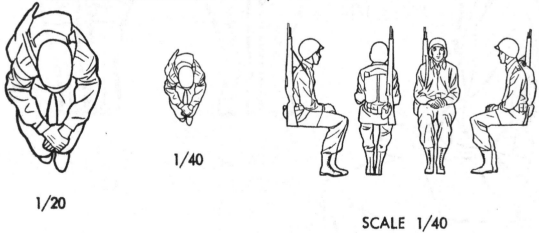

1/20

1/40

SCALE 1/40

Figure 3.23 Scaled Views of Sitting, Combat Ready Troops

3.3.2 Seating Layouts, Seats and Restraint Systems

An important decision in laying out the overall seating arrangement in passenger airplanes is that of x-seats-abreast versus y-seat-rows. This choice has a significant impact on:

1. cabin length and cabin width
2. fuselage weight and drag
3. future growth potential for the airplane
4. passenger appeal and therefore market acceptance

Sub-sub-section 3.3.2.1 provides information on seating arrangements and seats for general aviation airplanes. Sub-sub-section 3.3.2.2 gives similar data for transport airplanes.

In many airplanes it is desirable and/or necessary to have restraint systems built in, to prevent injury in the case of a crash. Sub-sub-section 3.3.2.3 deals with the layout of restraint systems.

3.3.2.1 Seating arrangements and seats for general aviation airplanes

Figures 3.48 through 3.52 in Section 3.4 provide dimensioned data on cabins and on seating layouts of existing general aviation airplanes.

In some types of general aviation airplanes it is necessary to carry a parachute. Parachutes can be worn as a back-pack or as a seat-pack. Figure 3.24 illustrates the two types with typical dimensions and weights. In the layout of seats, these dimensions need to be accounted for.

3.3.2.2 Seating arrangements and seats for transports

The passenger cabin should be laid out so that in cruise flight the cabin floor is level. If this criterion is not satisfied, cabin service and moving about in the cabin are made much more difficult. The level cabin floor requirement is linked directly to the choice of wing incidence. This is explained further in Chapter 4.

Figure 3.25 shows a statistical correlation between cabin length, seat pitch, total number of seats and seats abreast. As long as the right number and size exits are provided (See sub-section 3.3.3) the seating arrangement is up to the designer. However, the seat pitch near

Seatpack parachute:

length 13 in.
width 15.5 in.
depth 8 in.
(These dimensions include a 2 in. thick cushion)

Backpack parachute:

length 23 in.
width 14.5 in.
depth 6 in.
(These dimensions include the back pad)

(Data from Ref.14, p.349)

Figure 3.24 Typical Parachute Dimensions

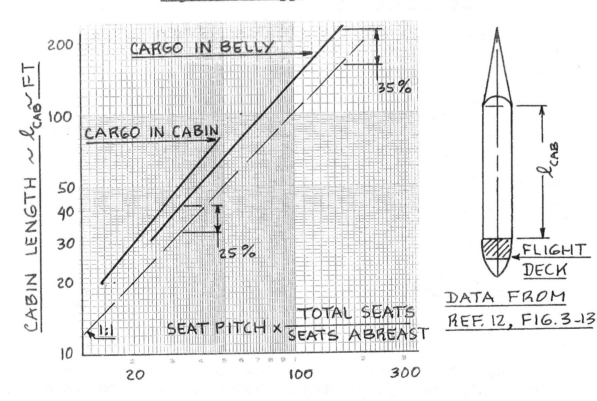

Figure 3.25 Statistical Relationship Between Cabin Length, Seat Pitch, Number of Seats and Number of Seats Abreast

emergency exits must meet the requirement stated in sub-section 3.3.3.

In laying out a proposed seating arrangement, remember that passengers (when given an equal choice) do not like three seats in a row. Nevertheless, many narrow body airplanes are configured in precisely this manner. Also remember that the FAR's prohibit more than three seats in a row, unless an extra aisle is included. The four-seat rows in the center of a B-747, flanked by two aisles are acceptable.

The following seat pitch values reflect industry practice:

```
seat pitch for: first class seating,  38 - 40 inches
                tourist/coach/econ.,  34 - 36 inches
                high density,         30 - 32 inches
```

Examples of transport seating arrangements are shown in Figures 3.26 and 3.27.

Note the cabin attendant (cabin crew) seats in the seating layouts of Figures 3.26 and 3.27. Requirements for cabin attendants as a function of the number of passengers carried are given in FAR 91. These requirements are summarized as follows:

<u>Light transports:</u> 1 attendant per 20 passengers (minimum)

<u>Large transports:</u> 1 attendant per 50 passengers (minimum)
1 attendant per 35 passengers (this reflects industry practice)

Table 3.1 relates seat classification (in terms of class of service) to seat dimensions. The seat dimension symbols are defined in Figure 3.28.

Finally, data on a number of seats available on the market are given in Figures 3.29a-c.

All seats must meet the requirements of FAR 23 and 25, parts 561 and 785. Table 3.2 presents the design limit load factors for seats with a nominal 170 lbs passenger according to FAR 25.785.

When preparing a Class II weight and balance statement (Step 21, page 19, Part II) it is useful to have actual data on seat weights. Table 3.3 provides these data for six seat types. Figures 3.29a-c contain additional seat weight data.

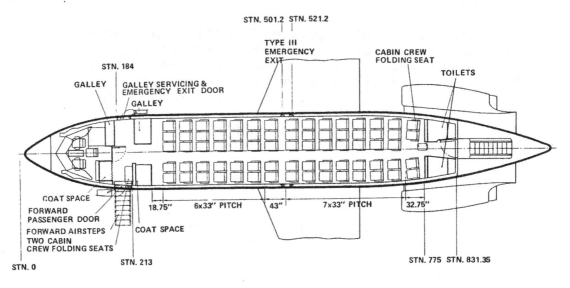

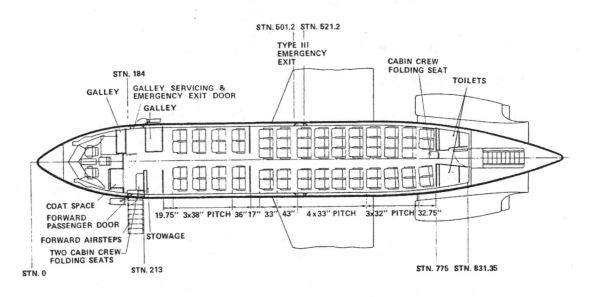

NARROW BODY   COURTESY: BAC

Figure 3.26 Example Seating Arrangements: BAC 111

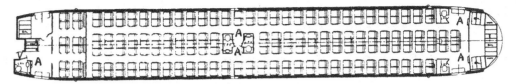

12 First Class (Total 212 Passengers)

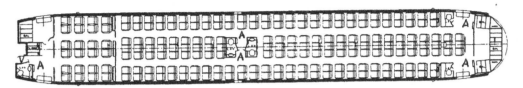

24 First Class (Total 210 Passengers)

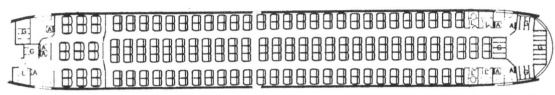

18 First Class-38 in. Pitch   213 Passengers   195 Tourist-34 in. Pitch

One and
One Half
Meal Service

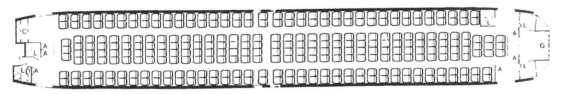

290 Passengers-31/30 in. Pitch

Very High
Density
Inclusive Tour
Two Overwing
Exits Per Side

A = ATTENDANT   L = LAVATORY
G = GALLEY

WIDE BODY   COURTESY: BOEING

Figure 3.27 Example Seating Arrangements: Boeing 767-200

Table 3.1 Seat Classification and Seat Dimensions
===

Note: see Figure 3.28 for definition of seat dimensions.

Seat Classification

Symbol	Unit	De Luxe	Normal	Economy
a	in.	20(18.5-21)	17(16.5-17.5)	16.5(16-17)
b	in.	47(46-48.5)	40(39-41)	39(38-40)
		for two seats per block		
b	in.	---	60(59-63)	57
		for three seats per block		
l	in.	2.75	2.25	2.0
h	in.	42(41-44)	42(41-44)	39(36-41)
k	in.	17	17.75	17.75
m	in.	7.75	8.5	8.5
n	in.	32(24-34)	32(24-34)	32(24-34)
p/p_{max}	in./in.	28/40	27/37.5	26/35.5
α/α_{max}	deg/deg	15/45	15/38	15/38

The data in brackets indicate the range of numbers found.

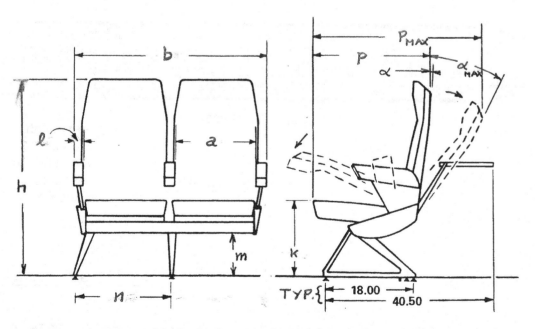

Figure 3.28 Definition of Seat Dimensions in Table 3.1

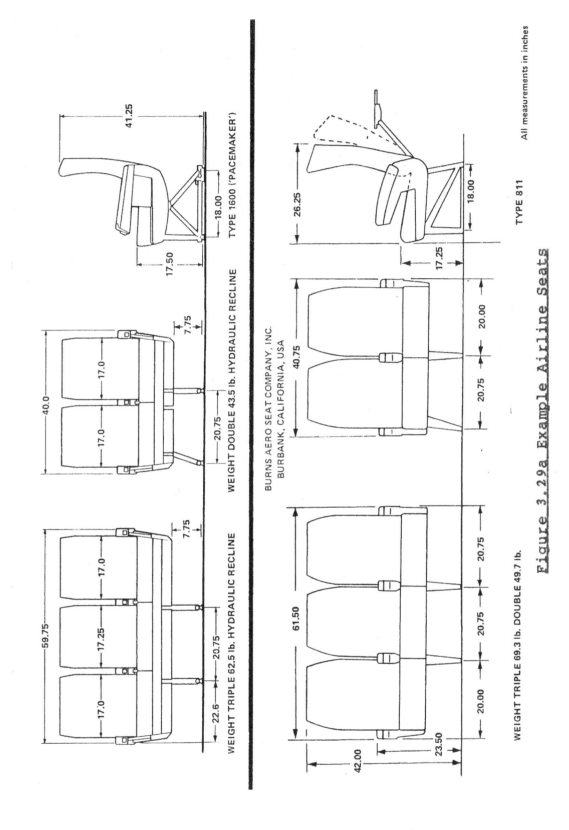

Figure 3.29a Example Airline Seats

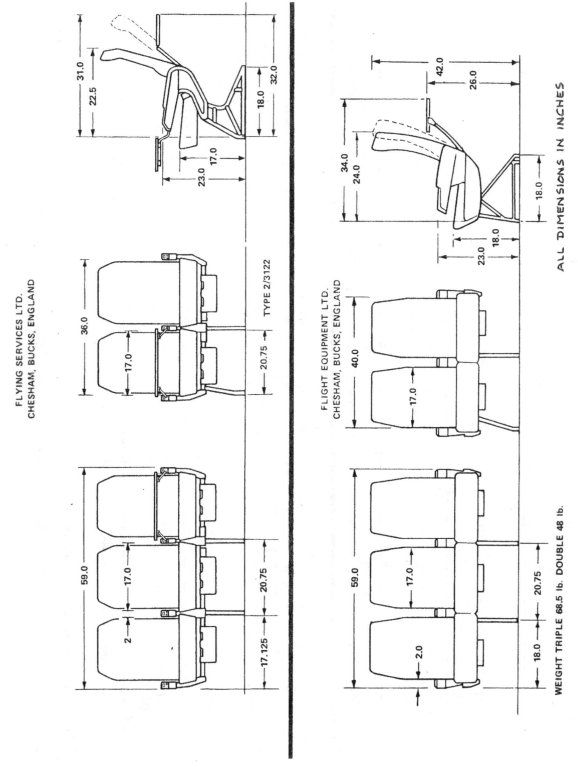

Figure 3.29b Example Airline Seats

Part III　　　　Chapter 3　　　　Page 64

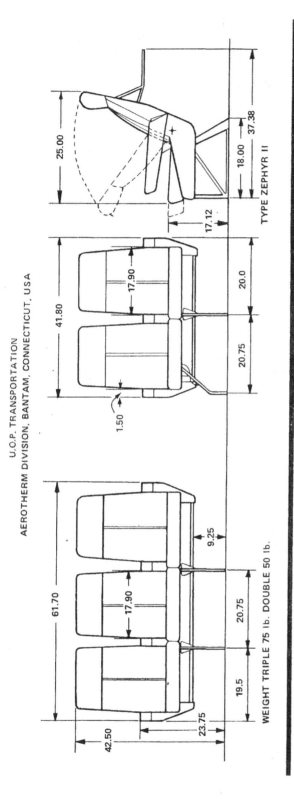

Figure 3.29c Example Airline Seats

Table 3.2 Design Limit Load Factors for Airplane Seats
==

Certif. Base	Forward	Rearward	Upward	Downward	Sideward
FAR 25.561	9.0	--	2.0	4.5	1.5
BCAR D3-8	9.0	1.5	4.5	4.0	2.25

The numbers reflect the amount of g's the seat must be able to withstand with a nominal 170 lbs passenger. Seat attachment fittings must be able to withstand an additional factor 1.33 according to FAR 25.785.

Table 3.3 Seat Weights for Commercial Airplanes
==

Seat Classification	Medium/Long Haul (lbs)	Short Haul (lbs)
De Luxe Single	47	40
Double	70	60
Normal Single	30	22
Double	56	42
Triple	78	64
Economy Single	24	20
Double	47	39
Triple	66	60
Commuter Single	--	17
Double	--	29
Light weight seats (civil and military)		14
Attendants seats	18	14
Executive seats	32 - 50	
Ejection seats (installed)	150	

Part III   Chapter 3   Page 66

3.3.2.3 Restraint systems

For protection of passengers and crew members in the case of flight through turbulence as well as in the case of a crash, restraint systems are required.

Reference 11, FAR 23 and 25 parts 561 and 785 define the design requirements for restraint systems. Seat belts are required for all seats. In addition, shoulder harnesses are required for crew members.

Reference 22 contains useful hints to airplane designers in the area of occupant injury prevention. Figure 3.30 shows the preferred geometry for shoulder harness installations according to Ref.23.

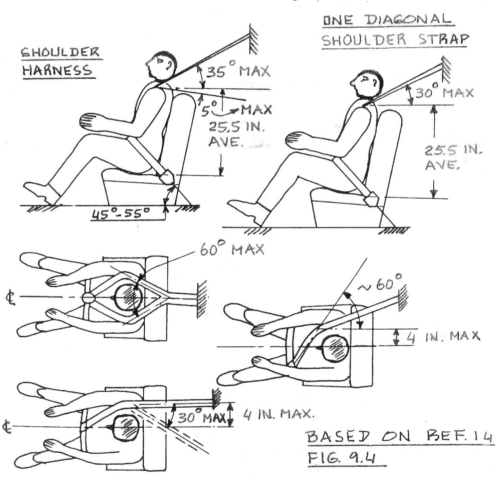

Figure 3.30 Geometry for Shoulder Harness Installation

3.3.3 Layout of Doors, Emergency Exits and Windows

All doors, exits and windows are potential sources for leaks, noise, drag and excess weight. Passenger comfort and emergency evacuation requirements demand a minimum number of as well as a minimum size for doors, exits and windows. Here is a clear conflict between requirements of safety, comfort and economics.

3.3.3.1 General aviation airplanes

FAR 23.807 (Ref.11) lists the requirements for emergency exits which apply to this airplane category. Consult this FAR before finalizing any door and exit layout.

Window layout is normally dictated by the seating arrangement in this type of airplane. The reader should look at the door and window layouts shown in the threeviews of Chapter 3, Part II.

Figures 3.48 through 3.52 contain information about door and exit layouts used in general aviation airplanes.

In many general aviation airplanes the wing attaches to the fuselage via two or more fuselage frames. In such cases the window spacing and sizing is dictated by structural considerations.

Refer to sub-sub-section 3.3.3.2 for more information about window location and window design.

3.3.3.2 Transport airplanes

Transport airplanes normally are required to have three types of doors/exits:

1. Passenger access doors   2. Service access doors
3. Emergency exits

<u>Passenger access</u> doors are normally located on the <u>port side.</u> <u>Servicing access</u> doors are normally located on the <u>starboard side.</u>

For airplanes carrying less than 80 passengers one passenger access door is normally sufficient. For airplanes carrying between 80 and 200 passengers at least two such doors should be provided. For airplanes carrying more than 200 passengers, the number of doors depends on the envisioned boarding scenarios.

Comfortably sized passenger access doors should be

6x3 ft. These dimensions are very difficult to achieve in smaller airplanes and a compromise is needed.

The reader must bear in mind that any door or exit represents a potential pressurization leak, a potential drag cause (because of seal deterioration) and a significant increment in weight. From a weight as well as from an economics viewpoint it makes sense to have as few doors and exits as possible. From an emergency evacuation viewpoint the opposite is true.

The number and the size of doors and emergency exits required in civil airplanes are defined in FAR 23 and 25 parts 807-813. The reader should consult these FAR's before starting the door layout process.

Table 3.4 defines the number and the type of required exits as a function of the number of passengers carried. Table 3.5 provides the minimum required dimensions for each type of exit. Figure 3.31 shows what the various exit types look like and where they are located.

Important notes:

1. FAR 25.807 also demands ventral and/or tailcone exits.

2. All emergency exits and doors must meet the 'unobstructed access' requirement. To satisfy this requirement the following dimensions are used:

   For Type I exits: 36 inches of access width
   For Type II exits: 20 inches of access width
   For Type III and IV exits: 18 inches of access width.

3. The unobstructed access width requirement affects the allowable seat pitch near emergency exits! Account for this in preparing a seating layout.

4. FAR 25.807 also requires escape chutes in some cases. Figure 3.32 shows an example for the Boeing 767-200.

5. Additional requirements apply to airplanes which are operated over water, to cope with emergency evacuation following a ditching.

The window pitch in a passenger transport is normally dictated by the 'frame spacing' requirement and not by the seating layout. Fuselage frames are typically spaced about 20 inches apart.

Windows should be shaped as circles, ovals or rec-

Table 3.4 Required Number of Exits per FAR 25
==

Number of Passenger Seats	Number of Required Exits on Each Side of the Fuselage			
	Type I	Type II	Type III	Type IV
1 - 10	none	none	none	1
11 - 19	none	none	1	none
20 - 39	none	1	none	1
40 - 59	1	none	none	1
60 - 79	1	none	1	none
80 - 109	1	none	1	1
110 - 139	2	none	1	none
140 - 179	2	none	2	none
more than 179	The FAA imposes special conditions			

Notes: 1. The BCAR requirements of Ref.23 are different.
2. Exits do not have to be located diametrically opposed to each other.
3. Instead of one Type III exit it is permissible to use two Type IV exits.
4. See Table 3.5 for dimensions of each exit Type.

Table 3.5 Minimum Dimensions for Exits of Table 3.4
==

Exit Type and Location	Dim. B	Dim. H	Dim. R	Maximum Step Height	
				Dim. h_1 inside	Dim. h_2 outside
I Floor level	24	48	8.0	not applicable	
II Floor level Above wing	20	44	6.7	not applicable 10	 17
III Above wing	20	36	6.7	20	27
IV Above wing	19	26	6.3	29	36

Notes: 1. See Figure 3.31 for explanation of dimensions.
2. All dimensions are in inches.

tangles with liberally rounded corners. This is to avoid unneccessary stress concentrations in the pressure shell.

Window tops should be located at eye level of a 90 percentile passenger.

Many dedicated cargo transports do not have windows to save weight!

In supersonic airplanes which fly at very high altitudes cabin windows should be as small as possible. The large pressure differentials encountered by such airplanes may dictate the elimination of passenger windows.

Examples of door and window layouts of passenger and cargo airplanes are given in Section 3.4.

3.3.3.3 Military airplanes

The door and exit requirements for military airplanes varies with the type of mission. The military RFP will normally include references from which the door and exit requirements can be distilled.

Figures 3.53 (AN-26 and Hercules) show examples of door and window layouts of typical military transports.

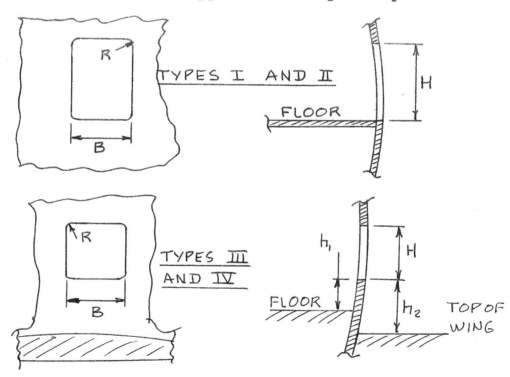

Figure 3.31 Definition of Exit Geometry

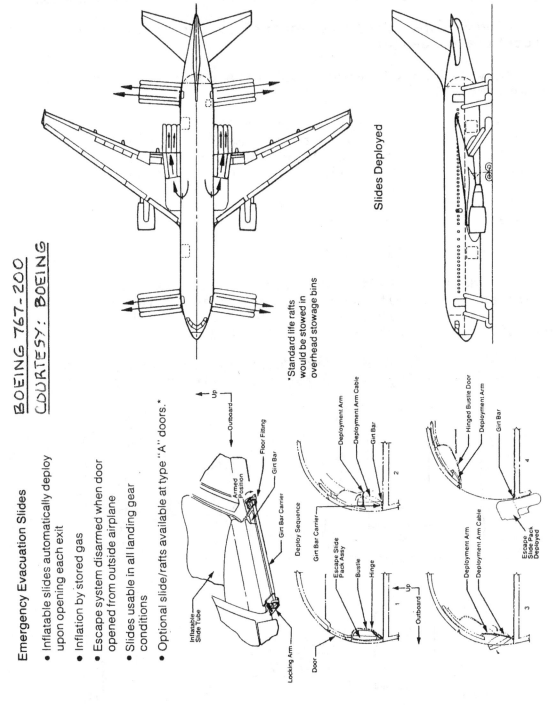

Figure 3.32 Example of Escape Chute Deployment

Part III　　　Chapter 3　　　Page 72

3.3.4 Galley, Lavatory and Wardrobe Layouts

Table 3.6 presents typical dimensions of galleys, lavatories and wardrobes as they are found in a number of airplane types. For smaller general aviation airplanes Figures 3.51 and 3.52 provide an indication of typical dimensions associated with these items.

Figures 3.33 and 3.34 show typical installations used in a narrow-body and in a wide-body transport.

Table 3.6 Typical Dimensions of Galleys, Lavatories and Wardrobes Used in Several Airplanes

Airplane Type	N_{pax}	Range	Galleys No.	Galleys Dim.	Lavatories No.	Lavatories Dim.	Wardrobes No.	Wardrobes Dim.
Business Jets								
HFB 320	7	1,000	1	24x24	1	30x26	1	24x15
Falcon 20F	10	1,500	1	27x18	1	44x30	1	51x25
BAe HS 125	8	1,450	none		1	35x28	1	24x12
Regional Turboprops								
Nord 262	29	400	1	23x20	1	41x28	1	40x24
Gulfstr. I	19	2,100	1	34x25	1	67x37	1	36x32
Bae HS 748	44	1,000	1	37x14	1	53x35	none	
Fokker F27	48	1,100	1	43x35	1	47x46	1	31x16
DHC-7	44	800	1	26x24	1	46x30	1	26x24
Electra	95	2,300	2	46x26	4	46x41	2	46x34
Jet Transports								
VFW 614	40	700	1	35x28	1	55x32	1	65x40
Fokker F28	60	1,025	1	44x25	1	58x25	1	25x21
BAC 111	74	900	2	49x22	2	65x35	1	49x22
McDD DC9-10	80	1,100	1	48x33	2	48x48	2	48x21
B 737-200	115	1,800	1	55x43	2	43x34	1	55x43
Caravelle	118	1,000	1	51x43	2	55x43	2	24x17
D Mercure	140	800	none		2	47x34	2	49x16
B 727-200	163	1,150	2	51x32	3	43x39	Overhead	
A 300 B4	295	1,600	3	?	5	59x35	Overhead	
L 1011	330	2,700	1	240x162*	7	45x36	Overhead	
McDD DC10	380	3,000	1	240x162*	9	40x40	2	76x22
B 747	490	5,000	4	79x25	12	40x40	2	71x28

*under floor galley

All dimensions in inches. Data based on Ref.12, page 80.

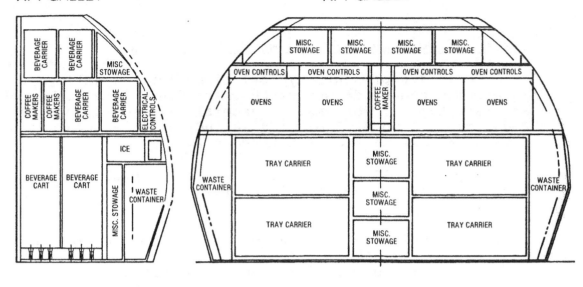

- Tourist class capacity
  - 2 beverage carts
  - 4 beverage carriers
- Main deck galleys
- Aft tray carrier service complexes
  - Galley complexes located at the aft end of the passenger cabin
  - Recessed floor gutter and door periphery connected to sump drain

- Tourist class capacity
  - 12 tray carriers-168 trays
  - 8 ovens-168 entrees

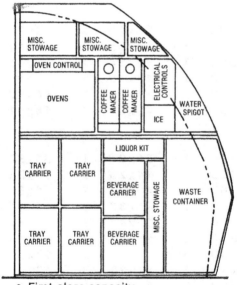

BOEING 757
COURTESY: BOEING

- First class capacity
  - 2 ovens-42 entrees
  - 4 tray carriers-56 trays

- Main deck galley
- Forward tray carrier service galley complex
  - Galley complex located at forward end of passenger cabin
  - Recessed gutters around galleys and periphery connected to overboard drain

Figure 3.33 Example of Narrow Body Galley Layout

Part III   Chapter 3   Page 74

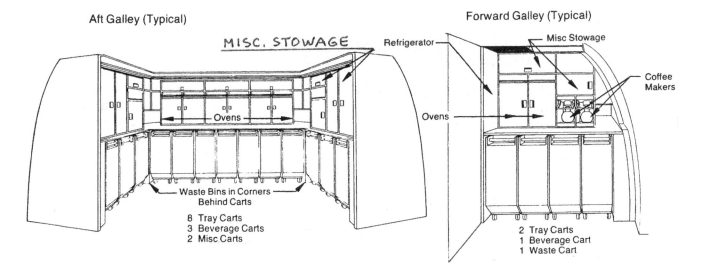

Passenger Services

Interior	Lavatory Ratio – Pass/Lav		Galley Volume – ft³/Pass (m³/Pass)		Coat Rod – in./Pass (cm/Pass)	Attendants
	F/C	T/C	F/C	T/C		
208 Mixed Class	18	48	7.72 (0.20)	2.10 (0.06)	0.16 (0.41)	6
211 Mixed Class	18	48	5.11 (0.14)	1.61 (0.05)	0.33 (0.84)	6
213 Mixed Class	18	49	5.11 (0.14)	2.04 (0.06)	0.33 (0.84)	6
216 Mixed Class	18	50	5.11 (0.14)	1.57 (0.04)	0.32 (0.81)	6
230 All Tourist		46		1.47 (0.04)		6
241 All Tourist		60		1.67 (0.05)		6
255 (7) All Tourist		64		1.85 (0.03)		6
255 (8) All Tourist		51		1.85 (0.03)	0.49 (1.24)	6
289(8) All Tourist		58		1.39 (0.03)		6

One Meal Service – 211 Passenger Arrangement (Typical)

Item	Total
Tray Carts 12 in. Wide (14 or 28 Tray Capacity)	10 *(252 Meals)
Bev Carts	4
Misc Carts	2
Waste Carts	1
Waste Bins	2
Coffee Makers	6
Ovens	7
Refrig. Compt.	5
Misc. Stowage	7 ft³

* 211 Meals Plus 19% Overage for Crew and Spare Meals

BOEING 767-200
COURTESY: BOEING
SEE P.61 FOR TYPICAL LOCATION OF PASSENGER SERVICES

Figure 3.34 Example of Wide Body Galley Layout

3.3.5 Layout of Cargo/Baggage Holds Including Data on Cargo Containers and Pallets

The amount of cargo which needs to be carried by a particular airplane depends strongly on the operator and his route system. Cargo is carried in passenger transports as well as in dedicated freighter type airplanes. In passenger transports the cargo is usually carried below the cabin floor. In dedicated freighter airplanes cargo is carried above the cabin floor as well as below the cabin floor. In addition there exist so-called 'quick-change' and 'combi' configurations of passenger transports where any mix of cargo and/or passengers can be carried.

Data on cargo and baggage volume requirements are presented in sub-sub-section 3.3.5.1. Shapes, sizes and weights of 'standard' containers and pallets are given in sub-sub-section 3.3.5.2.

Sub-sub-section 3.3.5.3 presents severel airplane configurations which have been used to allow for easy loading and unloading of cargo and luggage.

3.3.5.1 Cargo and baggage volume requirements

For cargo a workable average density is: 10 lbs/ft^3.

Loading efficiency for cargo varies from 75 to 100 percent depending on the method used in carrying the cargo.

For passenger baggage (luggage) the following numbers represent typical averages:

Baggage weight per passenger is:

    30-35 lbs for short haul flights
    40-45 lbs for long haul flights

Baggage density averages to: 12.5 lbs/ft^3.

Loading efficiency for baggage is 85 percent. This means that 15 percent of the design baggage volume is in fact lost!

Many passengers (particularly business people) insist on carrying their own luggage on board. Although the airlines have established limits on the amount of 'carry-on' luggage, they also have provided convenient overhead storage bins in most larger transport airplanes.

Figure 3.35 shows an example of such overhead storage facilities.

The cargo/baggage hold in the belly of transport airplanes should have an effective (i.e. usable) height of at least 35 inches if loading personnel are required to move about the belly holds. If no personnel need to be in the belly holds, a height of 20 inches may suffice. The need for a workable height in belly holds may force a non-circular cross section on the designer, such as a so-called 'double-bubble' cross section (See Fig. 3.36).

In laying out the belly hold it is essential that the impact of all possible loading scenarios on center of gravity location be accounted for. Chapter 10 of Part II contains a method for determining the 'weight and balance' consequences of different loading scenarios.

Many airplanes have belly holds forward of the wing as well as behind the wing for precisely this reason.

3.3.5.2 Data on standard containers and pallets

The use of pallets and containers greatly reduces the loading time of baggage and cargo. On nearly all medium to long haul operations, cargo as well as luggage is preloaded in containers and/or on pallets. The industry has developed a range of so-called standard containers and pallets as shown in Figure 3.37. Ref.12, p.85 contains data on dedicated freight containers and pallets.

3.3.5.3 Typical loading/unloading configurations

Figures 3.47 and 3.54 show typical dimensions of belly holds and freight floor layouts used in different airplane types. Examples of the door and ramp configurations which have been used in a number of airplanes are given in Figures 3.38 through 3.42.

It is essential that a certain amount of clearance be 'designed into' any freight/cargo door. Adequate clearance means a minimum of _five_ inches in height and a minimum of _nine_ inches in width.

To facilitate loading and unloading many floors are equipped with roller systems. The 747F of Figure 3.38 is a typical example.

Freight floors need to be equipped with tie-down provisions to prevent the cargo from sliding and thereby

changing the c.g. It is also necessary to build in
provisions to prevent cargo from sliding into passengers
or crew in the case of a crash. A 9g limit load
requirement is therefore imposed on all tie-down or
catch-net restraint systems.

Typical structural arrangements of freight floors
are given in Section 3.5.

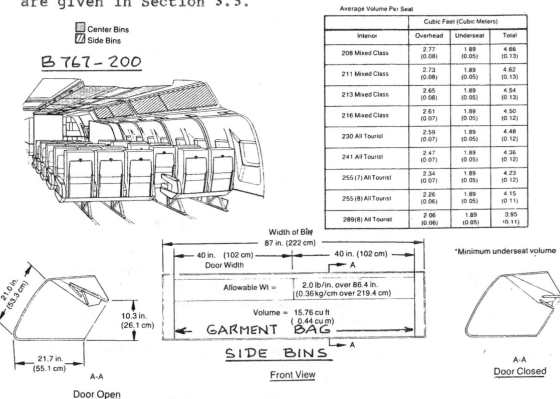

Figure 3.35 Example of Overhead Storage

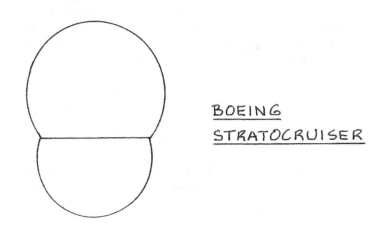

Figure 3.36 Example of a Double-Bubble Cross Section

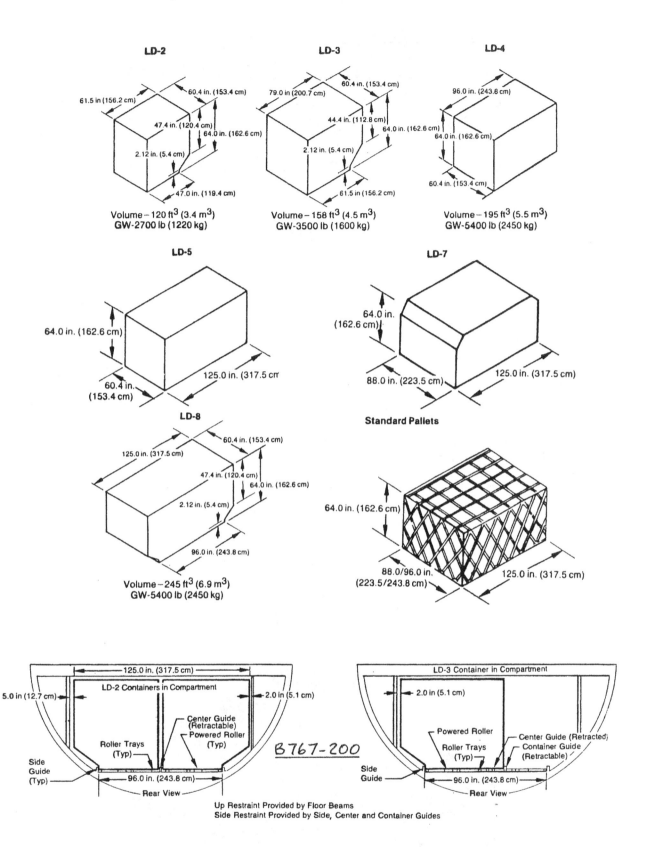

Figure 3.37 Typical Pallet and Container Sizes

Part III Chapter 3 Page 79

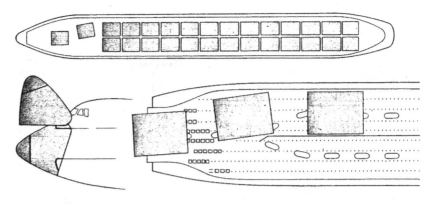

Figure 3.38 Cargo Deck Arrangement for the Boeing 747F

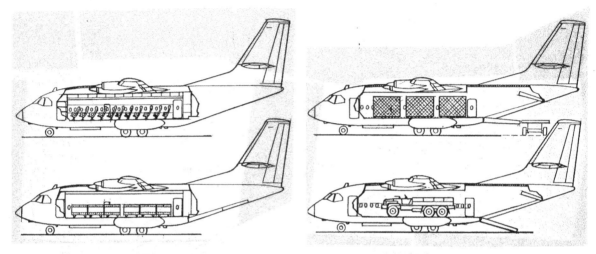

Figure 3.39 Passenger or Cargo Arrangement CASA 401

Figure 3.40 Loading/Unloading of a Fairchild C119

Figure 3.41 Example of a Detachable Pod Concept

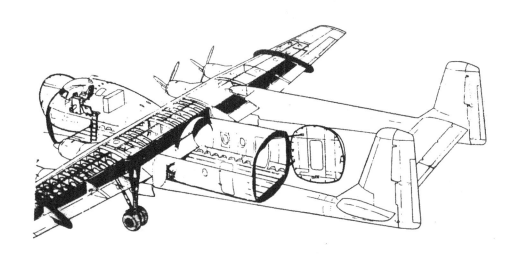

Figure 3.42 Rear and Front Loading Doors: AW650 Argosy

3.3.6 Inspection, Maintenance and Servicing Considerations

It is essential that those areas of the fuselage which require frequent access for inspection, replacement of parts or repairs be easily accessible. Design engineers need to work with airplane mechanics to find out the conditions under which typical inspection, maintenance and servicing procedures are being carried out!

The author has witnessed mechanics in freezing conditions trying to open an undersized access cover. Having finally achieved this after having to take off their gloves (!) they discovered that it was impossible to look inside without putting down their parka top, exposing their faces to sub-zero conditions.

ENGINEERS: design for maintenance and inspectability under climatic conditions which can be expected to prevail.

In transport operations it is essential that the 'turn-around' time be minimized as much as possible. This means that a large number of vehicles need to have simultaneous access to the airplane when parked at the gate. Typical of the services which need to be performed on an airplane simultaneously are:

1. load and unload passengers
2. refuel and reoil
3. replenish potable water
4. clean airplane cabin
5. remove food and beverage left-overs and load fresh food and beverage supplies
6. service lavatories

The required trucks and other servicing vehicles, loading and unloading ramps, stairs etc. must not interfere with each other or with protruding components of the airplane (for example: wing tip booms are an invitation to collisions). Careful attention needs to be given to the layout of all service doors, cargo doors and access covers.

Figures 3.43 and 3.44 present diagrams used to demonstrate to potential customers that an airplane design satisfies these requirements.

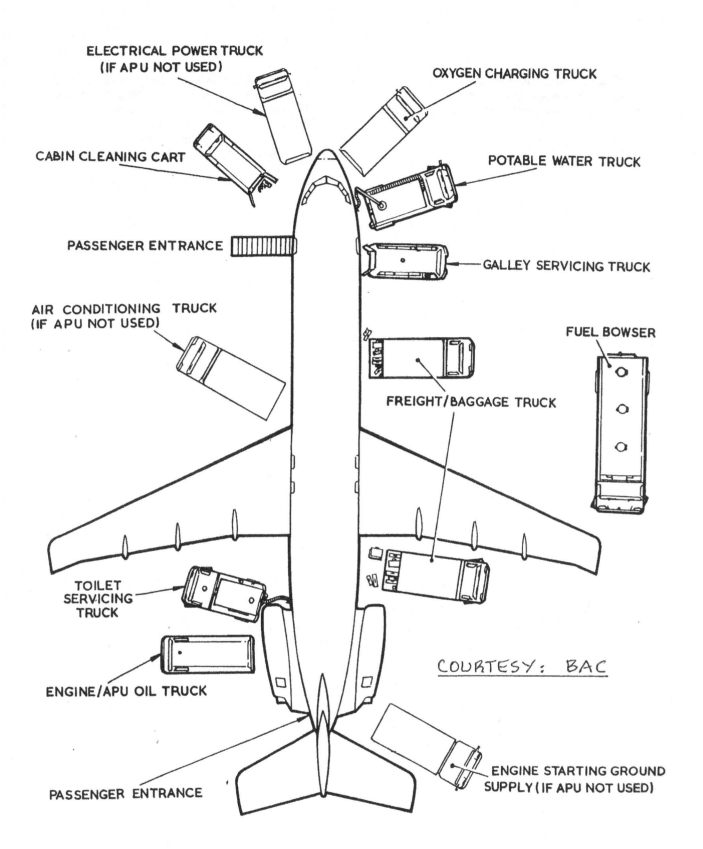

Figure 3.43 Terminal Servicing of a BAC 111

1. Radar Antenna, Glide Scope
2. Nose Gear Hinge Attach Bolt
3. Air Conditioner Condenser Blower and Relay
4. Avionics Compartment
5. Rudder Bell Crank
6. Rudder Tab Actuator
7. Elevator Bell Crank
8. Rudder Control Horn, Rudder Stops
9. Landing Gear Doors
10. Control Cables and Pulleys below Pedestal
11. Aft Fuselage Access, Outflow and Safety Valves
12. Oxygen Filler Gage and Service Valve
13. Air Conditioner Condenser
14. Horizontal Stabilizer Attachment
15. Elevator Push Rods, Pulleys, Tab Cables
16. Horizontal Stabilizer Attachment
17. Elevator Tab Actuators and Stops

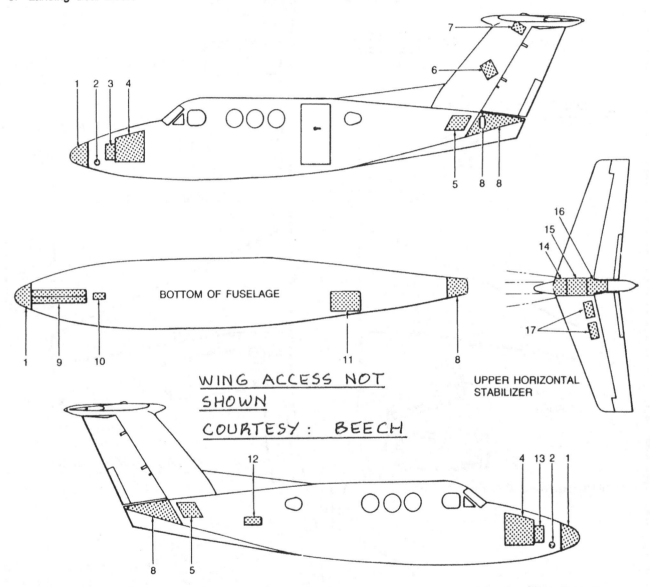

WING ACCESS NOT SHOWN

COURTESY: BEECH

Figure 3.44 Access Diagram for the Beech King Air F90

3.4 DESIGN DATA FOR FUSELAGE CROSS SECTIONS, CABIN AND CARGO HOLD LAYOUTS, WINDOW AND DOOR LAYOUTS

During the fuselage layout design process decisions need to be made regarding the basic fuselage cross section, the overall cabin and cargo hold arrangement, and the window and door layout. Steps 4.1, through 4.6 of Chapter 4, Part II deal with these decisions. To arrive at these decisions it is desirable to have available a set of design data showing solutions arrived at by various airframe manufacturers. The purpose of this section is to provide these design data.

The data are organized as follows:

For fuselage cross sections:

Figure 3.45:	Fuselage Cross Sections for Business Jets
Figure 3.46:	Fuselage Cross Sections for Regional Turboprops
Figures 3.47a-c:	Fuselage Cross Sections for Jet Transports

For cabin and cargo hold layouts:

Figures 3.48a-d:	Cabin and baggage hold dimensions for single engine prop. driven airplanes
Figure 3.49:	Cabin dimensions for single engine piston/propeller driven airplanes
Figure 3.50:	Cabin dimensions for twin engine piston/propeller driven airplanes
Figures 3.51a-f:	Cabin dimensions for twin engine turbo-propeller driven airplanes
Figures 3.52a-e:	Cabin dimensions for business jets
Figures 3.53a-c:	Fuselage and cargo hold dimensions for turbopropeller driven transports
Figures 3.54a-l:	Fuselage and cargo hold dimensions for jet transports

For window and door layouts:

Figures 3.45 through 3.54 all show window, door (and emergency exit) arrangements.

NOTE: The design information in this section is based on data accumulated over several years from the following magazines:

1. Flight International (British weekly)
2. Business and Commercial Aviation (US monthly)

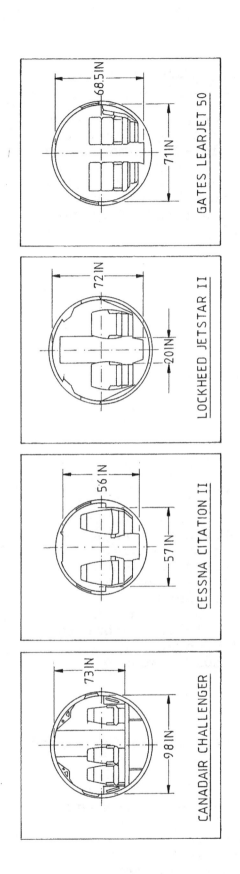
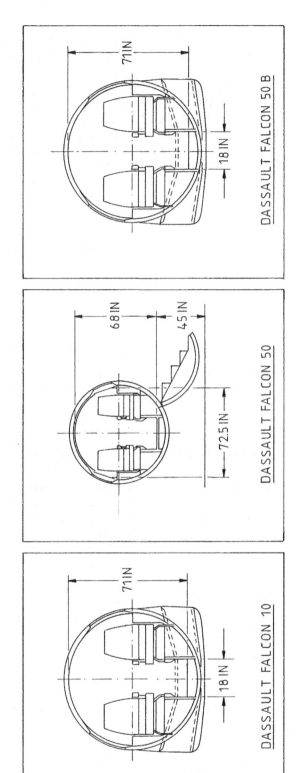

Figure 3.45 Fuselage Cross Sections for Business Jets

Part III Chapter 3 Page 86

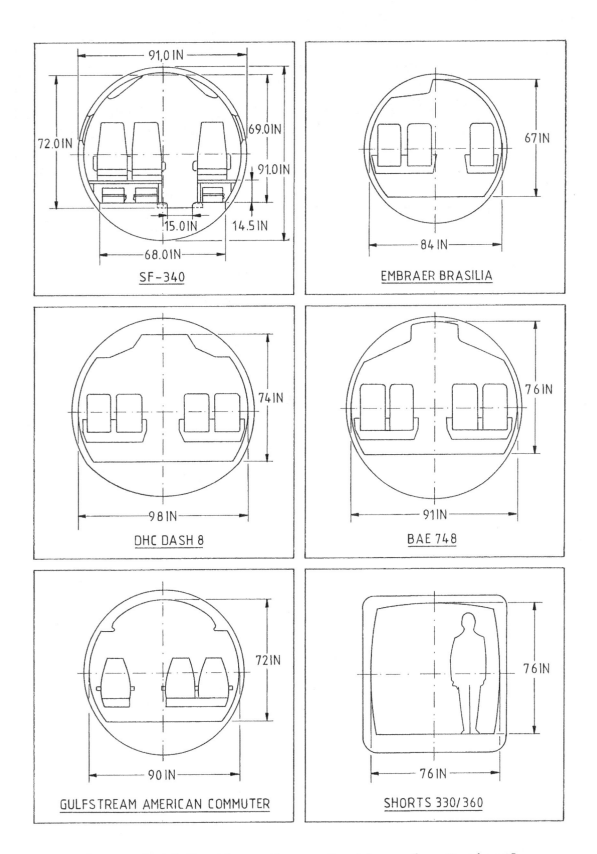

Figure 3.46 Fuselage Cross Sections for Regional Turboprops

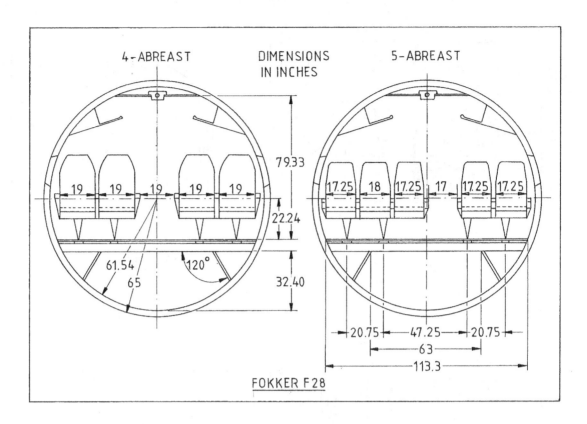
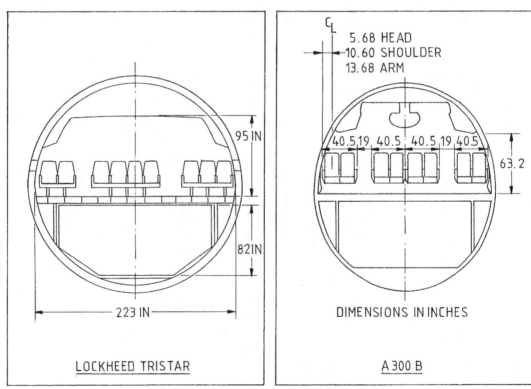

Figure 3.47a Fuselage Cross Sections for Jet Transports

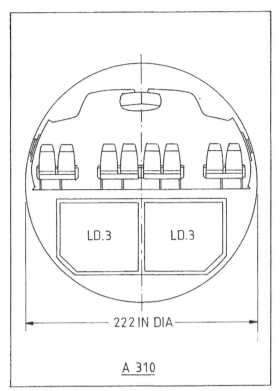
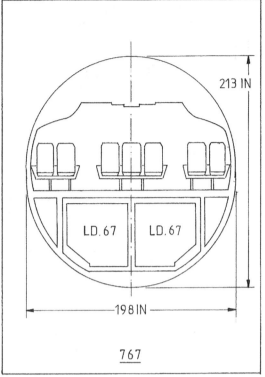
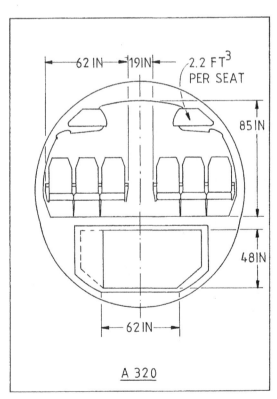
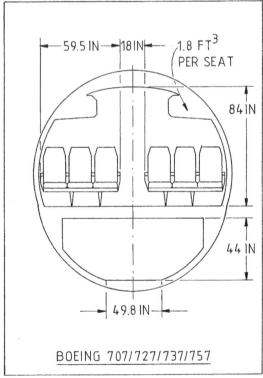

Figure 3.47b Fuselage Cross Sections for Jet Transports

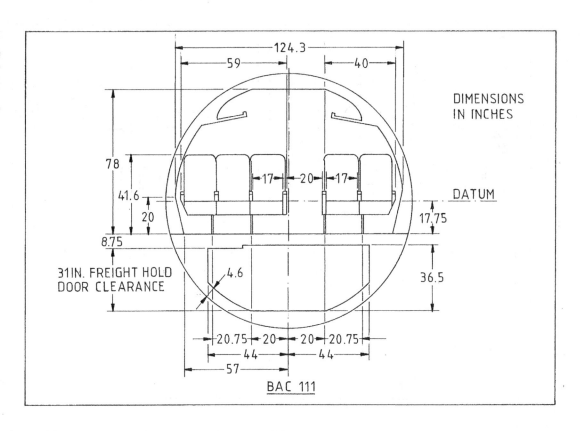

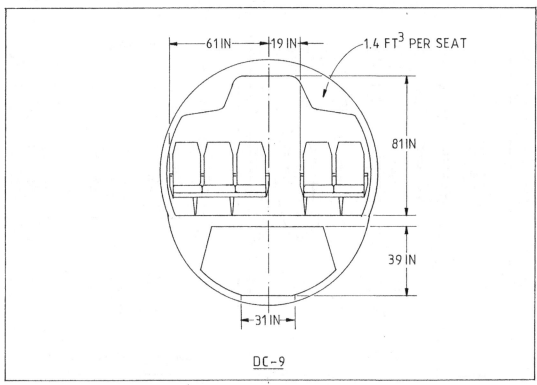

Figure 3.47c Fuselage Cross Sections for Jet transports

Part III   Chapter 3   Page 90

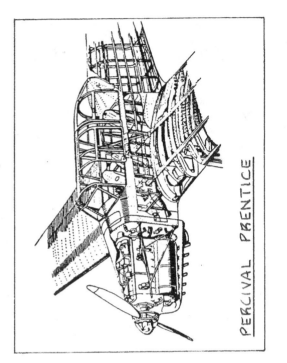
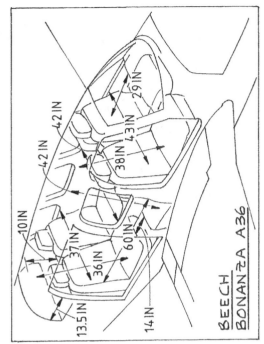
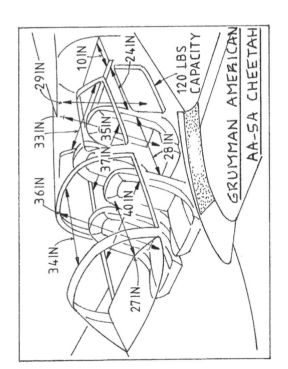
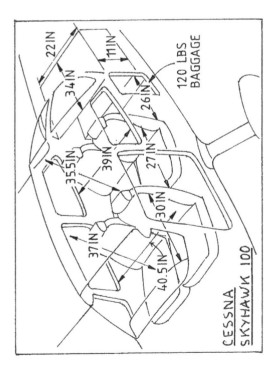

Figure 3.48a Cabin and Baggage Hold Dimensions for Single Engine Piston/Propeller Driven Airplanes

Part III　　　　　　　　Chapter 3　　　　　　　　Page 91

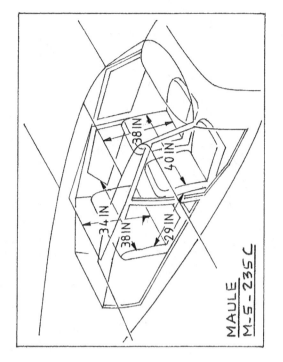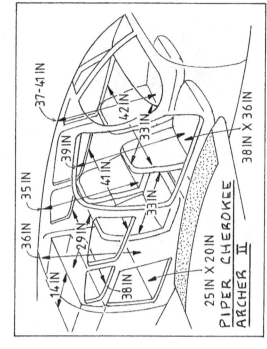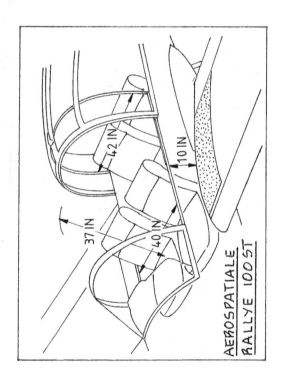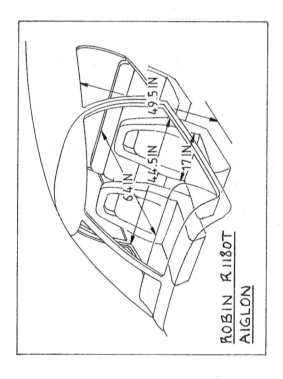

Figure 3.48b Cabin and Baggage Hold Dimensions for Single Engine Piston/Propeller Driven Airplanes

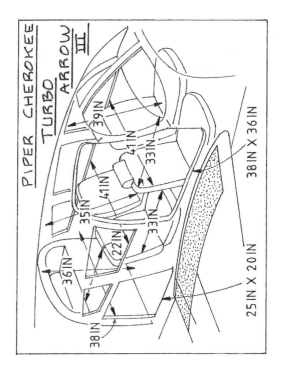
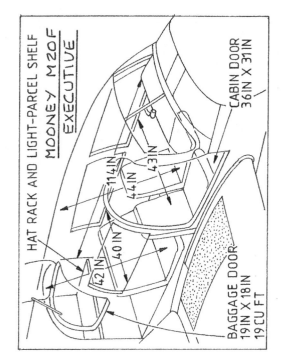
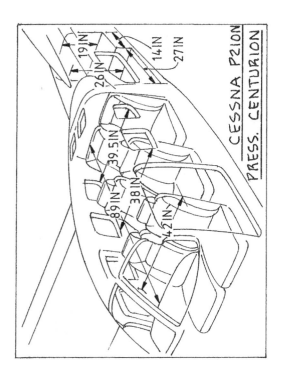
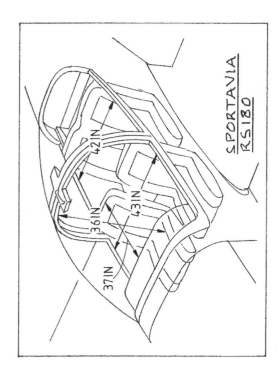

Figure 3.48c Cabin and Baggage Hold Dimensions for Single Engine Piston/Propeller Driven Airplanes

Part III　　　　Chapter 3　　　　Page 93

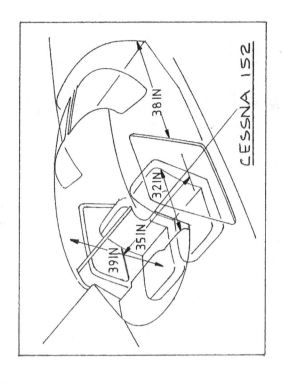
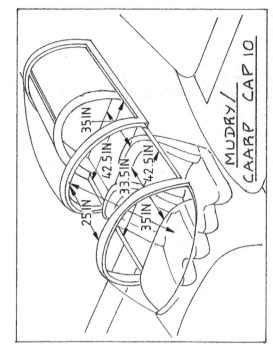
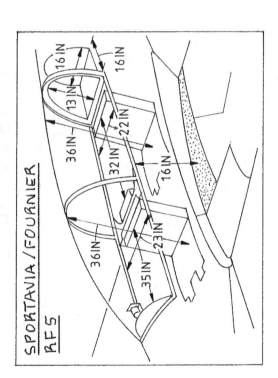
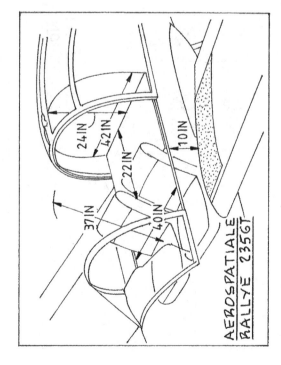

Figure 3.48d Cabin and Baggage Hold Dimensions for Single Engine Piston/Propeller Driven Airplanes

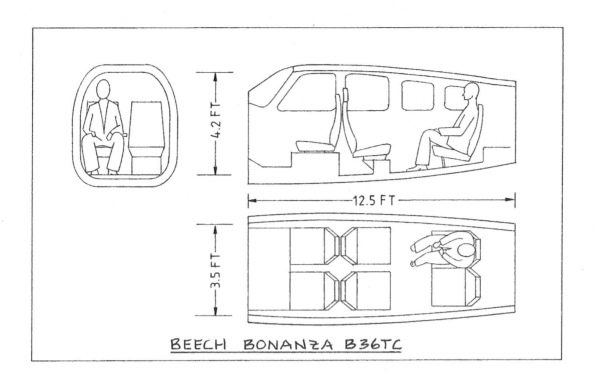

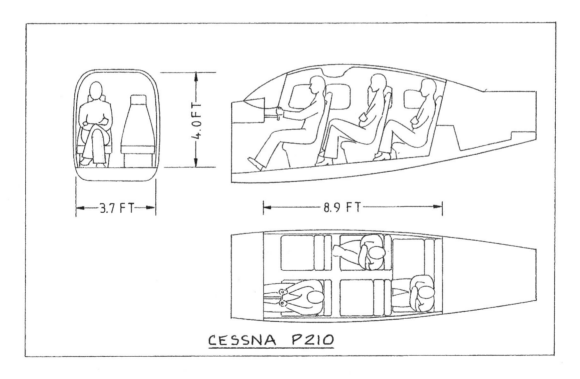

Figure 3.49 Cabin Dimensions for Single Engine Piston/Propeller Driven Airplanes

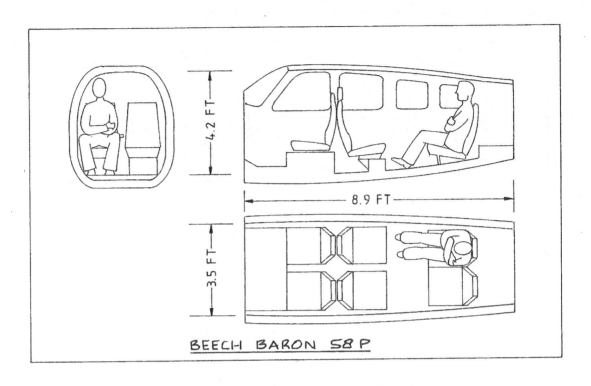

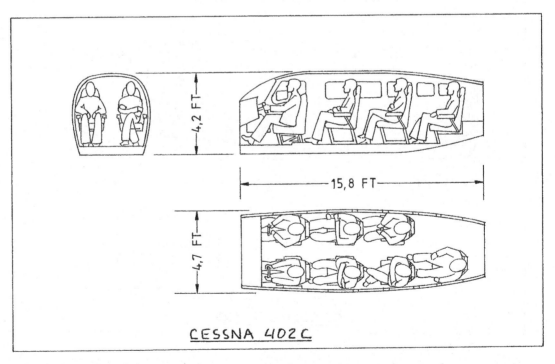

Figure 3.50 Cabin Dimensions for Twin Engine Piston/Propeller Driven Airplanes

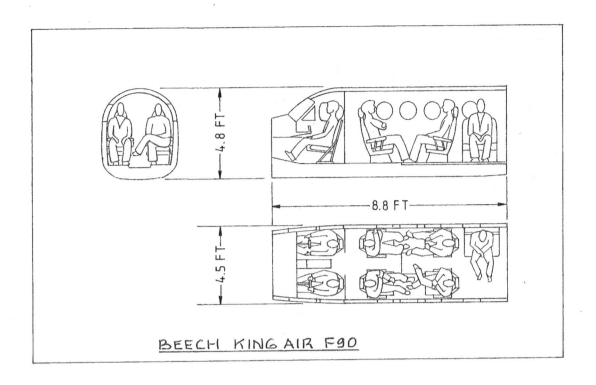

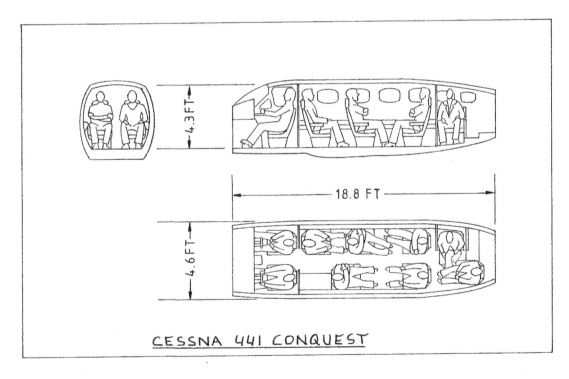

Figure 3.51a Cabin Dimensions for Twin Engine Turbo-Propeller Driven Airplanes

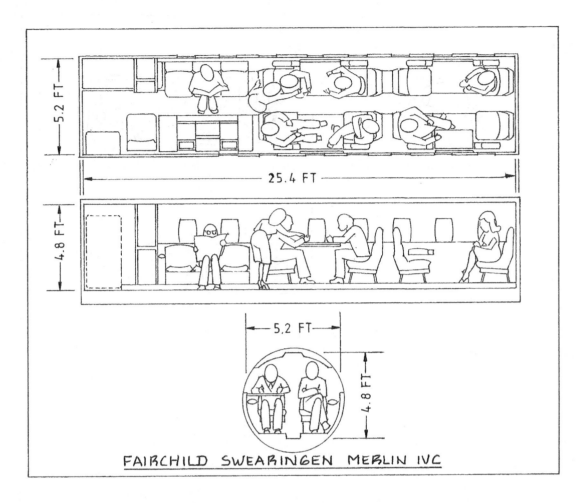

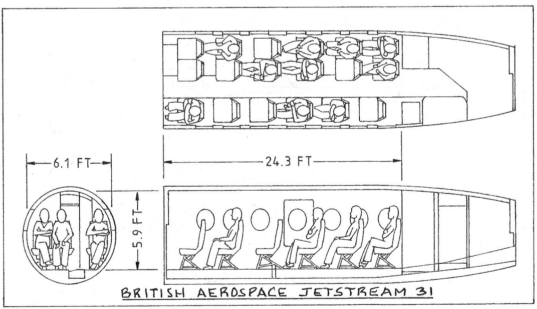

Figure 3.51b Cabin Dimensions for Twin Engine Turbo-Propeller Driven Airplanes

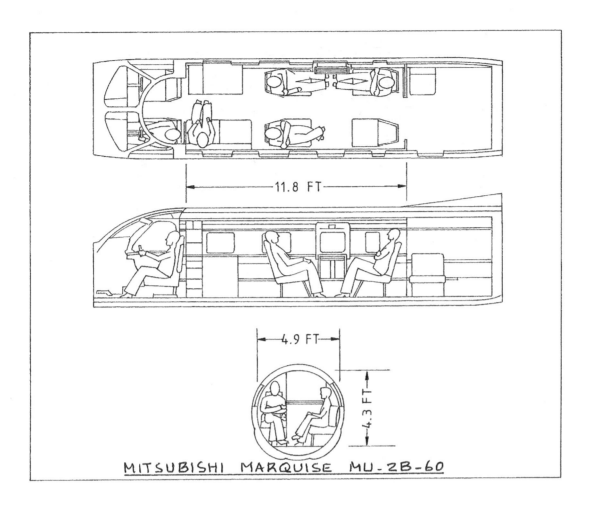

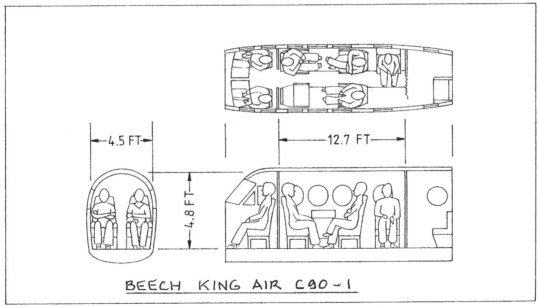

Figure 3.51c Cabin Dimensions for Twin Engine Turbo-Propeller Driven Airplanes

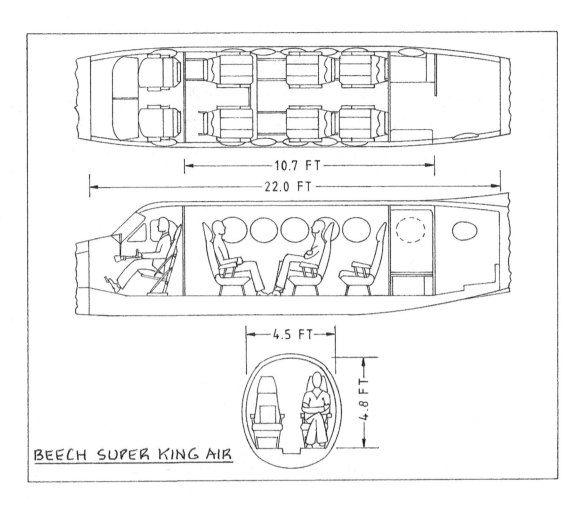

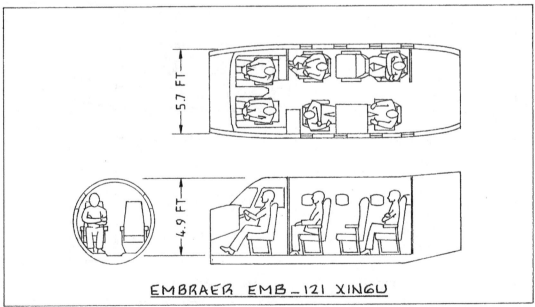

Figure 3.51d Cabin Dimensions for Twin Engine Turbo-Propeller Driven Airplanes

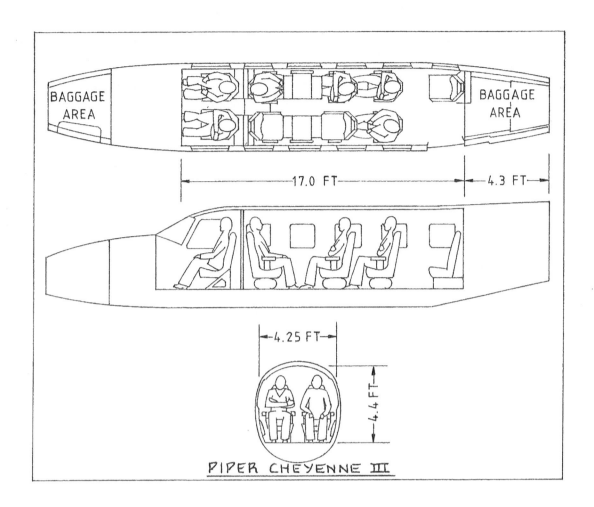

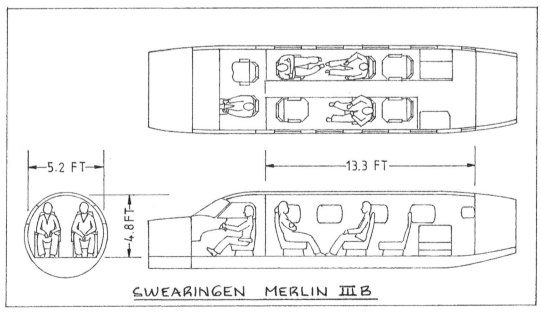

Figure 3.51e Cabin Dimensions for Twin Engine Turbo-Propeller Driven Airplaes

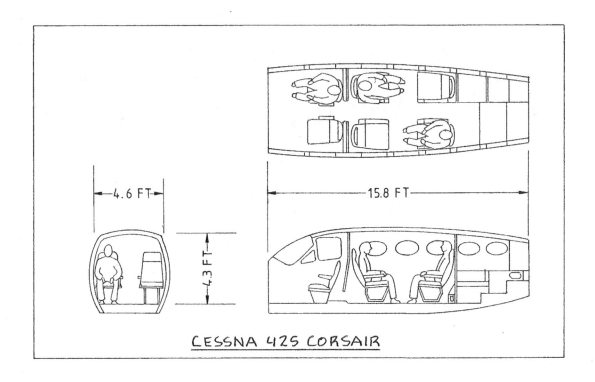

CESSNA 425 CORSAIR

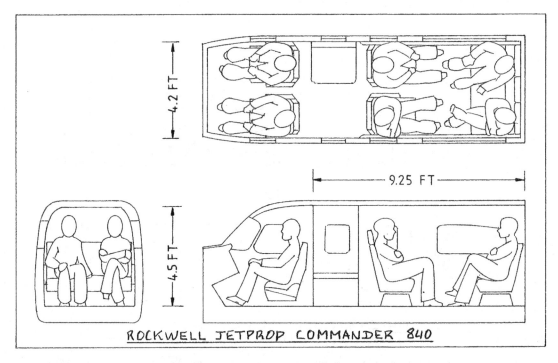

ROCKWELL JETPROP COMMANDER 840

Figure 3.51f Cabin Dimensions for Twin Engine Turbo-Propeller Driven Airplanes

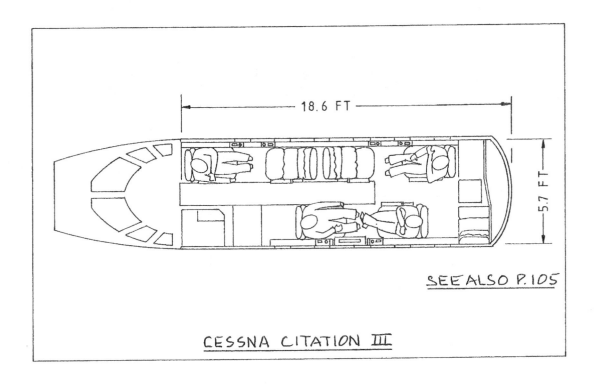

Figure 3.52a Cabin Dimensions for Business Jets

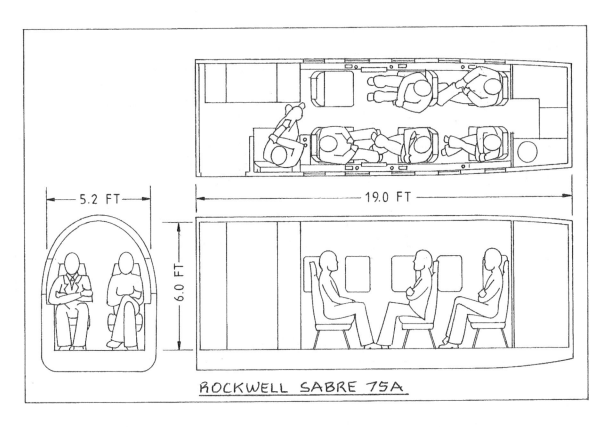

Figure 3.52b Cabin Dimensions for Business Jets

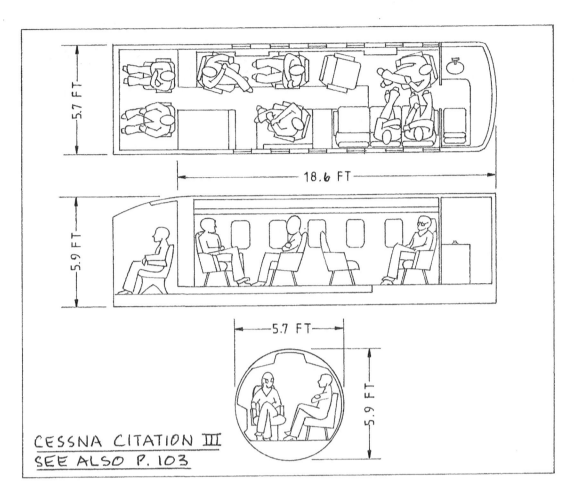

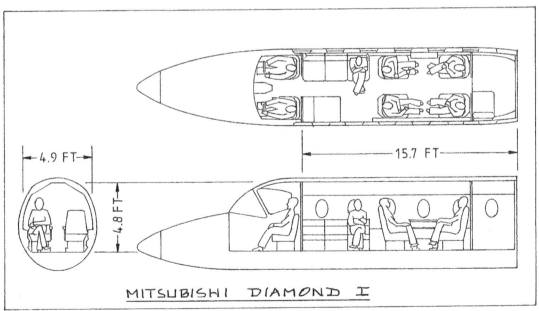

Figure 3.52c Cabin Dimensions for Business Jets

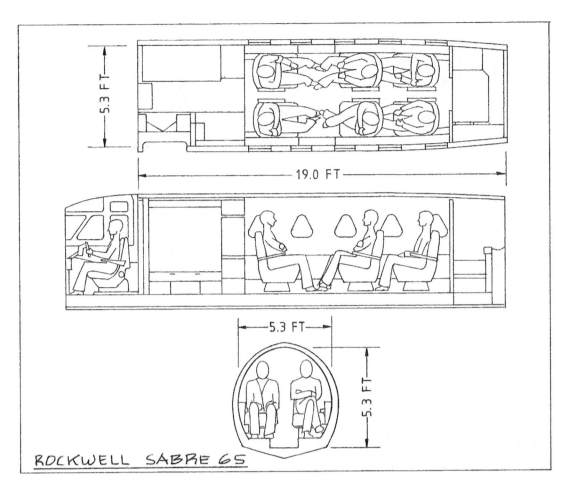
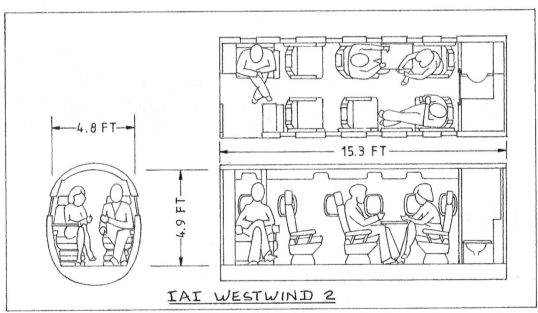

Figure 3.52d Cabin Dimensions for Business Jets

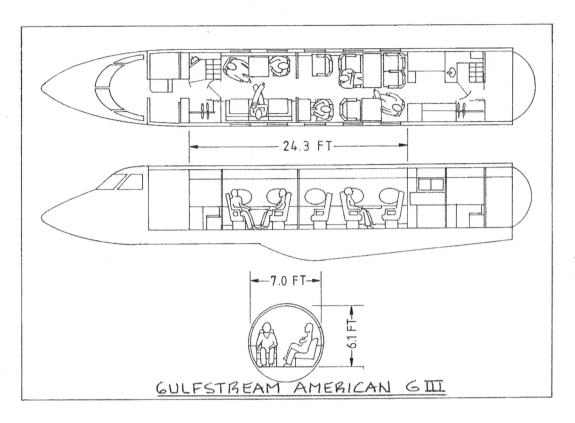

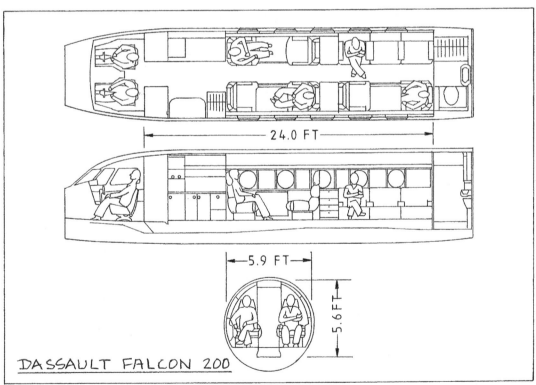

Figure 3.52e Cabin Dimensions for Business Jets

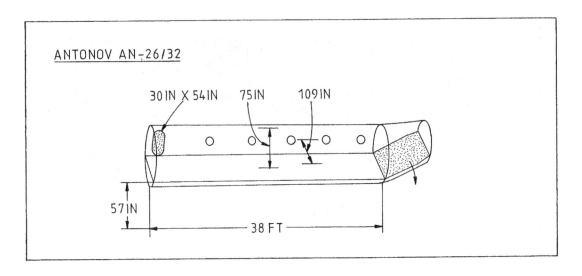

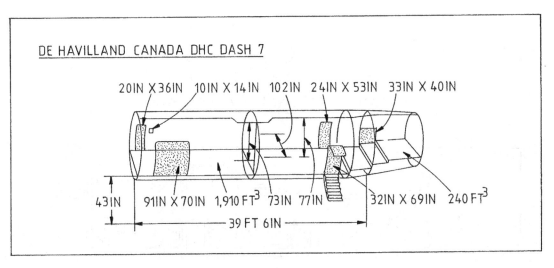

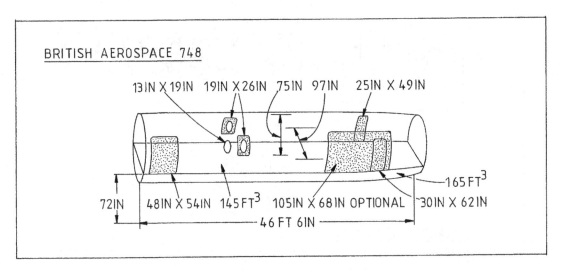

Figure 3.53a Fuselage and Cargo Hold Dimensions for Turbopropeller Driven Transports

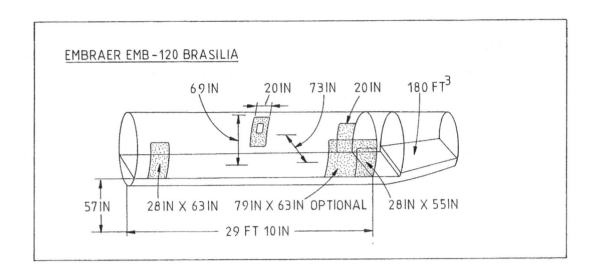
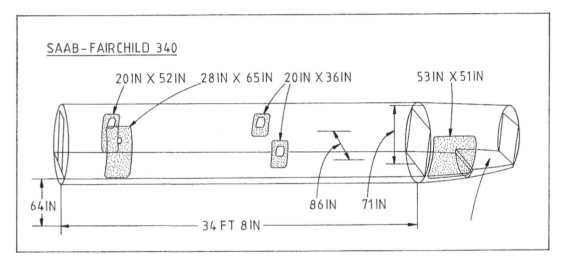
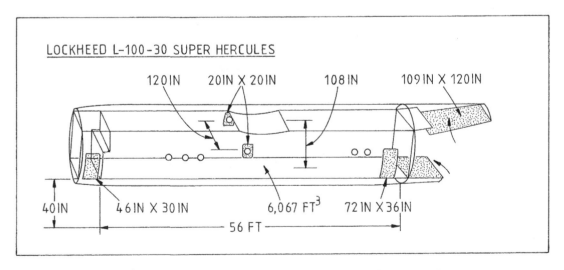

Figure 3.53b Fuselage and Cargo Hold Dimensions for Turbopropeller Driven Transports

Part III　　　　　　　　　　Chapter 3　　　　　　　　　　Page 109

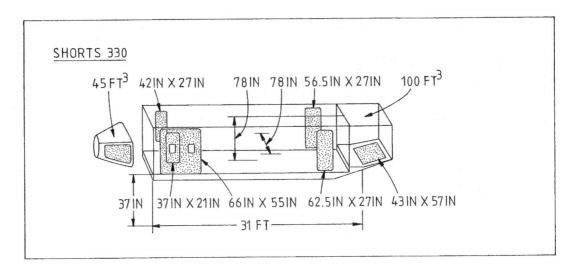
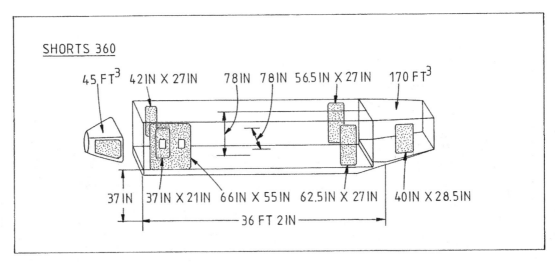
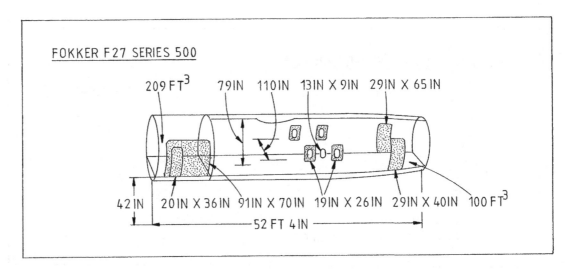

Figure 3.53c Fuselage and Cargo Hold Dimensions for Turbopropeller Driven Transports

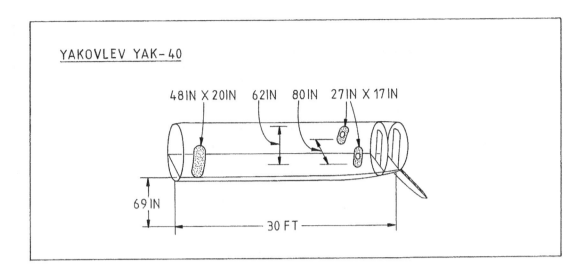

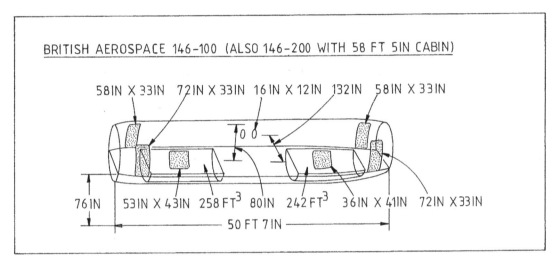

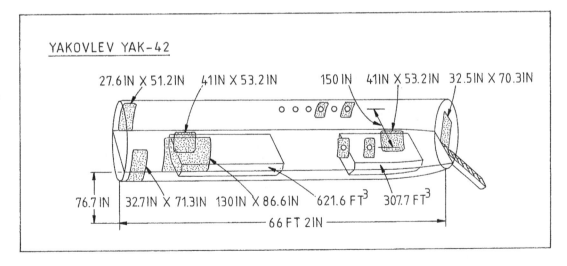

Figure 3.54a Fuselage and Cargo Hold Dimensions for Jet Transports

Part III   Chapter 3   Page 111

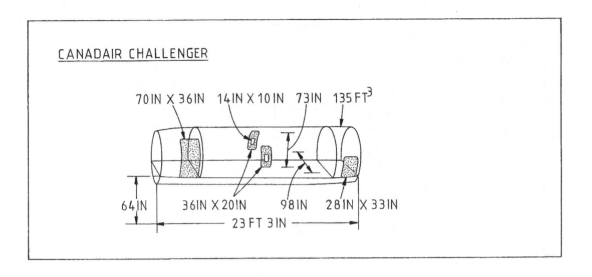

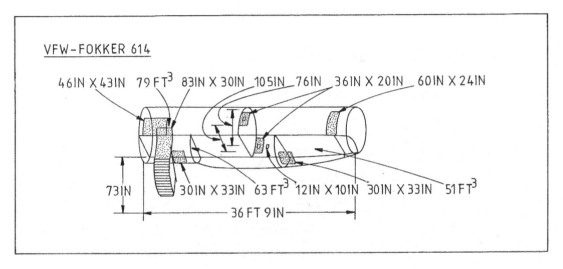

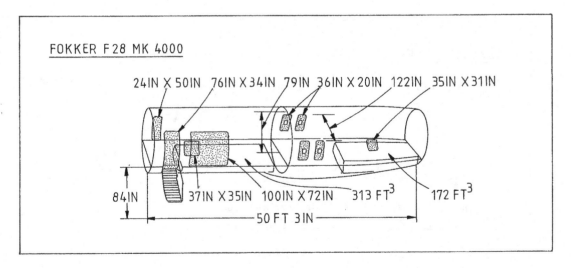

Figure 3.54b Fuselage and Cargo Hold Dimensions for Jet Transports

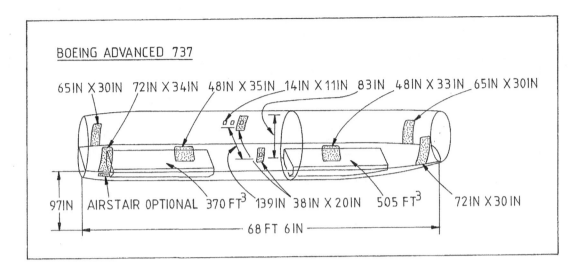

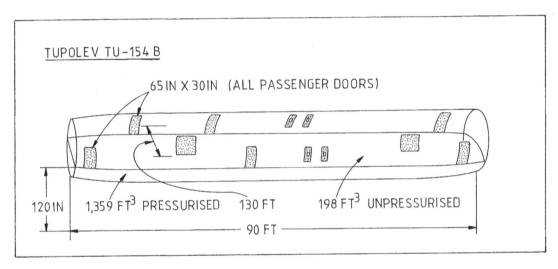

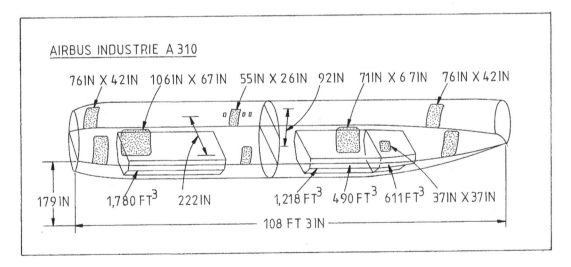

Figure 3.54c Fuselage and Cargo Hold Dimensions for Jet Transports

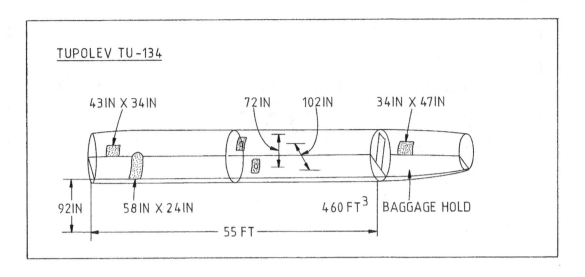

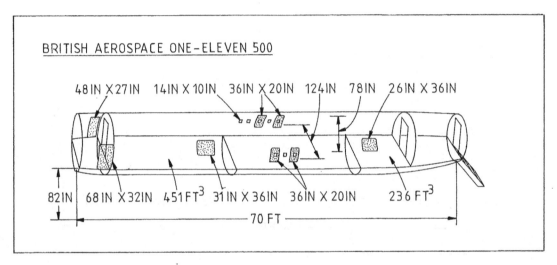

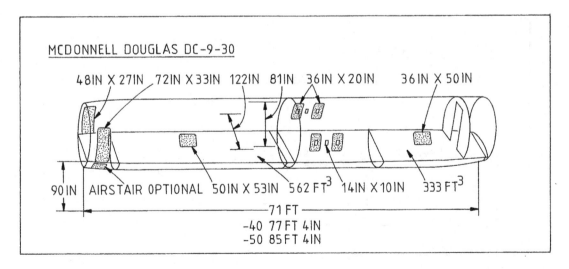

Figure 3.54d Fuselage and Cargo Hold Dimensions for Jet Transports

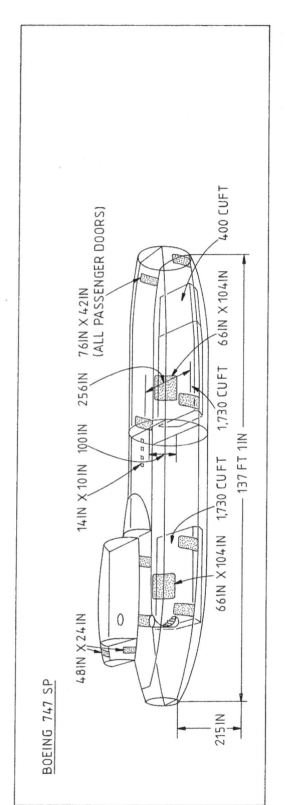
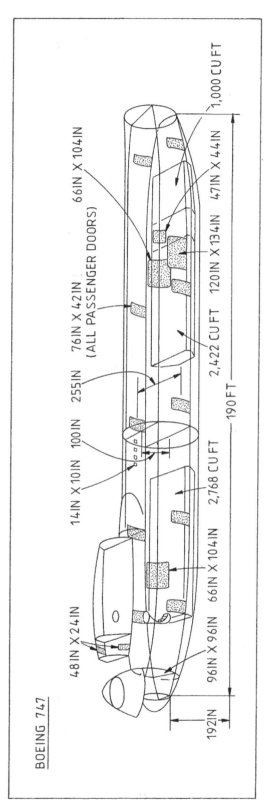

Figure 3.54e Fuselage and Cargo Hold Dimensions for Jet Transports

Part III   Chapter 3   Page 115

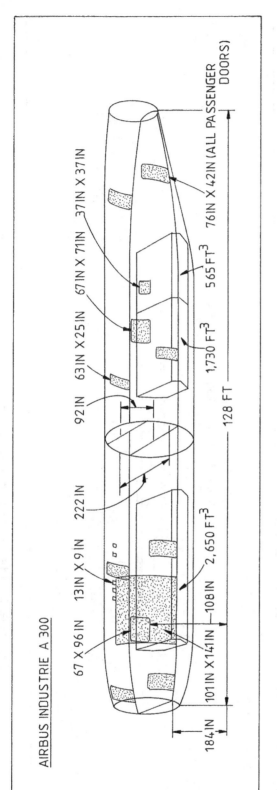
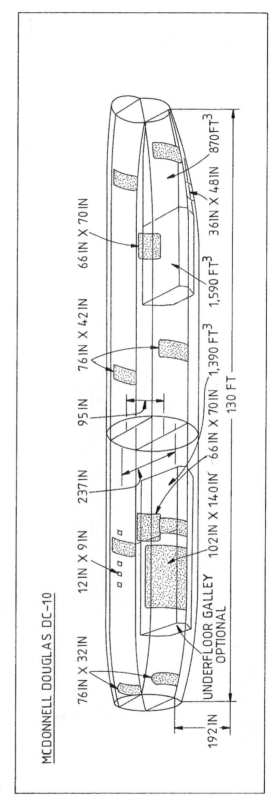

Figure 3.54f Fuselage and Cargo Hold Dimensions for Jet Transports

Part III　　　　　　　Chapter 3　　　　　　　Page 116

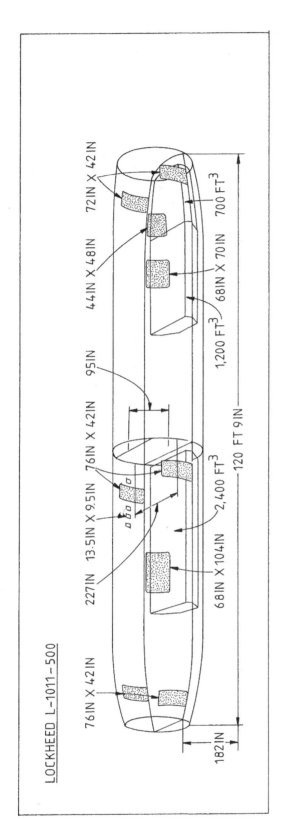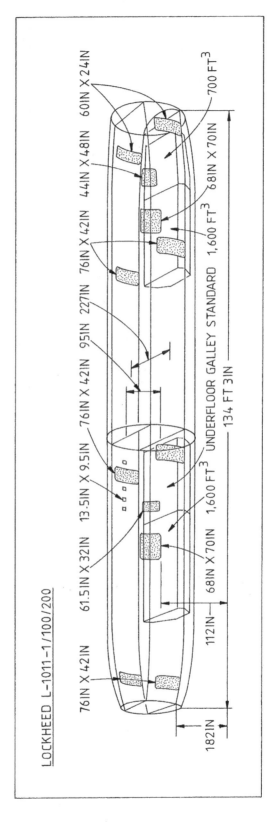

Figure 3.54g Fuselage and Cargo Hold Dimensions for Jet Transports

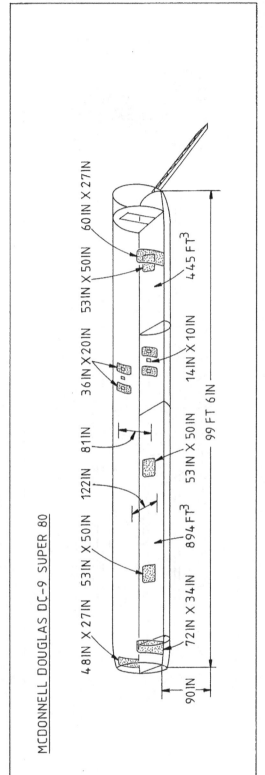
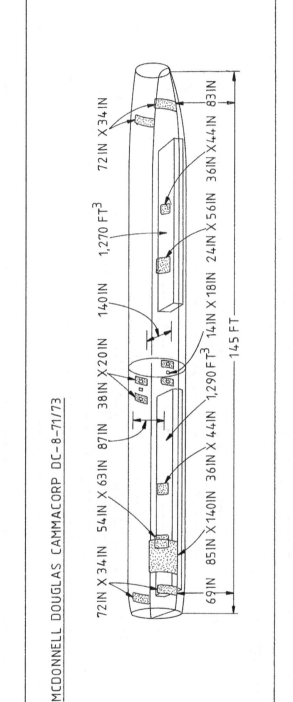

Figure 3.54h Fuselage and Cargo Hold Dimensions for Jet Transports

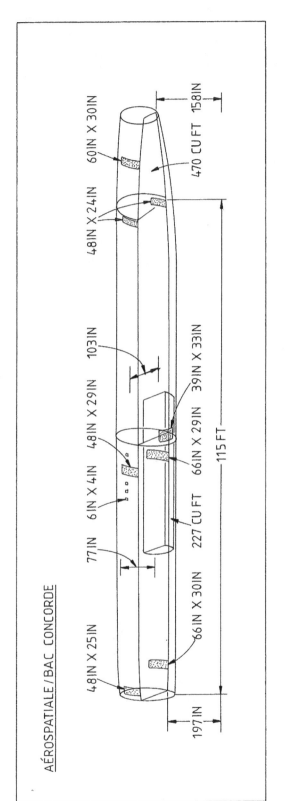
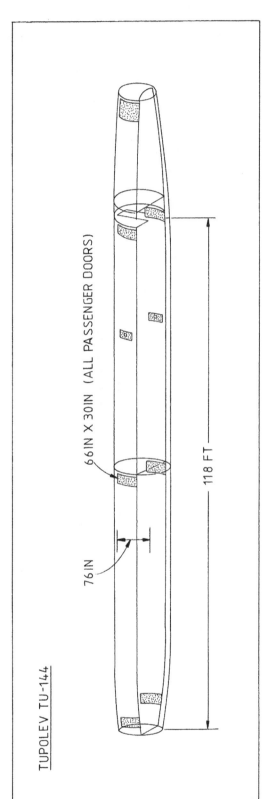

Figure 3.54i Fuselage and Cargo Hold Dimensions for Jet Transports

Part III Chapter 3 Page 119

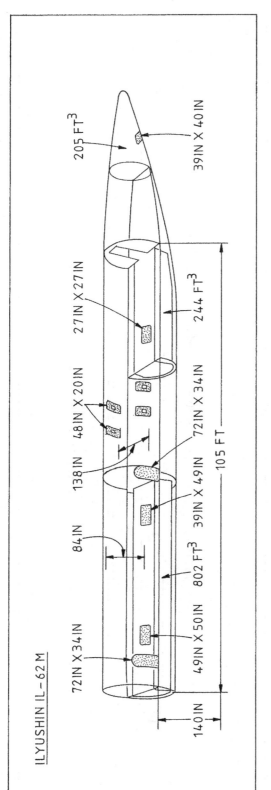
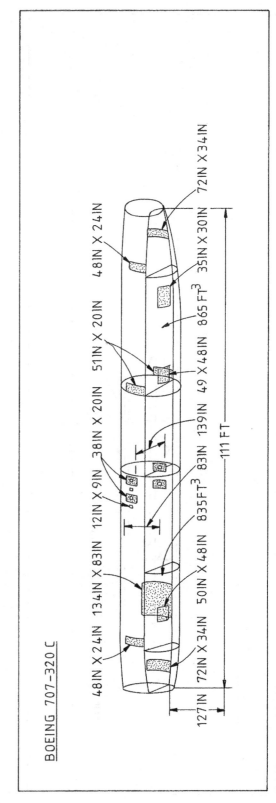

Figure 3.54j Fuselage and Cargo Hold Dimensions for Jet Transports

Part III　　　　　　Chapter 3　　　　　　Page 120

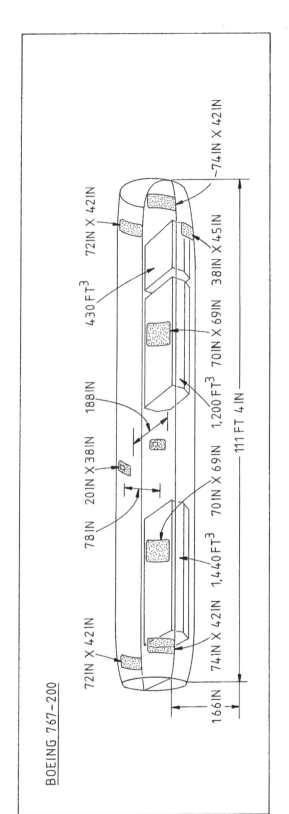 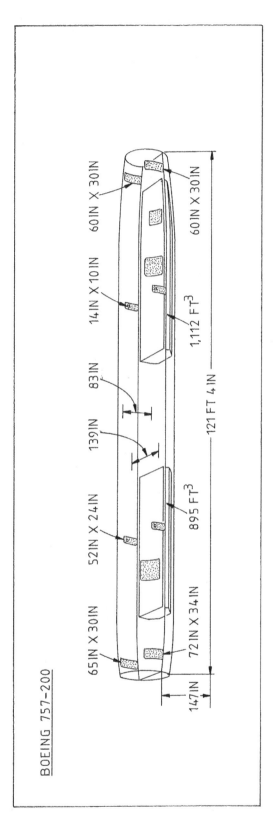

Figure 3.54k Fuselage and Cargo Hold Dimensions for Jet Transports

Part III Chapter 3 Page 121

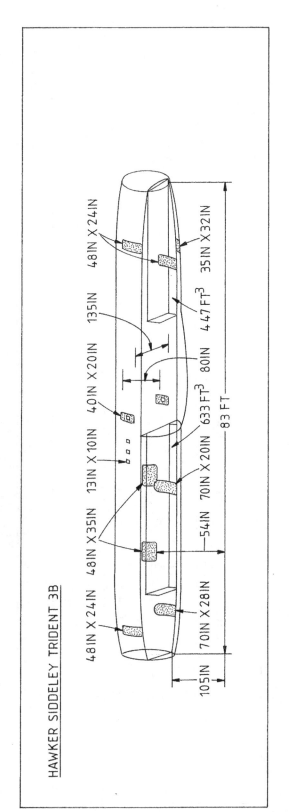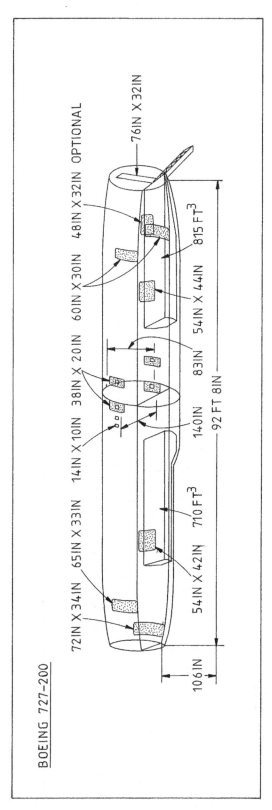

Figure 3.541 Fuselage and Cargo Hold Dimensions for Jet Transports

3.5 STRUCTURAL DESIGN CONSIDERATIONS AND EXAMPLES OF STRUCTURAL LAYOUT DESIGN OF FUSELAGES

From a structural design viewpoint the fuselage of most airplanes can be viewed as that component to which the wing, the empennage and in some instance the landing gear and the nacelles are attached.

The fuselage structure therefore must be designed so that the following types of load can be taken without major structural failures and without major structural fatigue problems:

1. Empennage loads due to trim, maneuvering, turbulence and gusts.

2. Pressure loads due to cabin pressurization.

3. Landing gear loads due to landing impact, taxiing and ground maneuvering.

4. Loads induced by the propulsion installation when the latter is attached to the fuselage.

References 11 and 19 define the loads for which airplanes must be designed. References 20 and 21 contain methods for structural sizing of structural components.

In 'survivable' crashes the fuselage must provide sufficient protection to prevent injuries to its occupants. References 22 and 24 through 27 contain information on the subject of 'design for reasonable crashworthiness'.

Cabin materials used for sound-proofing, decorative panels, seats, trays and carpets must not generate toxic fumes when exposed to fire. Ref.11 contains the regulations which must be observed in this regard.

Once the initial fuselage layout is completed (Chapter 4, Part II) and the dimensioned threeview of Step 15 (Part II) has been drawn up, it is possible to prepare an initial structural arrangement for the fuselage. Chapter 7 gives a step-by-step method for preparing the overall preliminary structural arrangement for an airplane.

The purpose of this section is to provide the reader with some ground rules for and some examples of typical structural arrangements employed in airplane fuselages.

The material is organized as follows:

3.5.1 Typical frame depths, frame spacings and longeron spacings
3.5.2 Examples of fuselage structural arrangements
3.5.3 Examples os fuselage shell layout
3.5.4 Examples of door and stair design
3.5.5 Examples of cockpit and cabin window design
3.5.6 Examples of floor design

3.5.1 Typical Frame Depths, Frame Spacings and Longeron Spacings

In laying out an initial structural arrangement for a fuselage the following design 'groundrules' are useful:

<u>Frame Depths:</u>

For small commercial airplanes: 1.5 inches.
For fighters and trainers: 2.0 inches.
For large transports: $0.02d_f + 1.0$ inches.

<u>Frame Spacings:</u>

For small commercial airplanes: 24 - 30 inches.
For fighters and trainers: 15 - 20 inches.
For large transports: 18 - 22 inches.

<u>Longeron Spacings:</u>

For small commercial airplanes: 10 - 15 inches.
For fighters and trainers: 8 - 12 inches.
For large transports: 6 - 12 inches.

Figure 3.55 defines these structural parameters. The actual numbers used for these structural design parameters depend on the skin thickness. In the case of composite structures these guidelines are not valid.

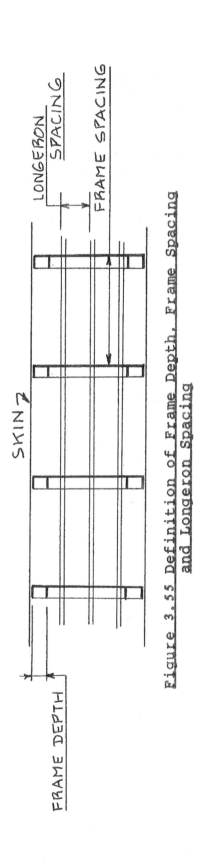

Figure 3.55 Definition of Frame Depth, Frame Spacing and Longeron Spacing

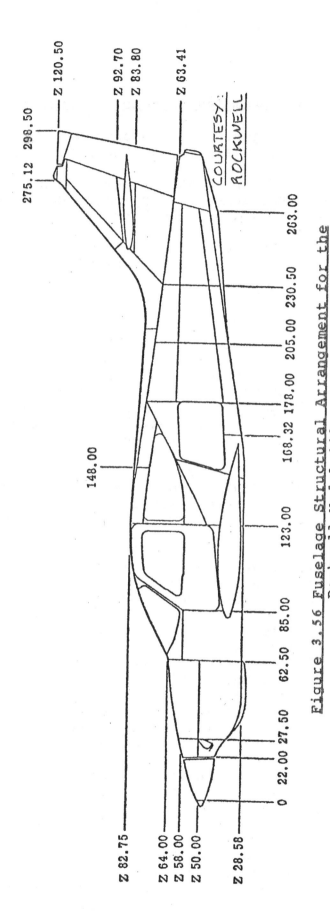

Figure 3.56 Fuselage Structural Arrangement for the Rockwell Model 112

3.5.2 Examples of Fuselage Structural Arrangements

Figures 3.56 through 3.60 show examples of structural arrangements for the fuselages of several airplanes.

Figure 3.56 represents a small, unpressurized, general aviation airplane. Note the wide spacing between frames and between longerons.

Figure 3.57 represents a small, unpressurized, twin engine turbopropeller airplane. There are no fuselage longerons. Their function has been taken over by bonded, corrugated skin panels.

Figure 3.58 represents a pressurized business jet. Note the tight spacing between fuselage frames and longerons. Also note that both pressure bulkheads are flat. Forward pressure bulkheads are normally small enough for the weight penalty due to flatness to be negligible. The nose gear is often attached to stiffeners placed on the forward pressure bulkhead. In that case any weight penalty due to flatness may disappear.

The rear pressure bulkheads are normally large. A significant weight penalty is incurred by making these flat. However, sometimes it is desirable to gain useful cabin volume by making the rear pressure bulkhead flat.

Figures 3.59 and 3.60 show the fuselage structural arrangements of typical jet transports. Both airplanes have flat forward pressure bulkheads which are also used to mount the weather radar disks.

Note that the 767 has a spherical aft pressure bulkhead while the DC10 has a flat one. In the latter, the aft pressure bulkhead is also used to mount the front spar of the vertical tail.

Figure 3.61 shows a typical structural arrangement for the fuselage of a fighter airplane. The irregular fuselage structure is mandated by the requirements for:

1. engine removal   3. nose wheel retraction
2. speed brakes   4. canopy

Note the tail hook attachment.

<u>Important observation:</u> The airplanes in Figures 3.56-3.58 violate the principle of p.238, Part II which requires the zero reference point to be well ahead of the nose! <u>Don't fall into this trap!</u>

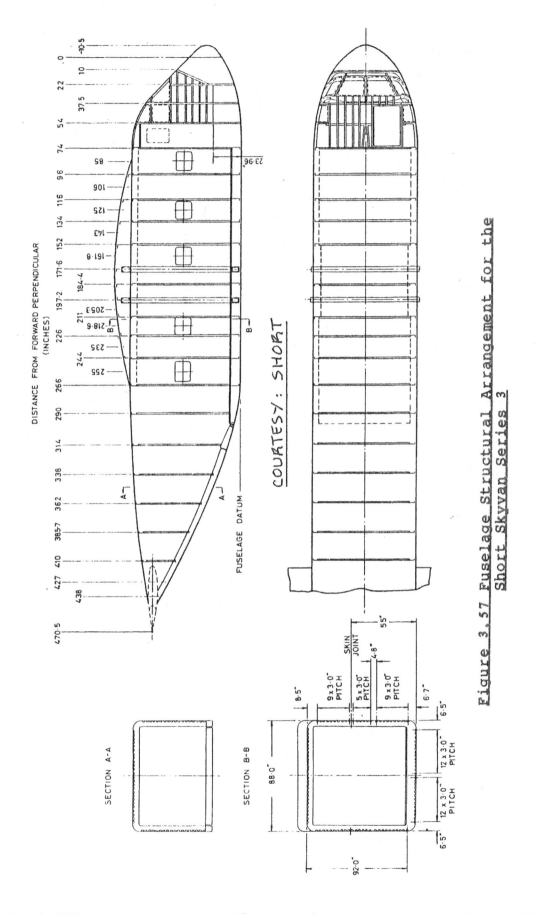

Figure 3.57 Fuselage Structural Arrangement for the Short Skyvan Series 3

Part III　　　　　Chapter 3　　　　　Page 127

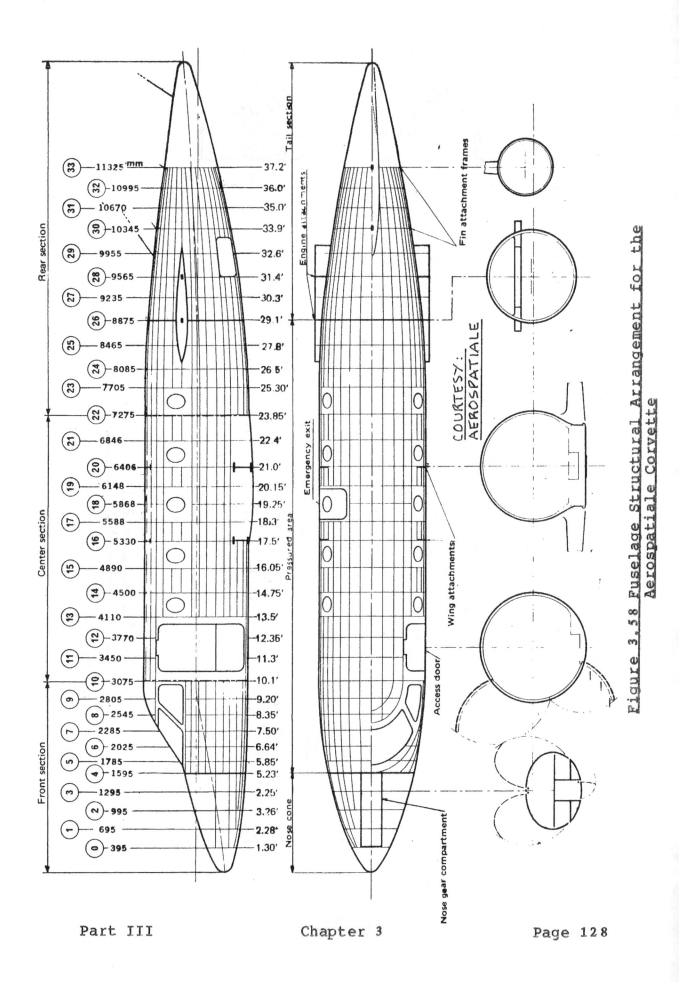

Figure 3.58 Fuselage Structural Arrangement for the Aerospatiale Corvette

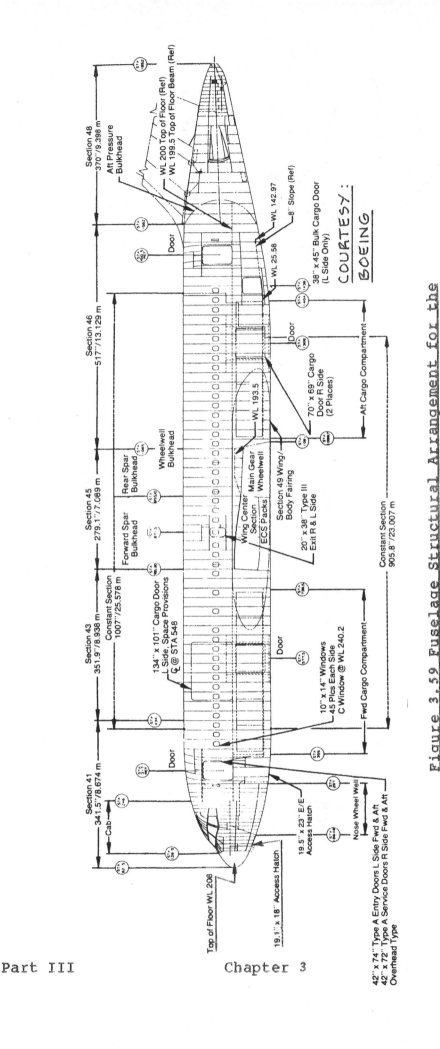

Figure 3.59 Fuselage Structural Arrangement for the Boeing 767-200

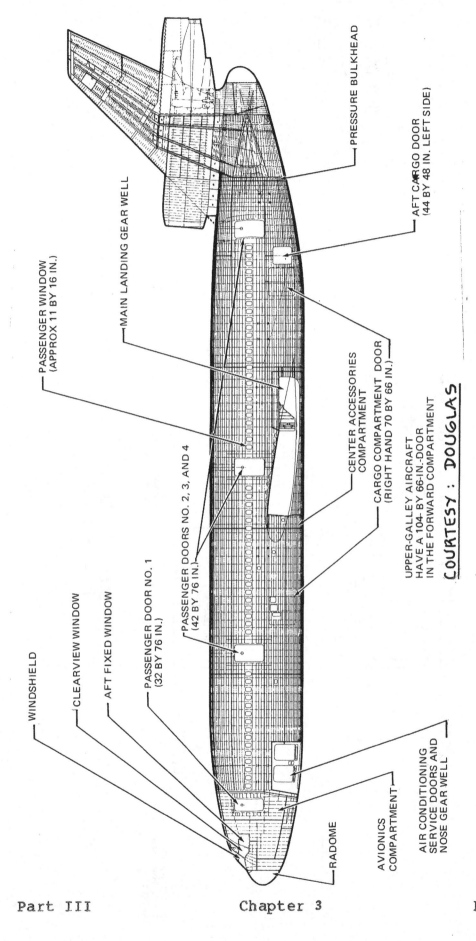

Figure 3.60 Fuselage Structural Arrangement for the McDonnell Douglas DC10

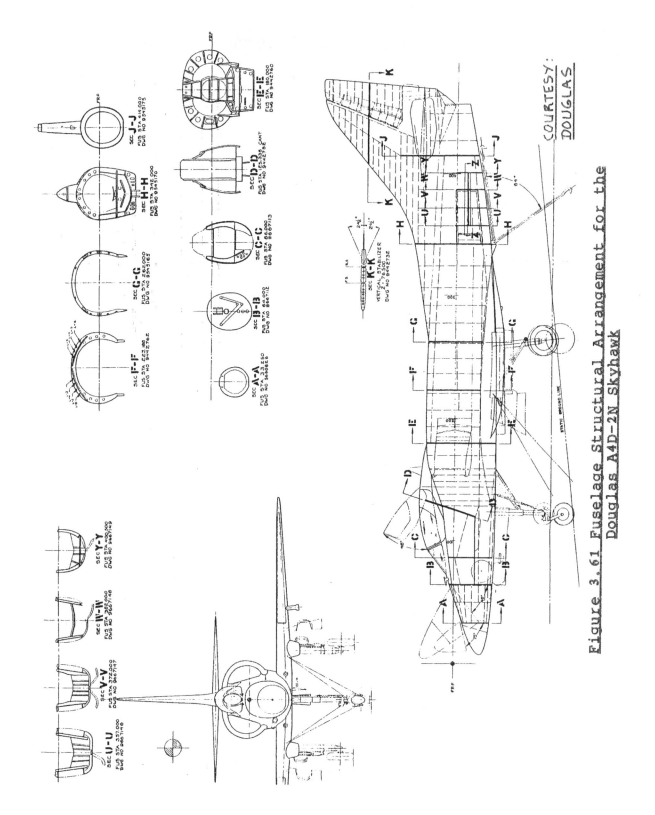

Figure 3.61 Fuselage Structural Arrangement for the Douglas A4D-2N Skyhawk

3.5.3 Examples of Fuselage Shell Layout

Figures 3.62 and 3.63 provide typical shell and skin layouts for general aviation airplanes. Note that relatively thin skin gauges are sufficient.

Figures 3.64 and 3.65 illustrate the shell layout used in a typical large jet transport. To facilitate manufacturing and final assembly, the fuselage and the skin are made in varying sections. This necessitates skin splicing. Figures 3.66 and 3.67 show typical skin splices used in such transports.

When using honeycomb panels for the fuselage skin, most frames and longerons are no longer needed. Fig.3.68 shows an example concept of such a shell layout, applied to a short haul airplane design.

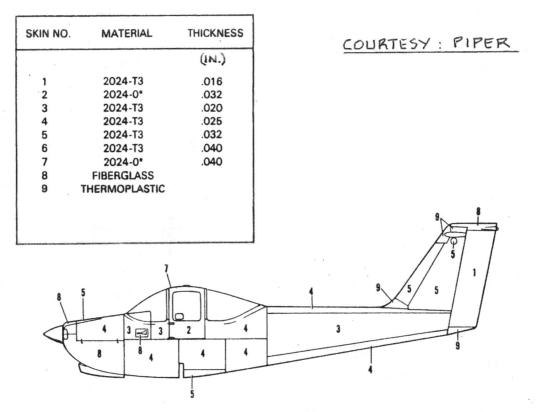

SKIN NO.	MATERIAL	THICKNESS (IN.)
1	2024-T3	.016
2	2024-O*	.032
3	2024-T3	.020
4	2024-T3	.025
5	2024-T3	.032
6	2024-T3	.040
7	2024-O*	.040
8	FIBERGLASS	
9	THERMOPLASTIC	

COURTESY: PIPER

Figure 3.62 Fuselage Shell and Skin Layout for the Piper PA-38-112 Tomahawk

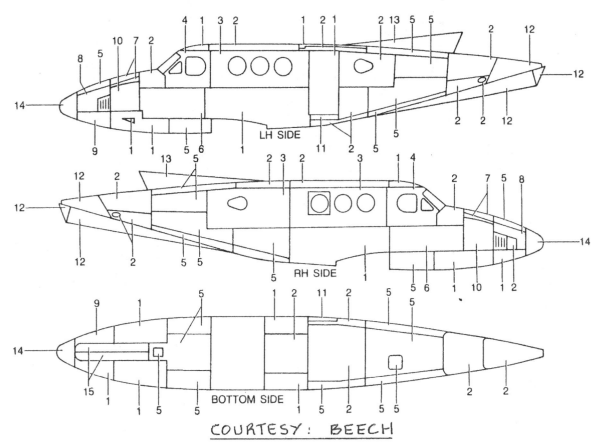

ITEM	MATERIAL	THICKNESS IN INCHES
1	2024T3	.040
2	2024T3	.050
3	2024T3	.063
4	2024T3	.071
5	2024T3	.025
6	2024T42	.040
7	2024T4	.025
8	2024T4	.050
9	2024T4	.032
10	6061T4	.040
11	2024T3	.080
12	6061T6	.025
13	Laminated No. 181 Glass Cloth per MIL-C-9084	.030
14	Laminated No. 181 Glass Cloth per MIL-C-9084	.035-.050
15	2024T3	.032

Figure 3.63 Fuselage Shell and Skin Layout for the Beech King Air F90

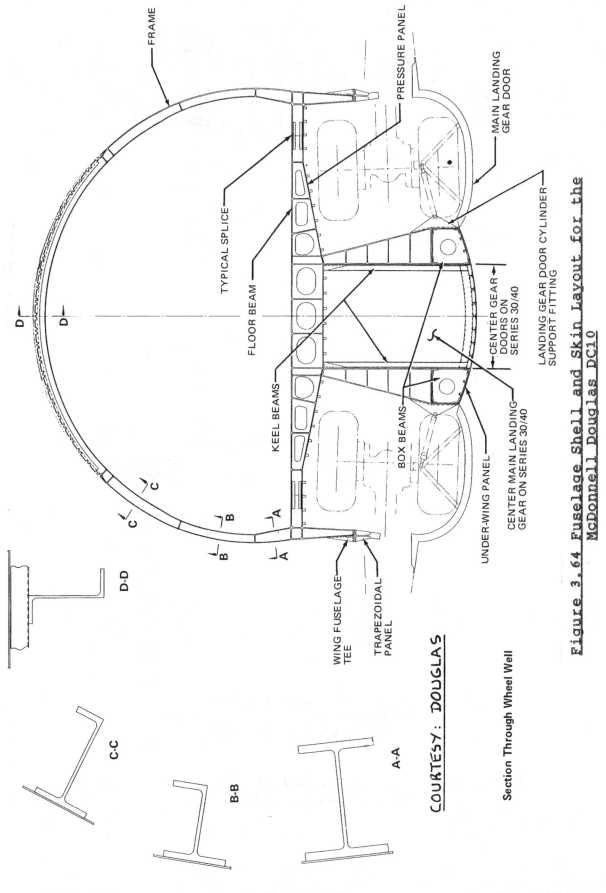

Figure 3.64 Fuselage Shell and Skin Layout for the McDonnell Douglas DC10

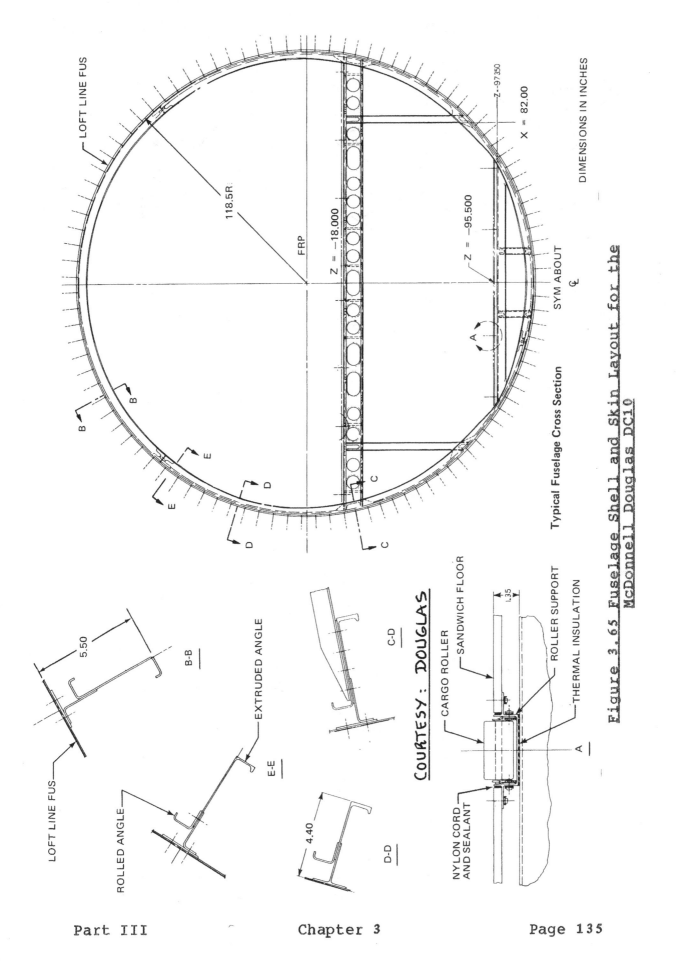

Figure 3.65 Fuselage Shell and Skin Layout for the McDonnell Douglas DC10

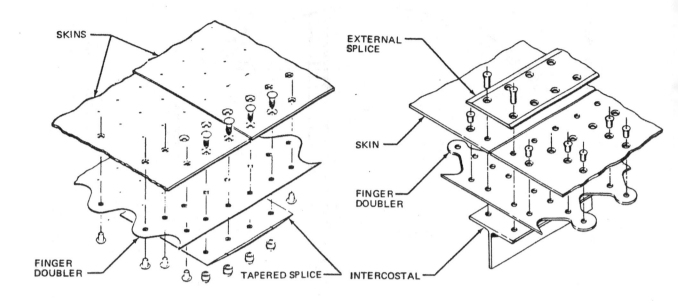

Figure 3.66 Typical Transverse Fuselage Skin Splice for the McDonnell Douglas DC10

Figure 3.67 Typical Longitudinal Fuselage Skin Splice for the McDonnell Douglas DC10

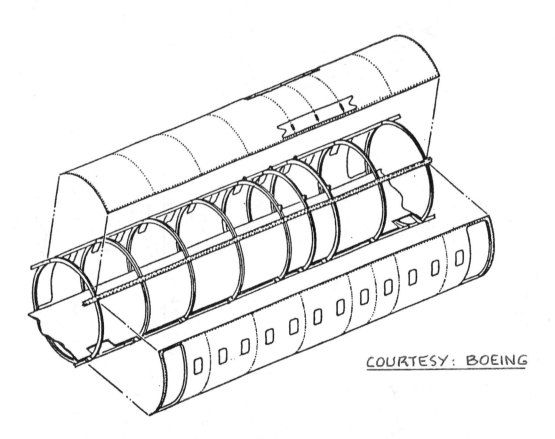

Figure 3.68 Example of a Fuselage with Honeycomb Skin Panels According to a Boeing Proposal

3.5.4 Examples of Door and Stair Design

Figure 3.69 shows a typical door installation used in a light, unpressurized airplane. Contrast this with the forward passenger door of Figure 3.70 for a jet transport! Many transports also have doors in their rear pressure bulkheads. Figure 3.71 shows an example of such a design.

When operating from fields without passenger loading ramps, transports need to carry 'built-in', retractable stairways. Figures 3.72 show an example of a ventral stairway mounted outside the pressure vessel.

Cargo compartments in pressurized transports must have special doors to allow for easy loading/unloading. These doors must be easy to open and close. They also must be able to hold the cabin pressure differential. Figure 3.73 shows an example of such a cargo door.

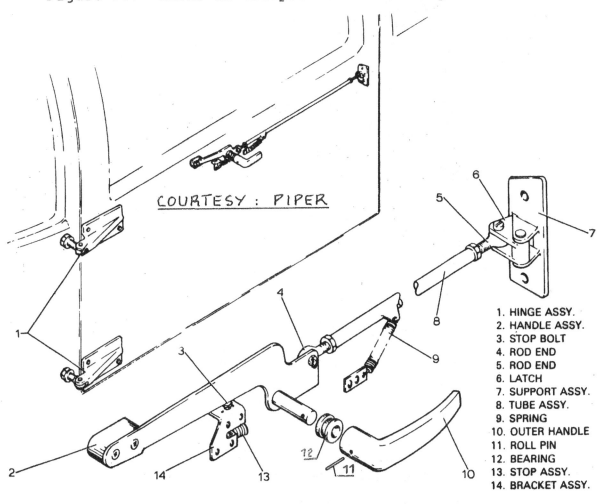

1. HINGE ASSY.
2. HANDLE ASSY.
3. STOP BOLT
4. ROD END
5. ROD END
6. LATCH
7. SUPPORT ASSY.
8. TUBE ASSY.
9. SPRING
10. OUTER HANDLE
11. ROLL PIN
12. BEARING
13. STOP ASSY.
14. BRACKET ASSY.

Figure 3.69 Cabin Door Installation for the Piper PA-38-112 Tomahawk

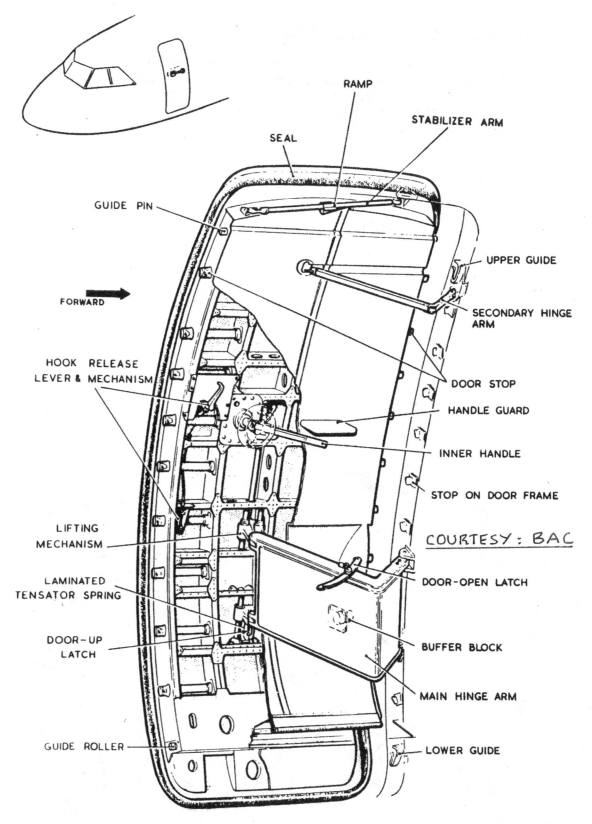

Figure 3.70 Passenger Door Installation for the BAC 111

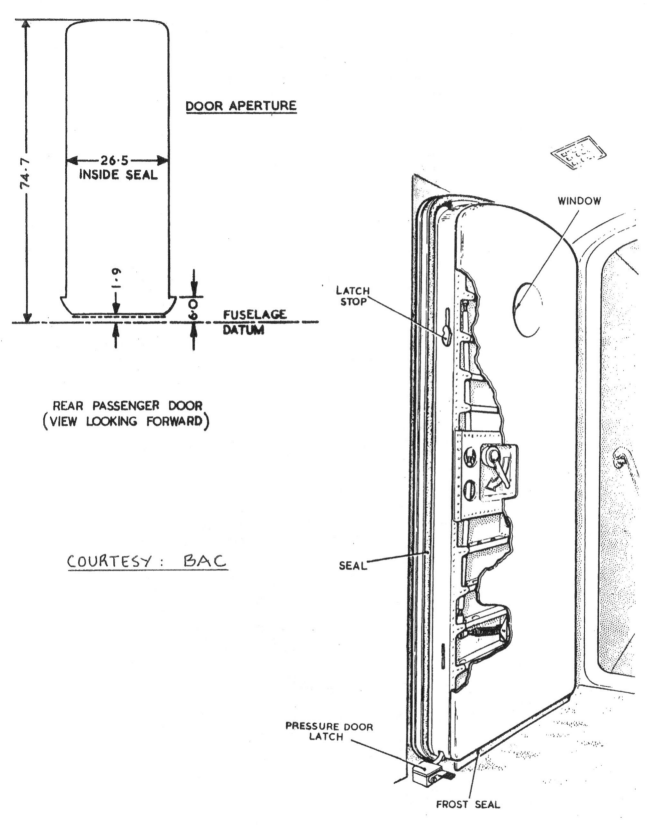

Figure 3.71 Pressure Bulkhead Door Installation for the BAC 111

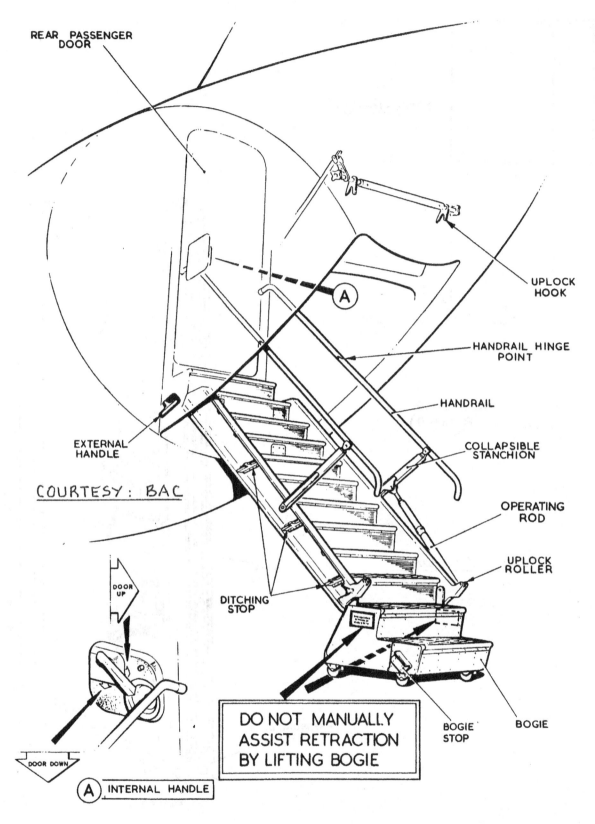

Figure 3.72a Ventral Stair Installation for the BAC 111

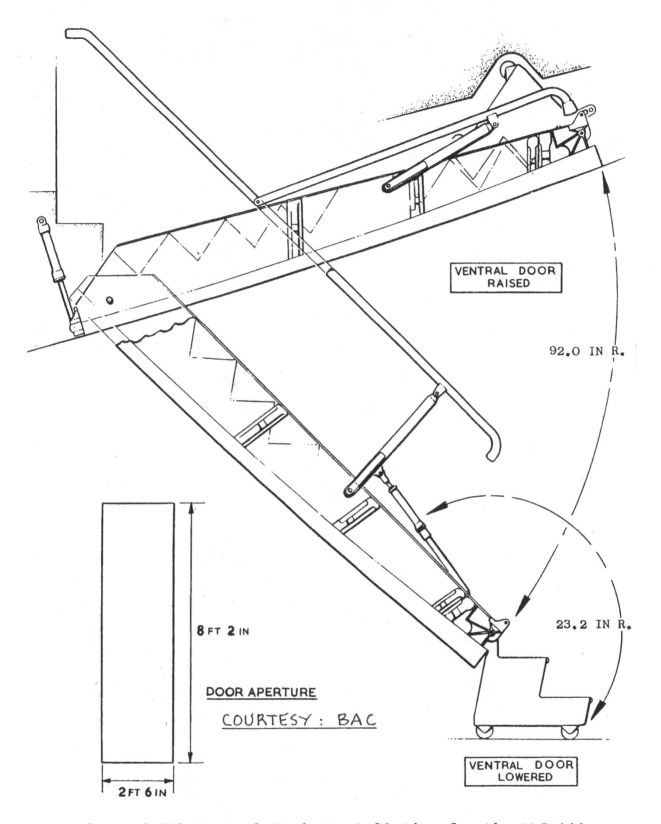

Figure 3.72b Ventral Stair Installation for the BAC 111

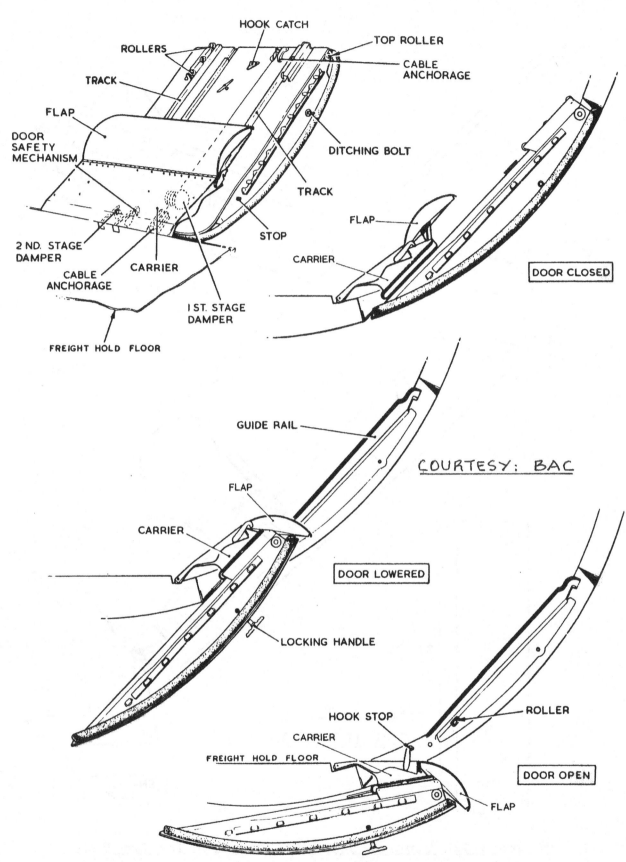

Figure 3.73 Cargo Compartment Door Installation for the BAC 111

3.5.5 Examples of Cockpit and Cabin Window Design

Cockpit windows must satisfy a number of conflicting design criteria:

1. they must be large enough to provide good visibility
2. they must have good optical qualities
3. they must be resistant to rain, hail and dust induced abrasion
4. they must cause little extra drag
5. for airplanes faster than 250 kts, they must meet the bird-strike requirement
6. they must be light

It is not easy to find a design solution which represents an acceptable compromise between these criteria.

Figure 3.74 shows an example of a typical light airplane windshield and window installation. Contrast the windshield of Figure 3.74 with that of Figure 3.75 for a business jet which must meet the bird-strike requirement. Note the complicated and heavy structure needed to meet this requirement.

In many business airplanes the cabin windows are arranged to polarize the incoming light. This is to prevent glare and still allow the passengers to look out the window. Figure 3.76 shows an example of such a polarized installation.

In jet transports the cabin windows are a potential source for leaks and for structural fatigue problems. The structure surrounding the windows is usually reenforced to prevent fatigue cracks from developing. Figure 3.77 shows an approach to this design problem: a forged (or cast) sub-frame is used to install each window. The window itself normally consists of three layers: two are redundant panes, each of which can hold the cabin pressure differential and an inner pane to protect the actual panes from passenger vandalism.

Figure 3.78 shows an example of a cabin window installation including the cabin wall trim. A cross section through the structure at the window is shown in Figure 3.79.

To cut noise coming through the wall, to provide thermal protection and to present a pleasant looking interior a series of provisions are made to the cabin inside. Figure 3.78 shows a typical trim installation. A cross section through a cabin wall is depicted in Figure 3.80.

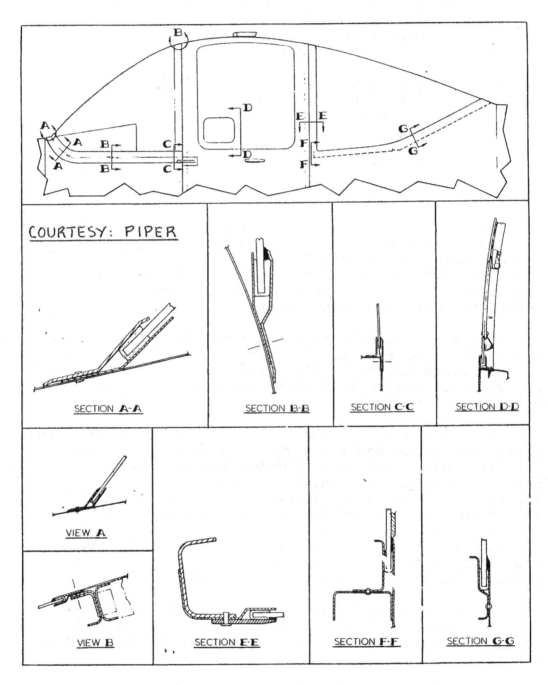

Figure 3.74 Windshield Installation for the Piper PA-38-112 Tomahawk

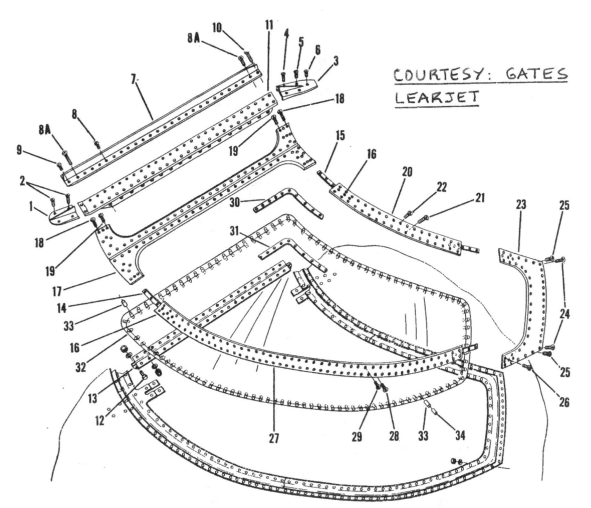

Figure 3.75 Windshield Installation for the Learjet 23

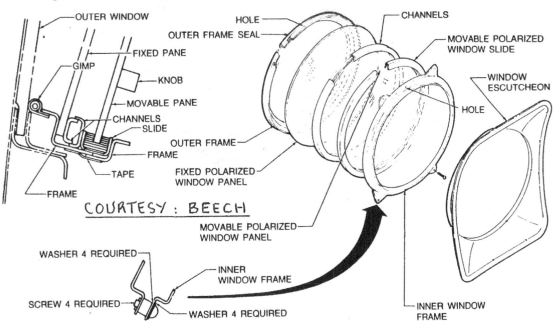

Figure 3.76 Polarized Window Installation for the Beech King Air F90

Part III Chapter 3 Page 145

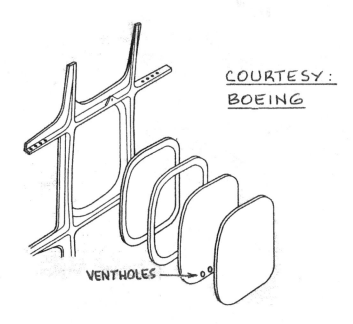

Figure 3.77 Typical Window and Window Frame Installation for a Jet Transport

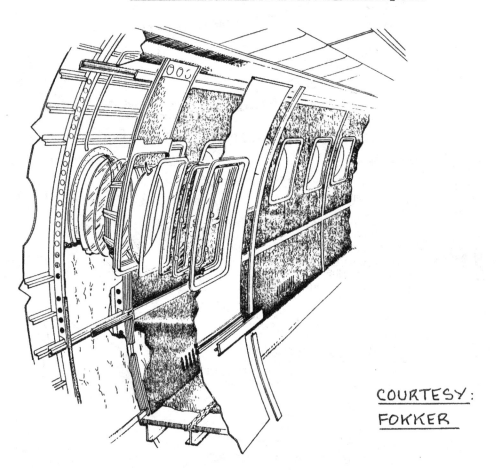

Figure 3.78 Cabin Window and Cabin Wall Trim Installation for the Fokker F28

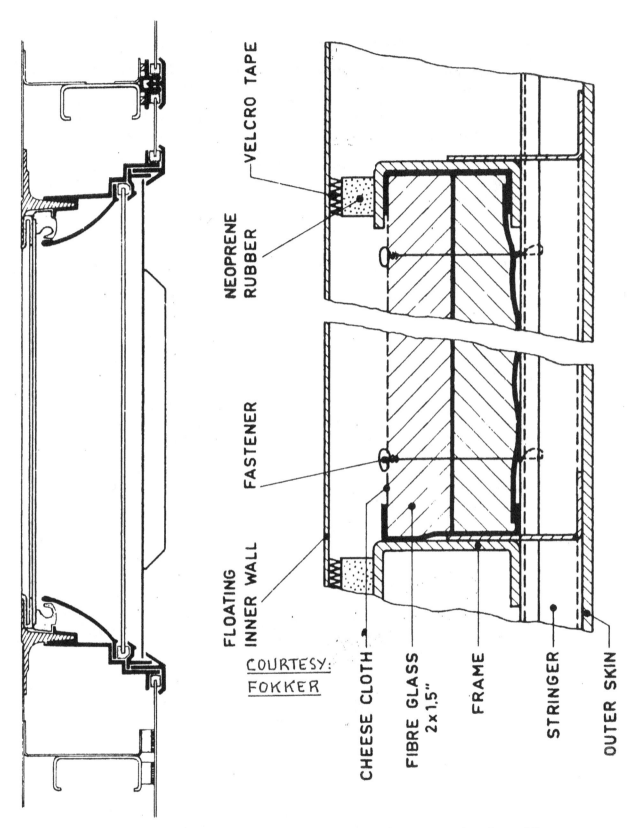

Figure 3.79 Typical Window Cross Section: Fokker F28

Figure 3.80 Typical Wall Cross Section: Fokker F28

Part III Chapter 3 Page 147

3.5.6 Examples of Floor Design

For new general aviation airplanes the author recommends a review of Ref.24 before committing to a specific floor design. Floors and the attached seat rails for crew and passenger seats can now be designed for maximum protection in the case of a crash.

For cargo carrying airplanes, floors need to be equipped with cargo restraint systems. Examples of tied down cargoes with appropriate restraints are depicted in Figures 3.81a-b.

In large cargo airplanes it is essential that the cargo can be easily moved to its proper position. This can be done with the help of self-driving roller systems. Figure 3.82 shows an example of a floor equipped in this manner.

The possibility of liquid spillage exists in passenger and in cargo airplanes alike. Floors need to be equipped with a drainage system. To be effective, the surrounding floor areas need to be sealed to prevent liquids running into areas where they could cause corrosion or malfunctioning of equipment. Figures 3.83 and 3.84 show typical drainage provisions.

Floors are ideal components for application of composite materials. Figure 3.85 show the extent to which composite floors are used in the Boeing 757.

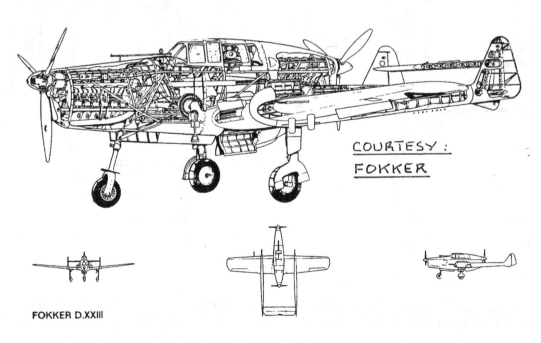

COURTESY: FOKKER

FOKKER D.XXIII

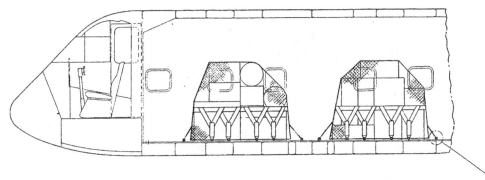
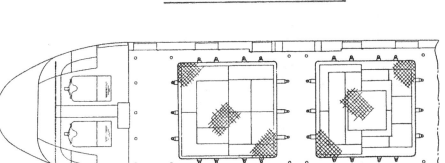

Figure 3.81a Palletized Cargo with Restraints in the Short Skyvan Series 3

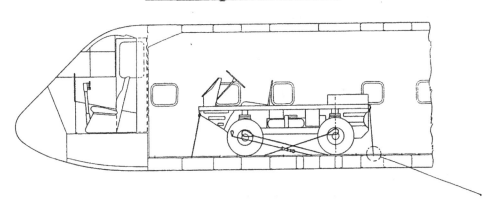

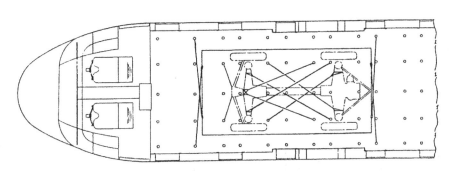

Figure 3.81b Vehicle with Restraints in the Short Skyvan Series 3

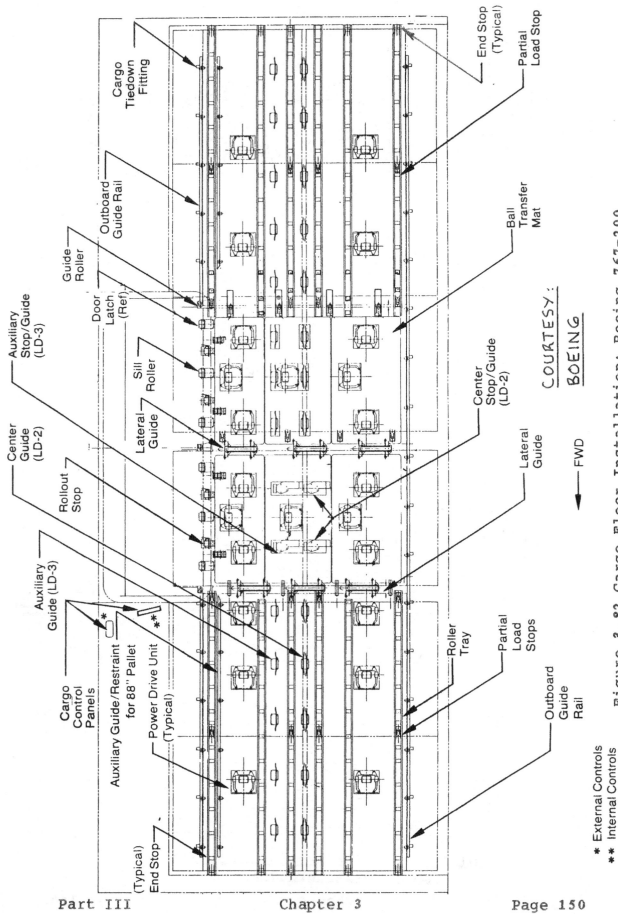

Figure 3.82 Cargo Floor Installation: Boeing 767-200

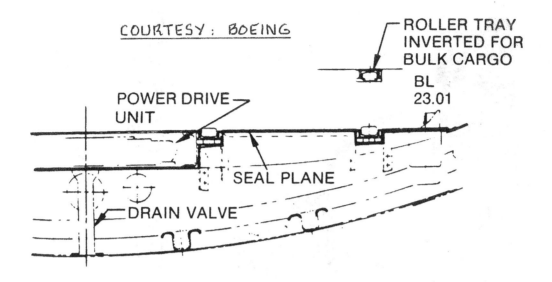

Figure 3.83 Cargo Floor with Drainage Provision: Boeing 757

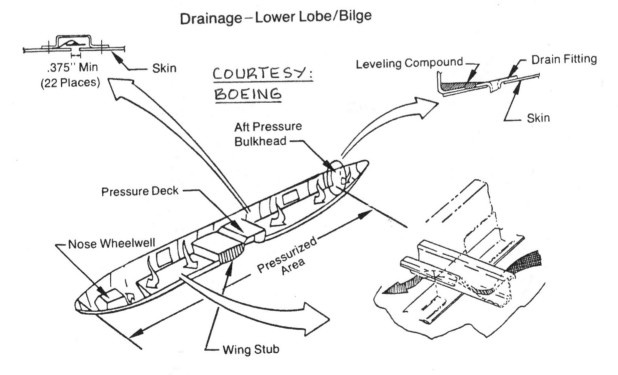

Figure 3.84 Drainage Provisions for the Boeing 767-200

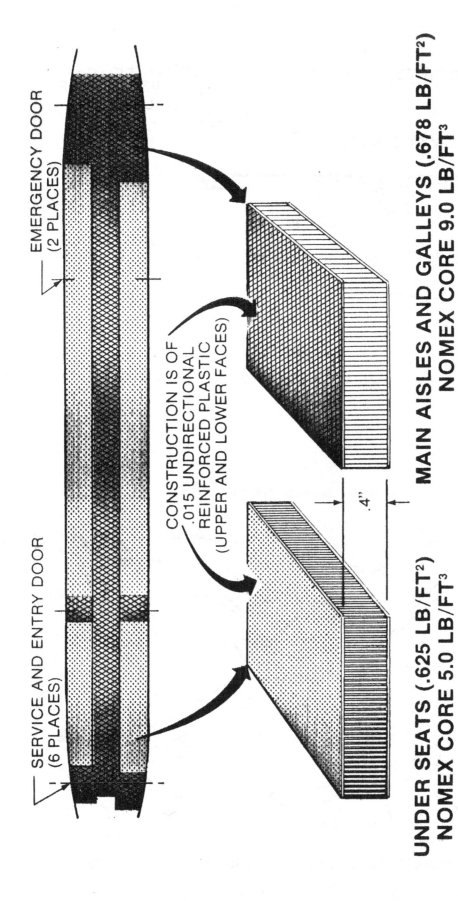

Figure 3.85 Composite Floor Application: Boeing 757

3.6 EXAMPLES OF INBOARD PROFILES

When the basic design decisions reflected by Sections 3.1 through 3.5 have been made it is useful to prepare a composite drawing showing the relative arrangement of important items. Such a composite drawing is called an 'inboard profile'. It is useful to include in the inboard profile all systems which are essential to the operation of the airplane. The latter implies that Step 4.7 of p.d. sequence I and Step 17 of p.d. sequence II have been completed.

The inboard profile serves primarily as an 'organizer' for the designer: it allows him to determine major conflicts. It also allows the designer to play the 'what-if' game, so essential to the ultimate flight safety of the proposed design. The 'what-if' game and its role in airplane design is discussed in Part IV.

Figures 3.86 through 3.94 provide examples of inboard profiles. The completeness of an inboard profile depends on the amount of detailed information available.

The inboard profiles are presented in the following sequence:

Figure 3.86 for a piston/propeller driven homebuilt: the Piel CP-80

Figure 3.87 for a jet powered homebuilt: the BD-5J

Figure 3.88 for a turbo/propeller driven twin: the Mitsubishi MU-2G

Figure 3.89 for a Jet Transport: the Boeing 767-200

Figure 3.90 for a military trainer/attack airplane: the SIAI-Marchetti S-211

Figure 3.91 for a military trainer: the Fairchild-Republic T-46A

Figure 3.92 for a fighter: the Macchi MB.339

Figure 3.93 for an experimental fighter: the Grumman X-29

Figure 3.94 for a turbo/propeller driven flying boat: the Shinmeiwa US-1

Figure 3.86 Inboard Profile: Piel CP-80

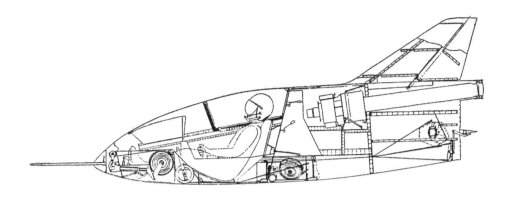

Figure 3.87 Inboard Profile: BD-5J

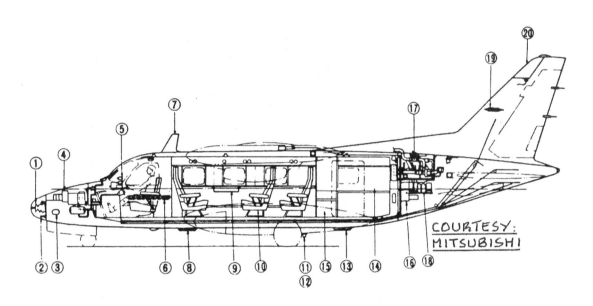

① Weather radar antenna
② Glide slope antenna
③ Landing and taxi light
④ Nose avionics compartment
⑤ Instrument panel
⑥ Circuit breaker panel
⑦ #1 VHF antenna
⑧ Marker beacon antenna
⑨ Table
⑩ Passenger seat

⑪ DME antenna (L.H)
⑫ ATC transponder antenna (R.H)
⑬ #1 ADF loop antenna
⑭ Baggage compartment
⑮ Toilet
⑯ Batteries
⑰ Air-conditioning system
⑱ Rear avionics compartment
⑲ VOR antenna
⑳ #2 VHF antenna

Figure 3.88 Inboard profile: Mitsubishi MU-2G

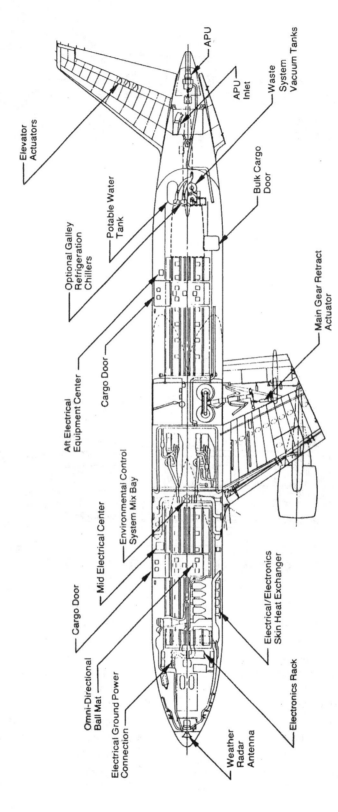

Figure 3.89a Inboard Profile Below Floor: Boeing 767-200

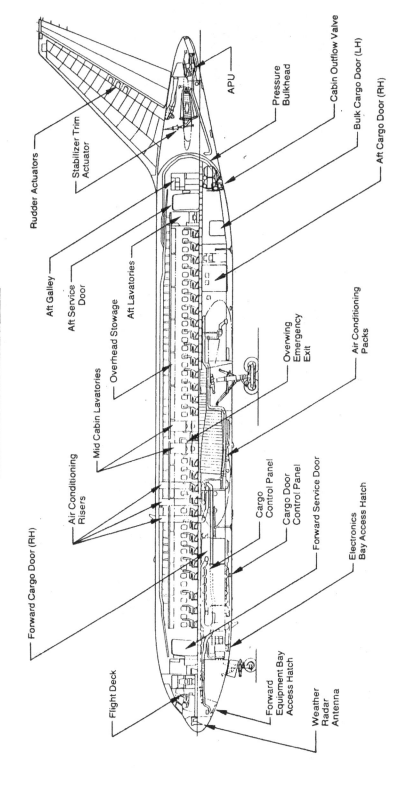

Figure 3.89b Inboard Profile: Boeing 767-200

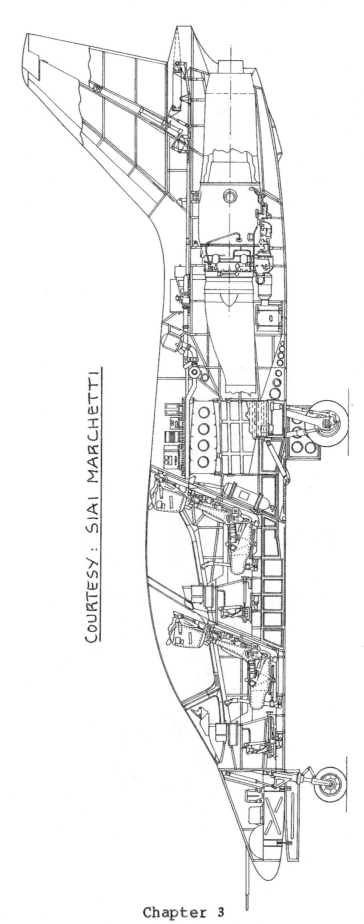

Figure 3.90 Inboard Profile: SIAI Marchetti S-211

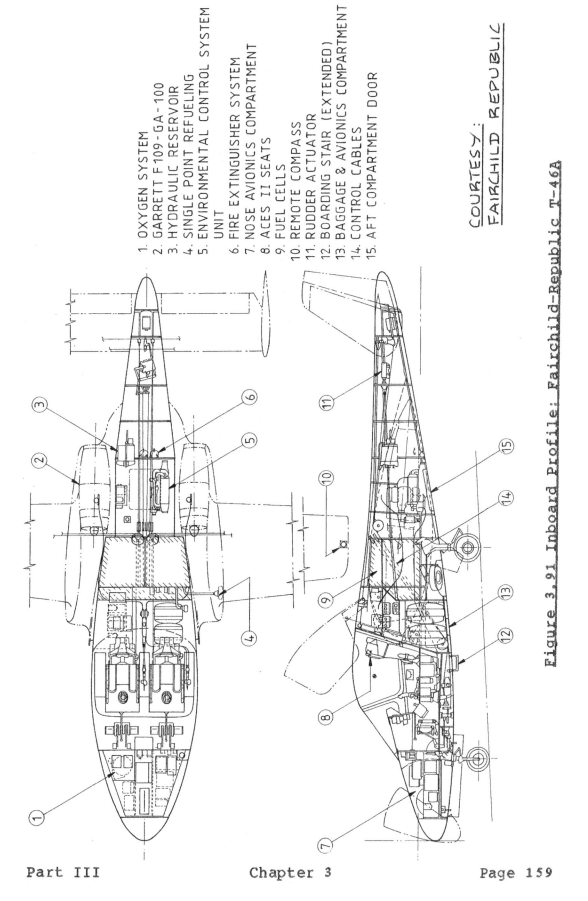

Figure 3.91 Inboard Profile: Fairchild-Republic T-46A

1. OXYGEN SYSTEM
2. GARRETT F109-GA-100
3. HYDRAULIC RESERVOIR
4. SINGLE POINT REFUELING
5. ENVIRONMENTAL CONTROL SYSTEM UNIT
6. FIRE EXTINGUISHER SYSTEM
7. NOSE AVIONICS COMPARTMENT
8. ACES II SEATS
9. FUEL CELLS
10. REMOTE COMPASS
11. RUDDER ACTUATOR
12. BOARDING STAIR (EXTENDED)
13. BAGGAGE & AVIONICS COMPARTMENT
14. CONTROL CABLES
15. AFT COMPARTMENT DOOR

COURTESY: FAIRCHILD REPUBLIC

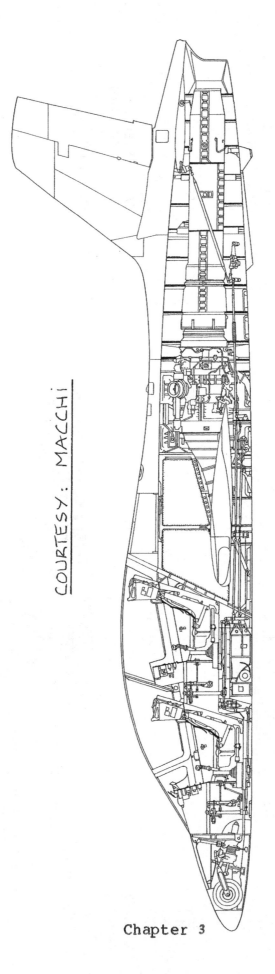

Figure 3.92 Inboard Profile: Macchi MB.339

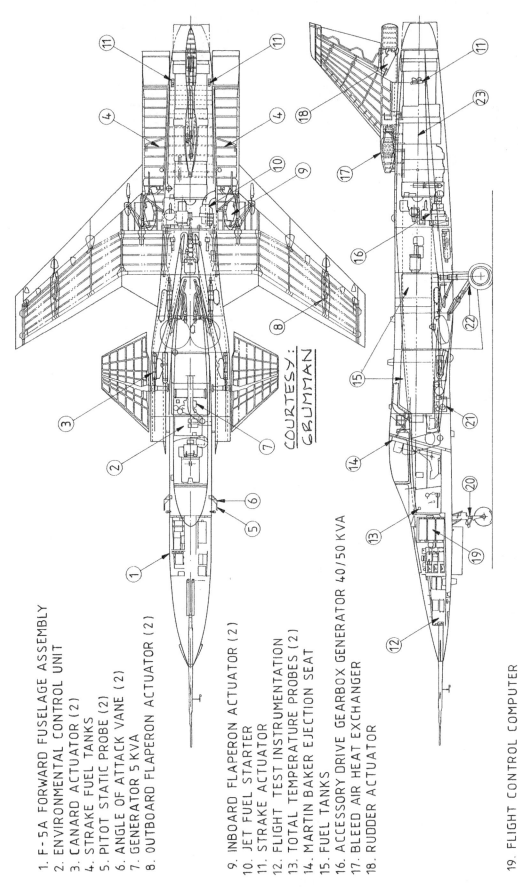

1. F-5A FORWARD FUSELAGE ASSEMBLY
2. ENVIRONMENTAL CONTROL UNIT
3. CANARD ACTUATOR (2)
4. STRAKE FUEL TANKS
5. PITOT STATIC PROBE (2)
6. ANGLE OF ATTACK VANE (2)
7. GENERATOR 5 KVA
8. OUTBOARD FLAPERON ACTUATOR (2)
9. INBOARD FLAPERON ACTUATOR (2)
10. JET FUEL STARTER
11. STRAKE ACTUATOR
12. FLIGHT TEST INSTRUMENTATION
13. TOTAL TEMPERATURE PROBES (2)
14. MARTIN BAKER EJECTION SEAT
15. FUEL TANKS
16. ACCESSORY DRIVE GEARBOX GENERATOR 40/50 KVA
17. BLEED AIR HEAT EXCHANGER
18. RUDDER ACTUATOR
19. FLIGHT CONTROL COMPUTER
20. F-5A NOSE LANDING GEAR
21. EMERGENCY POWER UNIT
22. F-16 MAIN LANDING GEAR
23. GENERAL ELECTRIC TURBOFAN F404-GE-400

Figure 3.93 Inboard Profile: Grumman X-29

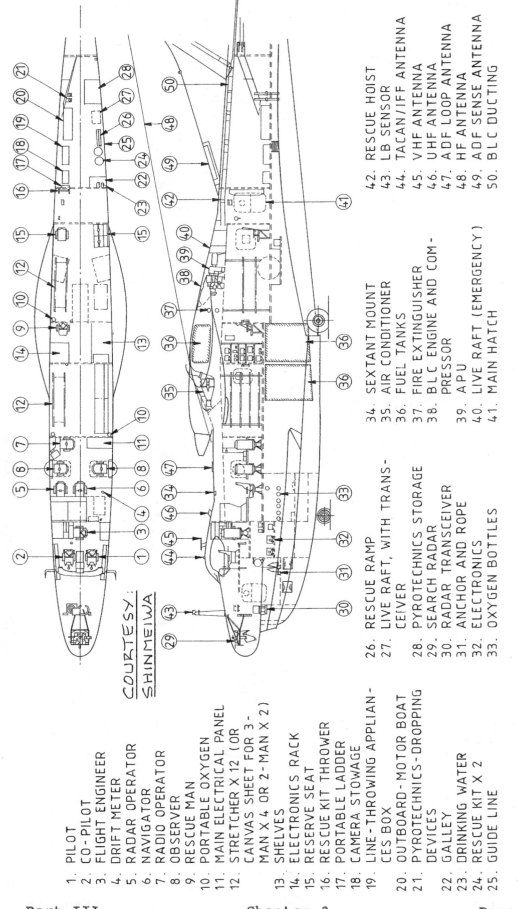

Figure 3.94 Inboard Profile: Shinmeiwa US-1

4. WING LAYOUT DESIGN

The purpose of this chapter is to provide design considerations, design data and design examples for the layout design of wings.

A step-by-step procedure for arriving at a satisfactory Class I preliminary wing layout was presented in Chapter 6 of Part II. That procedure is meant to be used in conjunction with p.d. sequence I of Chapter 2 in Part II. During the next phase of wing design (Class II in p.d. sequence II as outlined in Chapter 2 of Part II) it is recommended that the reader use the same procedure, but now augmented with the broad range of wing design considerations presented in this chapter.

The reader should also review the large number of wing/airplane configurations presented in Chapter 3 of Part II. It is always useful to determine what has been done by various manufacturers.

Section 4.1 presents a general discussion of wing configuration design aspects: aerodynamic as well as operational.

Section 4.2 contains a discussion of wing design integration considerations. Guidelines for the structural design of wings are presented. In addition, the structural integration of the wing into the fuselage is discussed with examples. Mechanizations of flaps and wing mounted lateral controls are also given.

Section 4.3 provides a discussion of several military/operational design considerations such as: stores, pivoting stores, wing folding, and wing pivot construction.

Section 4.4 contains examples of the overall structural arrangement of wings for a number of airplanes.

4.1 WING CONFIGURATION: AERODYNAMIC AND OPERATIONAL DESIGN CONSIDERATIONS

An overview of airplane configurations including discussions of wing configurations is presented in Chapter 3 of Part II. A step-by-step procedure for arriving at a satisfactory preliminary wing layout is contained in Chapter 6 of Part II.

The purpose of this section is to present additional design information relative to the choice of the wing configuration.

References 12, 13, 14 and 29 should be consulted for additional information on wing design.

The following wing configuration aspects will be discussed:

4.1.1   Wing size: large or small? Or,
       Wing loading: low or high?
4.1.2   High, mid or low wing?
4.1.3   Forward sweep, no sweep or aft sweep?
4.1.4   Variable sweep: one pivot or two?
4.1.5   Bi-plane, braced wing or joined wing?
4.1.6   Wing aspect ratio: high, low and/or winglets?
4.1.7   Wing thickness ratio: large or small?
4.1.8   Wing taper ratio: high or low?
4.1.9   Straight taper or variable taper?
4.1.10 Wing twist: how much?
4.1.11 Wing dihedral angle: how much?
4.1.12 Wing incidence on the fuselage: how much?
4.1.13 Variable camber (MAW = Mission Adaptive Wing)?
4.1.14 Wing leading edge strakes (Lexes)
4.1.15 Planform tailoring: why and when?
4.1.16 Area ruling: when is it required?
4.1.17 Wing span: when is it too high?
4.1.18 Aerodynamic coupling?
4.1.19 Flaps: what size and which type(s)?
4.1.20 Lateral controls: what size and which type(s)?

Finally, a review of wing drag contributions is presented:

4.1.21 Review of wing drag contributions.

The reader is referred to Tables 6.1 through 6.12 and 8.1 through 8.12 in Part II for detailed wing airfoil and wing planform design information.

4.1.1 Wing Size: Large or Small? Or, Wing Loading: Low or High?

The question of wing size (or wing loading) depends mostly on performance oriented objectives. Chapter 3 of Part I has addressed the general question of wing and thrust sizing to a wide range of performance objectives.

Wing size or wing loading primarily affects the following characteristics:

1. Take-off/landing field length
2. Cruise performance (L/D)
3. Ride through turbulence
4. Weight

Two examples will be discussed of trade studies showing the effect of wing loading on fieldlength performance and on cruise efficiency. The reader can develop similar relations for other performance objectives such as: climb performance, maximum speed performance and maneuvering performance. During Class II wing design it is essential to carry out these individual trade studies to ensure the proper sizing of the wing.

1. Take-off/landing field length: To achieve short field lengths, large wings (low wing loading) are better than small wings (high wing loading). The wing can be kept small by using flaps. Flaps provide the possibility to obtain high values of $C_{L_{max}}$.

Figure 4.1 illustrates the trends based on the following equations which are valid for sealevel only:

For landing field length at sealevel standard:

$$s_L = 429(W/S)_L/C_{L_{max_L}} \tag{4.1}$$

For take-off field length at sealevel standard:

$$s_{TOFL} = 37.5(W/S)_{TO}/(T/W)_{TO}C_{L_{max_{TO}}} \tag{4.2}$$

These equations are approximations of FAR 23 and 25 fieldlength Eqns. (3.8), (3.14) and (3.16) of Part I.

The strong influence of maximum lift coefficient in the landing configuration and of wing loading on required landing field length is apparent.

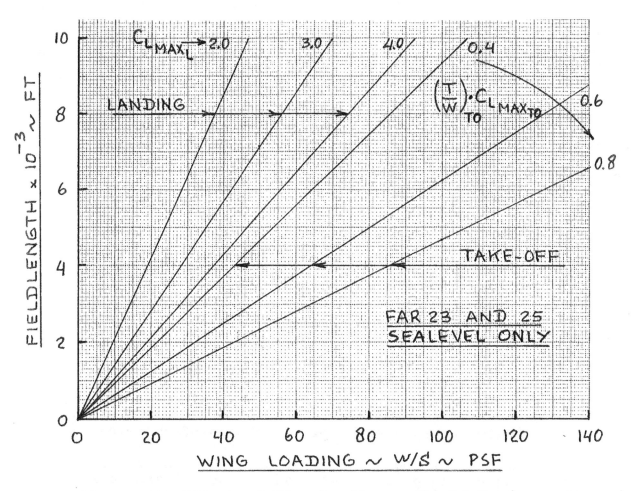

Figure 4.1 Effect of Wing Loading on Field Performance

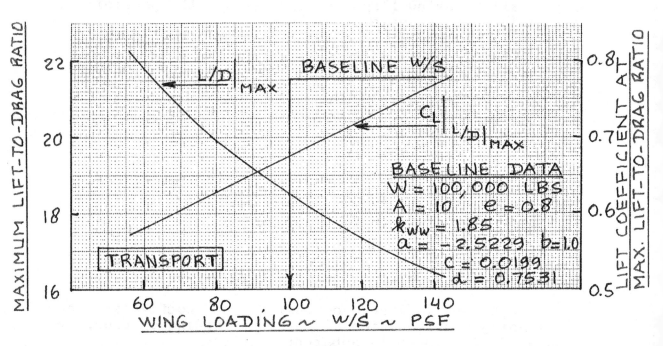

Figure 4.2 Effect of Wing Loading on Cruise Parameters

2: Cruise performance (L/D): To achieve cruise flight close to $(L/D)_{max}$ a high wing loading is needed, so that the cruise lift coeffcient can be close to that at $(L/D)_{max}$. On the other hand, if flight at extremely high altitudes and moderate speeds is required, a large wing area may be essential (U-2 and Canberra).

An appreciation for the effect of wing loading on $(L/D)_{max}$ can be obtained as follows:

$$(L/D)_{max} = (\pi A e / 4 C_{D_0})^{1/2} \quad (4.3)$$

with:

$$C_{D_0} = (1/S)\text{invlog}_{10}[a + b\log_{10}\{(\text{invlog}_{10}(c + d\log_{10}W_{TO})) + k_{ww}(S - S_{baseline})\}] \quad (4.4)$$

Equations (4.3) and (4.4) can be used to study the effect of varying wing area (wing loading) on $(L/D)_{max}$ for a given type airplane. To perform such a study the following input information is required:

* Aspect ratio, A   * Oswald's Efficiency Factor, e

* Regression constants a, b, c and d: (See Tables 3.4 and 3.5 in Part I)

* Take-off weight, W_{TO} * Baseline wing loading, $(W/S)_{baseline}$

* k_{ww} = 1.85 (approximately)

The constant k_{ww} accounts for that part of the wing which is 'buried' in the fuselage and which therefore does not contribute to wetted area.

The value for $S_{baseline}$ in Eqn.(4.4) follows from:

$$S_{baseline} = W_{TO}/(W/S)_{baseline} \quad (4.5)$$

The lift coefficient at $(L/D)_{max}$ follows from:

$$C_{L_{(L/D)_{max}}} = (C_{D_0} \pi A E)^{1/2} \quad (4.6)$$

Figure 4.2 illustrates the results of a trade study of the effect of wing loading on these cruise performance parameters for a transport type airplane. By varying the input data in the appropriate manner similar results can be generated for any type airplane. The reader is encouraged to do this for any new design.

WARNING: do not assume that an airplane will always be able to cruise at the lift coefficient for $(L/D)_{max}$ as given by Eqn.(4.6). Verify the actual cruise lift coefficient from:

$$C_{L_{cruise}} = W_{cruise}/\bar{q}S = W_{cruise}/1482\delta M^2 \qquad (4.7)$$

It will be found that most airplanes cruise at significantly lower values of lift coefficient than the one indicated by Eqn.(4.6).

3. **Ride through turbulence:** Wing loading also has a significant effect on the ride quality of an airplane through turbulence. Ride response to turbulence is proportional to the parameter n_α:

$$n_\alpha = \bar{q}C_{L_\alpha}/(W/S) \qquad (4.8)$$

Eqn.(4.8) clearly shows that airplanes with low wing loading have high values of n_α which translates into 'poor' ride qualities.

4. **Weight:** The larger the wing area, the greater the weight of the wing and therefore the weight of the airplane. The wing weight equations in Part V allow for the calculation of wing weight as a function of wing size (area).

Table 4.1 summarizes the range of typical take-off wing loadings found in twelve types of airplanes. The reader is reminded of the fact that additional data on airplane wing loadings may be found in Ref.30.

Table 4.2 summarizes the effect of wing loading on a number of design characteristics.

Table 4.1 Typical Values For Take-off Wing Loadings
==

Note: Ranges for take-off wing loadings are in psf.

Airplane Type	$(W/S)_{TO}$	Airplane Type	$(W/S)_{TO}$
1. Homebuilts	5 - 15	9. Fighters	
2. Single Engine Prop. Driven	10 - 25	Jets	70 - 140
		Props	40 - 70
3. Twin Engine Prop. Driven	20 - 45	10. Mil. Patrol, Bomb and Transport Airplanes	70 - 120
4. Agricultural	15 - 30	11. Flying Boats, Amphibious and Float Airplanes	
5. Business Jets	40 - 80	Jets	50 - 90
6. Regional TBP	30 - 55	Props	30 - 60
7. Transport Jets	80 - 120	Floats	20 - 50
8. Mil. Trainers		12. Supersonic Cruise Airplanes	80 - 120
Jets	40 - 80		
Props	20 - 40		

Table 4.2 Summary of the Effect of Wing Loading
==

Item	Effect of Wing Loading on Item	
	High W/S	Low W/S
Stall speed	High	Low
Fieldlength (landing and take-off)	Long	Short
Max. lift-to-drag ratio	High	Low
Ride quality in turbulence	Good	Bad
Weight	Low	High

4.1.2. High, Mid or Low Wing?

The vertical location of the wing on a fuselage affects the following characteristics:

1. Drag   2. Dihedral effect

3. Operational considerations

To a large extent the choice of high, mid or low wing configuration depends on operational considerations associated with the mission of the airplane. Therefore, which type of airplane is being considered plays an important role in deciding the vertical location of the wing.

Cargo Transports: For cargo transports, the requirement for easy loading and unloading via self-contained ramps, virtually dictates a high wing configuration. Floor levels must be close to the ground. This would not be possible with the wing carry-through structure passing below the floor in a low wing configuration.

Figure 4.3 illustrates these points.

Passenger Transports: For passenger transports a high wing configuration offers good visibility for all passengers. The Fokker F27 and the BAe 146 (Figures 3.18a and 3.21d in Part II) are examples of such configurations. If baggage and cargo needs to be carried below the floor of the passenger cabin, this results in a very heavy landing gear if the gear is wing mounted (F27). If the gear is fuselage mounted (BAe 146), this results in draggy bulges and/or marginal lateral stability during ground operations.

In the case of small high wing passenger transports it is possible to allow passenger loading without the help of external loading ramps and without the penalty self-contained stairs impose.

A synergistic advantage of a swept, high wing layout is that it allows the use of negative dihedral (anhedral) to obtain better dutch roll damping. In a low wing transport this is difficult to do because it results in long landing gear legs. The BAe 146 and the SAAB 105XT (Fig.4.4) illustrate these points.

In the case of large passenger transports, efficient access to baggage and cargo space is possible only with

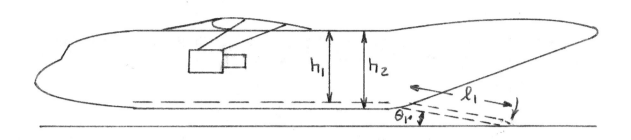

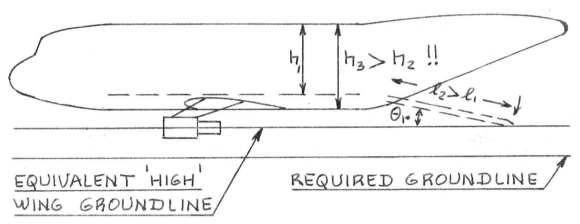

Figure 4.3 Effect of Wing Location on Ground Clearance

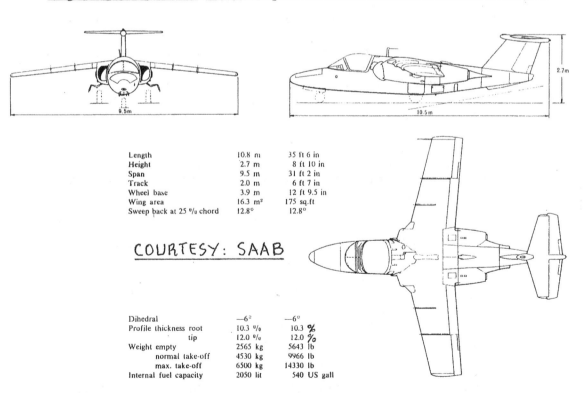

Figure 4.4 Use of Anhedral on the SAAB 105XT

the cargo holds close to the ground. This leads to the choice of a low wing configuration.

Both low and high wing configurations must carry fairings to smooth out the flow at the wing-fuselage intersection. If this is not done the interference drag becomes too high. Such fairings can be very large and cause an increase in weight and in manufacturing cost. Figures 3.30a and 3.15b in Part II illustrate these points.

<u>Light Airplanes:</u> For light airplanes, the choice of wing location seems to be more a matter of company tradition than anything else. A solid case pro or con either location is difficult to make.

A considerable aerodynamic (drag) advantage is associated with the mid wing configuration. Such a wing arrangement tends to minimize interference drag. A structural problem is the requirement for a carry-through structure passing through the middle of the fuselage. This makes it necessary to keep the passenger cabin either forward or behind the wing carry-through. Examples of this wing configuration are shown in Fig. 4.5 and in Fig. 3.47 of Part II.

Placing the passenger cabin forward of the wing usually results in large c.g. excursions which require special attention in empennage design. By utilizing canard and/or three-surface layouts it is possible to overcome these trim problems caused by large c.g. excursions. Figures 3.42 and 3.47 of Part II are examples of such layouts.

<u>Fighters and Trainers:</u> In fighter and trainer airplanes the mid wing configuration is often selected because of the favorable drag effect. No passengers need to be carried so the c.g. excursion argument does not apply in these cases.

Mid to high wing layouts in fighter airplanes also lead to easier loading and unloading of underwing stores.

<u>Flying Boats, Amphibians and Float Airplanes:</u> In this airplane category the high wing layout is often dictated by water clearance requirements. In float equipped airplanes the low wing layout has also been used.

Table 4.3 summarizes the effect of wing location on the fuselage.

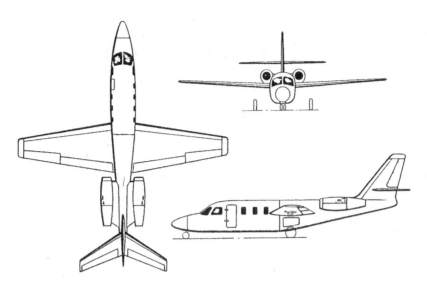

Figure 4.5 Use of Mid Wing Location on the Jet Commander

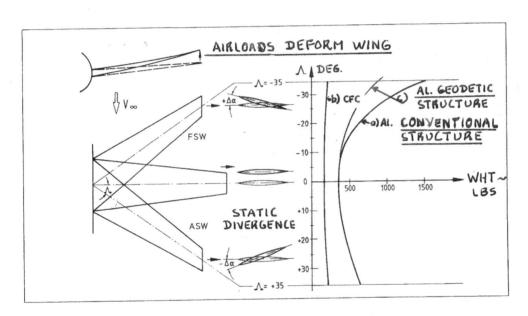

Figure 4.6 Effect of Sweep on Wing Weight

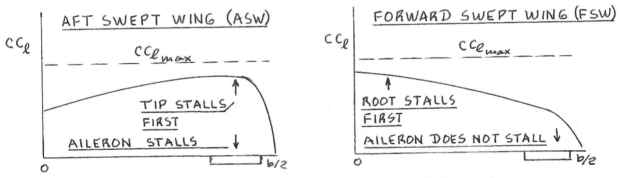

Figure 4.7 Effect of Sweep on Stall Behavior

Part III   Chapter 4   Page 173

Table 4.3 Summary of the Effect of Wing Location on the Fuselage

Item	Effect of Wing Location on Item		
	High	Mid	Low
Interference Drag	Poor	Good	Poor
Dihedral Effect	Negative	Neutral	Positive
Passenger Visibility	Good	Good	Poor for some
Landing Gear:			
Wing mounted		Long/heavy	Short/light
Fuselage mounted		Possibly draggy	
Loading and unloading	easy	easy	need stairs

Table 4.4 Summary of the Effect of Wing Sweep

Item	Effect of Increased Wing Sweep on item		
	Forward	None	Aft
Lift-curve Slope	Low	High	Low
Pitch attitude in low speed, level flight	High	Low	High
Ride through turbulence	Good	Poor	Good
Asymmetric stall	Best	Good	Poor
Lateral control at stall	Best	Good	Poor
Compressibility drag	Low	High	Low
Wing weight	Highest	Low	High

Tables 6.1 through 6.12 in Part II provide data on overall wing type and wing location for a wide range of airplanes.

4.1.3 Forward Sweep, No Sweep or Aft Sweep?

Adding sweep to a wing has important consequences to the following characteristics:

1. Compressibility drag 2. Weight 3. Stall behavior

4. Balance 5. Pitch attitude and ride 6. Good looks

There is a substantial weight penalty associated with the addition of sweep to a wing. Figure 4.6 illustrates this point for aft and for forward sweep.

1. Compressibility drag: It is clear from Fig. 6.1 in Part II that sweep has a very favorable effect on compressibility drag. Note that the sign of the sweep angle is not influenced by this: aft sweep and forward sweep yield similar reductions in compressibility drag.

2. Weight: Adding sweep to a wing increases wing weight substantially. The wing weight equations of Part V demonstrate this quantitatively for aft swept wings. Figure 4.6 illustrates this point for both aft swept and forward swept wings.

Figure 4.6 shows that aft sweep does translate into a weight advantage compared with forward sweep. The reason is the structural divergence phenomenon associated with forward sweep. By tailoring the ratio of bending to torsion stifness it is possible to make the weight penalty associated with forward swept wings quite acceptable. Such tailoring is inherently possible with composite structures. However, by using tailored machined skins (geodetic structure) and ribs it is possible to achieve similar results with metal wings. References 31 and 32 discuss these concepts in some detail.

3. Stall behavior: A significant advantage of forward sweep over aft sweep is the superior stall characteristics. Figure 4.7 illustrates the potential advantage of forward sweep. An important consequence of the superior stall behavior is the fact that outboard mounted lateral controls maintain their effectiveness well into the stall. Aft swept wings need to be twisted to protect against roll-off in the stall.

4. Balance: If a new design of an airplane with an unswept wing runs into minor trouble with its weight and balance (for example the c.g. is a bit too far aft) then it may be possible to 'fix' the problem by a slight amount of aft sweep. This has the effect of moving the airplane a.c. aft faster than the c.g. thereby bringing the airplane back into balance. This is in fact the reason for the fairly pronounced sweep of the Douglas DC2! The wing sweep feature of the DC2 was inherited by the DC3 and became one of its recognition features.

Adding aft sweep to a flying wing has a beneficial effect on the inherent longitudinal damping characteristics of such a configuration.

Adding aft sweep to a wing will also create the necessary moment arm for the addition of directionally stabilizing surfaces. The Beech Starship I of Fig. 3.42 (Part II) is a case in point.

5. Pitch attitude and ride: Sweep angle also has a significant effect on the lift-curve slope, C_{L_α} and therefore on:

1. Pitch attitude at low speed and therefore on runway visibility
2. Ride characteristics: high sweep improves ride and low sweep worsens the ride through turbulence

Figures 4.8 and 4.9 illustrate point 1. Point 2 can be visualized with Equation (4.8) which allows the calculation of the 'change in load factor with angle of attack derivative. This equation shows that lowering C_{L_α} by increasing sweep angle lowers the load factor reaction to turbulence induced angles of attack. Note that Eqn.(4.8) also shows that increased wing loading improves the ride!

6. Good looks: The 'good looks' aspects of sweep selection should not be dismissed as entirely frivolous. There is plenty of evidence that 'good looks' helps sell airplanes!

Table 4.4 summarizes the effect of wing sweep angle on a number of important characteristics.

The reader should refer to Tables 6.1 through 6.12 in Part II for data on wing sweep angles used on a wide range of airplanes.

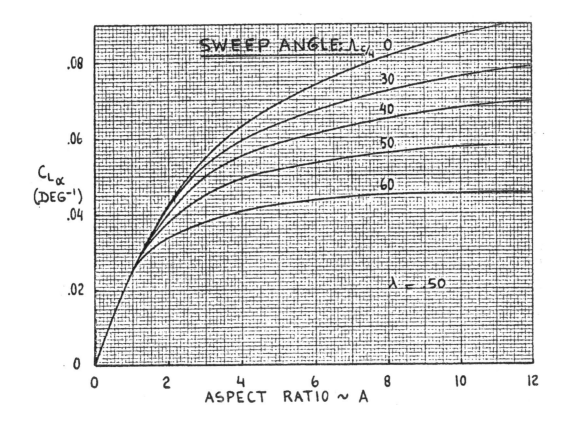

Figure 4.8 Effect of Sweep on Lift Curve Slope

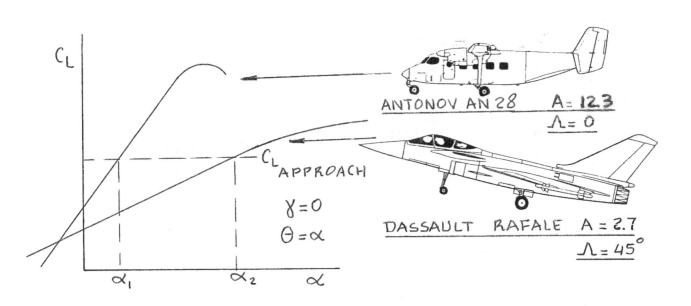

Figure 4.9 Effect of Sweep on Approach Attitude

4.1.4 Variable Sweep: One Pivot or Two?

The primary reasons for using variable sweep are:

1. for good take-off and landing performance as well as good handling characteristics very low sweep angles are best.
2. for low drag and good ride characteristics at high speed, high sweep angles are best.
3. to obtain optimum values of L/D throughout a wide performance envelope, variable sweep is best.

1. Take-off/landing and handling: Sub-section 4.11 shows that a high value of maximum lift coefficient is required to achieve good fieldlength performance. Sub-section 4.1.3 shows that at high sweep angles asymmetric stall problems may arise which compromise good handling qualities.

High sweep angles cause a drop in maximum lift coefficient according to the so-called cosine rule:

$$C_{L_{max_\Lambda}} = C_{L_{max_{\Lambda=0}}} \cos\Lambda \qquad (4.9)$$

This cosine rule applies below 35 degrees of sweep.

2. Low drag and good ride: The dragrise data of Fig.6.1, Part II and Eqn.(4.8) demonstrate the correctness of statement 2.

3. Lift-to-drag ratio: Figure 4.10 illustrates the possibilities with regard to L/D thereby showing the correctness of statement 3.

Evidently, with variable sweep all these things are possible. However, at a price: the variable sweep wing results in a significant weight penalty due to the need for a pivot structure and a system to change wing sweep angle in flight. A major design question with a variable sweep wing is: where to place the pivot(s)?

The chordwise pivot location is fairly obvious: close to the mid-chord position of the wing torque box. The spanwise pivot(s) position depends on whether one or two pivots are used.

If two pivots are used (symmetrical variable sweep) the weight penalty can be lessened by using outboard pivot locations. Outboard pivot locations also result in a reduction of a.c. travel due to sweep which in turn

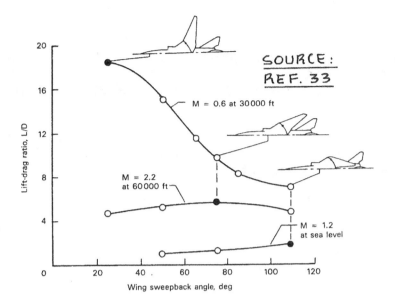

Figure 4.10 Effect of Sweep on Lift-to-Drag Ratio

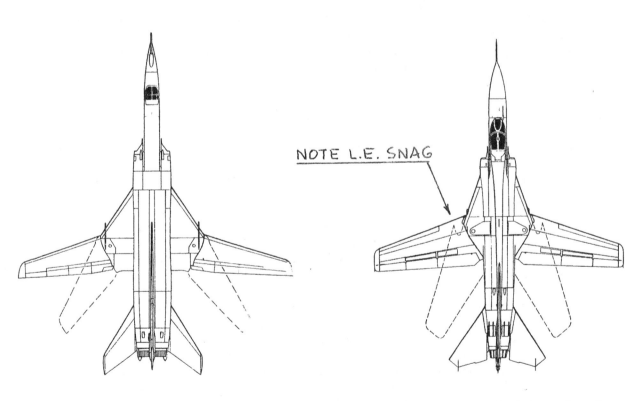

TU-22M BACKFIRE
32% PIVOT LOCATION

MIG-23 FLOGGER-G
21% PIVOT LOCATION

Figure 4.11 Example of Inboard and Outboard Wing Pivot Locations

results in less trim drag penalties. Outboard pivot locations result in large highly swept strakes which in turn result in formidable 'pitch-up' behavior at high angles of attack.

With inboard pivot locations the 'pitch-up' problem is less dominant. Weight is larger and aeroelastic effects become more important.

Figure 4.11 illustrates both inboard and outboard pivot locations.

Variable aft sweep has been used on airplanes such as: XF10F, F14, F111, Tornado, B-1, Backfire, Flogger and other Soviet airplanes.

Variable forward sweep has not yet been used operationally despite its obvious advantages.

Asymmetrical variable sweep (oblique wing) has been demonstrated in theory and in the windtunnel to result in considerable drag reductions, particularly in the transonic speed range. Because of the presence of only one pivot, a weight reduction relative to conventional VSW airplanes may be possible. The divergence behavior of the forward panel will require some increase in weight and partially off-set the pivot weight advantage.

Figure 4.12 shows why the oblique wing is (in principle) lighter than a conventional variable sweep wing. A flight test program on the NASA AD-1 (Fig. 4.13) has shown that the handling characteristics of the oblique wing are quite acceptable even without stability augmentation systems. With the advent of multiple surface, digital control systems the handling qualities issue against the oblique wing has become a non-issue.

Figure 4.14 shows a high speed application of an oblique wing to an F-8 fighter airplane.

A more radical approach to the problem of variable sweep is the so-called 'slewed wing' concept advanced by Handley Page. Figure 4.15 shows a potential application to a transport airplane. With the use of multiple trailing edge mounted flight control surfaces and modern control system design the controllability of this type of configuration is beyond question.

Table 4.5 summarizes the pros and cons of fixed versus variable sweep wing configurations, including the location of the pivot.

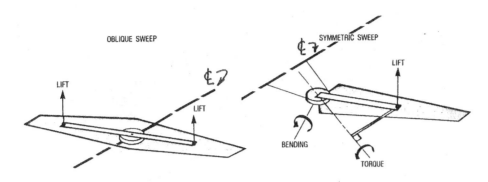

Figure 4.12 Load Comparison of Oblique and Symmetrically Variable Sweep Wings

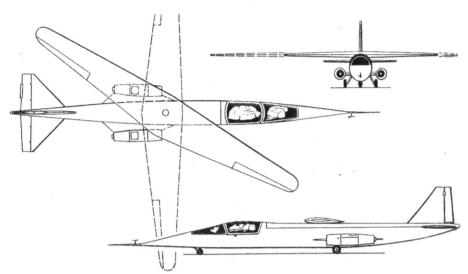

Figure 4.13 Experimental Oblique Wing Airplane: NASA AD-1

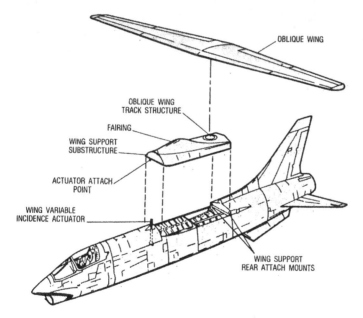

Figure 4.14 Experimental Oblique Wing Airplane: NASA F-8

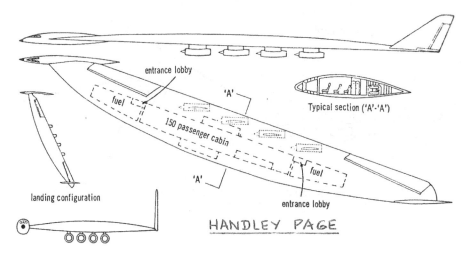

Figure 4.15 Proposed Slewed Wing Configuration

Figure 4.16 Bi-plane Example: Beech Model 17 Staggerwing

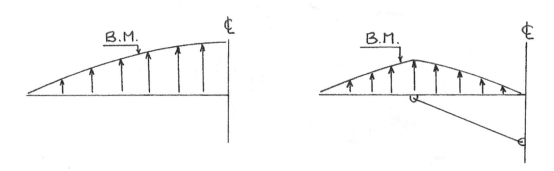

Figure 4.17 Comparison of Bending Moment Distributions of Braced and Cantilever Wings

Table 4.5 Summary of the Effect of Wing Pivot Location
==

Item	Effect of Wing Pivot Location		
	Two Pivots		One Pivot
	Inboard	Outboard	Oblique
Wing weight	Highest	High	Low
Aeroelastic effects	Severe	Moderate	Moderate
Aerodynamic center shift	Large	Small	Little
Pitch-up at high alpha	Moderate	Severe	Moderate

Table 4.6 Summary of Effect of Monoplane/Bi-plane/Joined
==

Item	Effect of Wing Configuration			
	Monoplane		Bi-plane	Joined
	Cantilever	Braced		
Wing weight	High	Low	Very Low	Low
Profile Drag	Low	High	Higher	Moderate
Interf. drag	Low	High	Higher	High

Table 4.7 Summary of the Effect of Aspect Ratio
===

Item	Effect of Aspect Ratio on Item	
	High	Low
Induced Drag	Low	High
Lift-curve slope	High	Low
Pitch attitude (approach)	Low	High
Ride in turbulence	Poor	Good
Wing weight	High	Low
Wing span	Large	Small

4.1.5 Bi-plane, Braced Wing or Joined Wing?

Bi-planes, compared to cantilever monoplanes have the following advantages, according to Reference 14:

1. Compactness: for the same required area they can have smaller spans which results in easier storage

2. For a given size airplane a bi-plane can carry more wing area which results in shorter field length requirements

3. Particularly the wire or strut-braced bi-planes are cheaper to build than comparable monoplanes, all else remaining equal.

For the low subsonic speed range it is not at all clear that a properly designed bi-plane should have poor L/D ratios. The 'old' Beech 17 Staggerwing of Fig. 4.16 sported a cruise speed of 202 mph and an $(L/D)_{max}$ of 11.7 (despite its suspension wires) according to Reference 33.

For fundamental aerodynamic data and theories on bi-planes the reader is referred to References 14 and 34.

Braced wings, compared with strutted wings have the advantage of lower structural weight. Figure 4.17 shows the reason: much less root-bending moment resulting in lower weight. Part V, p.70 suggests a factor of <u>30 percent wing weight advantage</u> for the strutted wing over the cantilevered wing!

Whether or not this difference in weight offsets the difference in drag (the strut can cause a significant increase in profile and interference drag) depends on detail design and on the required speed range. Detailed trade studies must provide the answer to this question.

It is possible to combine the weight advantage of the braced wing with the drag advantage of the cantilever wing by adapting the so-called joined wing. Figures 3.44 and 3.45 in Part II show examples of such an arrangement. Ref.35 claims significant weight and drag advantages for such a wing arrangement. A problem is that no such airplane has yet been built and certified. Therefore, many designers are weary of this possibility.

Table 4.6 summarizes the pros and cons of cantilever versus braced wing configurations.

4.1.6 Wing Aspect Ratio: High, Low and/or Winglets?

Wing aspect ratio, $A = b^2/S$ affects the following characteristics:

1. Induced drag   2. Lift-curve-slope   3. Weight   4. Span

1. Induced drag: High aspect ratio wings tend to have lower induced drag and therefore larger values of $(L/D)_{max}$. However, to utilize this advantage the associated larger value of lift coefficient requires flight at higher altitudes or at lower speeds. Equations (4.3), (4.6) and (4.7) demonstrate these points.

2. Lift-curve-slope: High aspect ratio wings tend to have high lift-curve slopes. Figure 4.8 illustrates this. High lift curve slopes have two consequences:

a) At low speed the approach attitude is conducive to good runway visibility from the cockpit. See Figure 4.9.

b) The ride through turbulence is rougher: Eqn.(4.8) shows this.

3. Weight: High aspect ratio wings tend to weigh more than low aspect ratio wings. The wing weight equations in Chapter 5 of Part V demonstrate this.

4. Span: High aspect ratio wings tend to have large spans. The definition of aspect ratio, $A = b^2/S$ shows this. Sub-section 4.1.17 presents a discussion of factors which can constrain wingspan.

Table 4.7 summarizes the effect of aspect ratio.

In some instances it is possible to increase the 'effective' aspect ratio of a wing by the use of winglets. Figure 4.18 shows an example application. The following question has been raised by designers: is it better to design a new wing with or without winglets? There still are no definitive answers to this question. Here are some opinions of the author:

a) New airplane design: A lower overall wing weight can be achieved by going to a higher aspect ratio instead of using a lower aspect ratio with winglets. However, if 'good looks' dictate the use of winglets, use them!

If, because of the configuration (aft located wing such as the Starship I) winglets are needed also for directional stability and control, it is advantageous to use winglets.

<u>b) Retrofit or growth of an existing airplane:</u> To improve cruise L/D of an existing airplane it is often lighter to add winglets than to extend the wingspan. In some cases a combination of the two works out best. An example of the latter case is the 747-400: See Fig.4.18.

There are situations where the choice of high aspect ratio leads to wing spans which are too high for reasons of operational constraints. Sub-section 4.1.17 addresses some of those constraints.

Table 4.7 summarizes the effect of wing aspect ratio on several important characteristics.

Refer to Tables 6.1 through 6.12 in Part II for values of aspect ratio used in a wide range of airplanes.

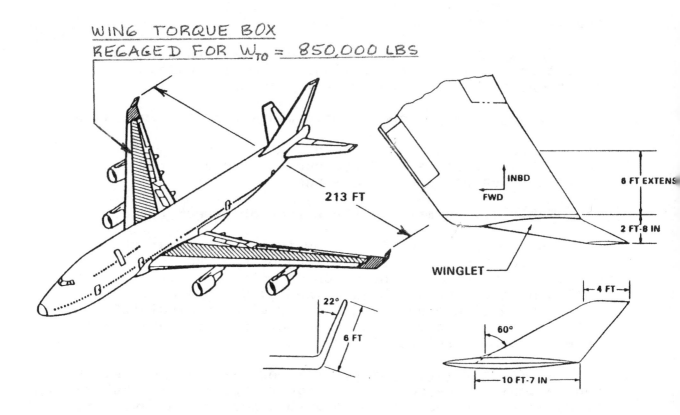

Figure 4.18 Winglet Example on the Boeing 747-400

4.1.7 Wing Thickness Ratio: Large or Small?

Wing (airfoil) thickness primarily affects the following characteristics:

1. Drag   2. Weight   3. Maximum Lift   4. Fuel Volume

1. Drag: Increased thickness means higher profile drag in the subsonic flight regime. It also means higher wave drag in the transonic and supersonic flight regime. In the transonic flight regime, Fig. 6.1 of Part II shows the large effect of thickness on compressibility drag. Use of super-critical airfoils allows designers to use larger thickness ratios and maintain fairly high subsonic Mach Numbers. Fig. 6.2 (Part II) shows this effect.

Reference 36 shows that wave drag is proportional to the parameter $(t/c)^2$ in the supersonic flight regime. Because of the very rapid increase of wave drag with t/c, the thickness of wing and empennage surfaces of supersonic airplanes must be very carefully selected.

2. Weight: Increased wing thickness means decreased wing weight since both bending and torsional stiffness increase with increasing thickness. The wing weight formulas in Chapter 5 of Part V indicate this trend.

3. Maximum Lift: Figure 7.1 in Part II shows that up to 12-14 percent thickness, maximum lift coefficients of airfoils tend to increase with increasing thickness.

4. Fuel Volume: Increased thickness translates into greater fuel volume. The effect of thickness ratio on fuel volume is given by Eqn.(6.3) in Part II. If possible fuel should be carried in the wing: this helps reduce wing weight (inertial relief) and tends to make the fuel system simpler as well.

Designers should always try to use as high a thickness ratio as possible, consistent with performance constraints. The weight decrease which results from this improves the operating economics of all airplanes.

Figure 4.19 illustrates the drag and weight trends.

Table 4.8 summarizes the primary effects of thickness ratio.

Tables 8.1 through 8.12 in Part II provide data on taper ratios used in a wide range of airplanes.

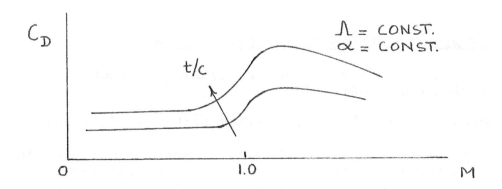

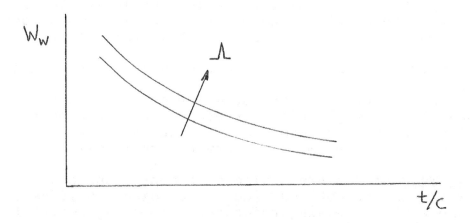

Figure 4.19 Effect of Thickness Ratio on Compressibility Drag and on Wing Weight

Table 4.8 Summary of the Effect of Thickness
===

Item	Effect of Thickness	
	Low t/c	High t/c
Wing weight	High	Low
Wing drag: subsonic	Low	High
supersonic	Acceptable	Very high
Wing fuel volume	Poor	Good
Maximum lift	Poor	Good
		Up to 12-14 percent depends on airfoil

Part III   Chapter 4   Page 188

4.1.8 Wing Taper Ratio: Large or Small?

Wing taper ratio: $\lambda = c_t/c_r$, tends to affect primarily the following items:

1. Weight 2. Tip stall 3. Fuel volume 4. Cost

1. Weight: Because wing lift distributions tend to zero at the wing tip the area near the tip of a wing is not very effective. A wing with $\lambda = 1$ will therefore 'waste' area and because of that weigh more than a wing with a smaller taper ratio.

2. Tip stall: Wings with small taper ratios tend to have small tip chords. This implies lower tip Reynold's numbers and therefore lower maximum lift coefficients. This is conducive to tip stall. An extreme example is a delta wing. Decreasing taper ratio also shifts the spanload distribution outboard : see Figure 4.20. This further aggravates tip stall.

3. Fuel volume: Large taper ratios mean more fuel volume. Eqn.(6.3) of Part II shows this effect.

4. Cost: Straight, untapered wings (no sweep, constant thickness ratio and $\lambda = 1.0$) allow for common wing ribs. This tends to lower manufacturing cost. Several light airplanes employ such wings as seen in Figures 3.4d and 3.6b,c in Part II.

Note: any value of taper ratio different from 1.0 will eliminate this advantage.

The reader may wonder if anyone has ever tried to build a wing with inverse taper? The answer is yes! An extreme example of such a wing is found in the Republic XF-91 Thunderceptor of Figure 4.21.

Reasons for using inverse taper (despite the wing weight penalty) were:

1. Improved tip stall. This was a real problem in early jet fighters. Stall protection by control-law algorithms did not exist at the time of the XF-91.

2. Improved cross sectional area distribution allowed the fuselage to be kept smaller, regaining some weight and drag.

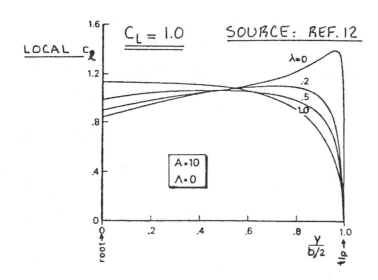

Figure 4.20 Effect of Taper Ratio on Span Loading

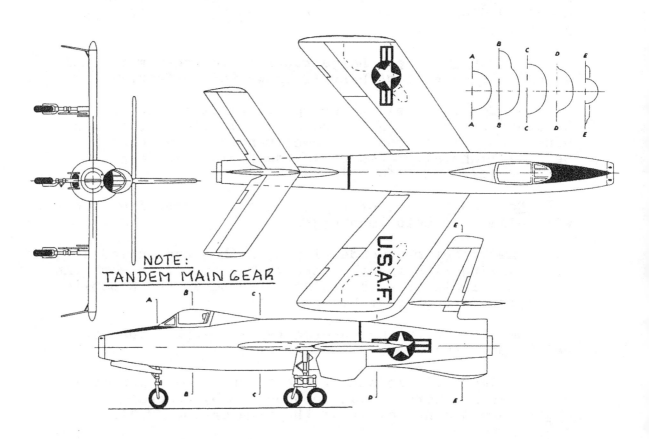

Figure 4.21 Republic XF-91 Thunderceptor: Example of an Inversely Tapered Wing

An interesting consequence of the inverse tapered wing was that the landing gear would no longer fit into the wing. To solve that problem the gear was configured as a tandem gear and retracted into the outboard wing. Note the large gear doors almost at the wingtips in Figure 4.21.

Other interesting features of the XF-91 were its variable incidence wing and its wing slats which gave it very good low speed characteristics. The XF-91 was the first USAF fighter to exceed Mach 1 on its first flight!

Table 4.9 summarizes the effect of wing taper ratio.

Tables 6.1 through 6.12 in Part II list values for wing taper ratio used in a wide range of airplanes.

4.1.9 Straight Taper or Variable Taper?

Wings with constant taper ratio are also called straight-tapered wings.

As the configuration drawings in Chapter 3 of Part II indicate, straight tapered wings are used in many airplanes. The taper ratio is then selected as a compromise between weight and good handling in the stall.

In many instance the use of 'broken' or 'curved' leading and trailing edge wings is advantageous. The following examples are offered:

1.) Cessna has been using planforms without taper inboard and straight taper outboard. Such wings have proved very effective in the low speed regime. Figures 3.5a and 3.5b in Part II are examples.

2.) In many transports, yehudis and/or gloves are used in the inboard wing section. The reasons for this are:

    a) to increase the root thickness which allows for a lighter wing.

    b) to decrease the root thickness ratio which allows for higher drag divergence Mach number.

    c) to increase the leading edge sweep angle at the root which allows for higher drag divergence Mach number (glove).

    d) to create room behind the wing spar for the moun-

Table 4.9 Summary of the Effect of Taper Ratio
===

Item	Effect of Taper Ratio	
	High	Low
Wing weight	High	Low
Tipstall	Good	Poor
Wing fuel volume	Good	Poor

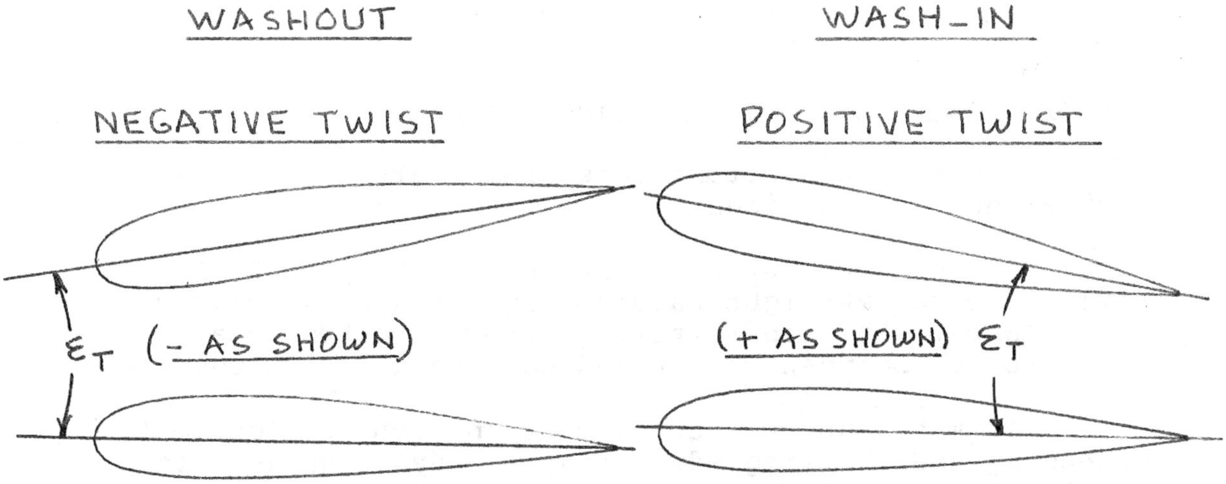

WASHOUT   WASH-IN

NEGATIVE TWIST   POSITIVE TWIST

ε_T (- AS SHOWN)   (+ AS SHOWN) ε_T

Figure 4.22 Definition of Wing Twist

Table 4.10 Summary of the Effect of Twist
===

Item	Effect of Twist Angle (Washout)	
	Large	Small
Induced drag	High	Small
Tipstall	good	poor
Wing weight	mildly lower	mildly higher

Part III   Chapter 4   Page 192

ting and retraction of the main landing gear. Gear retraction can thus be done without interfering with the flaps (yehudi).

Particularly Boeing has used the glove/yehudi combination with success: Figures 3.19 in Part II show examples.

4.1.10 Twist: How Much?

The wing twist angle affects primarily:

1. Wing tip stall   2. Induced drag   3. Wing weight

Figure 4.22 defines the wing twist angle ε_t: it is referred to as wing washout as drawn.

1. Wing tip stall: Many wings are designed with built-in twist so that the wing incidence angle actually decreases in the outboard direction. The reason for this is to delay tip stall. Tip stall is generally felt to be undesirable because it inevitably occurs in an asymmetrical manner causing serious roll control problems when approaching stall or when in a stall. Particularly aft swept wings must be twisted to prevent tip stall from happening.

2. Induced drag: A penalty which is caused by twist is an increase in induced drag. Forward swept wings (see 4.1.3) have an advantage in this regard since they can be built without twist.

In supersonic applications the use of conical camber was introduced on the F106 and B58 airplanes. The reason for this form of twist was a reduction in drag at high speed.

3. Weight: Washout as defined by Figure 4.22 tends to decrease the aerodynamic loading at the tip. This shifts the center of pressure inboard and results in a decrease in wing root bending moment. In turn this results in lower weight.

Table 4.10 summarizes the effect of twist.

Tables 6.1 through 6.12 in Part II contain typical values of wing incidence angles at the root and at the tip from which the twist angle (in case of linear twist distribution) may be deduced as:

$$\varepsilon_t = i_{w_{tip}} - i_{w_{root}} \qquad (4.10)$$

4.1.11 Wing Dihedral: How Much?

The choice of wing geometric dihedral affects the following characteristics:

1. Spiral stability 2. Dutch Roll stability
3. Ground and water clearance

Figure 4.23 defines positive and negative dihedral.

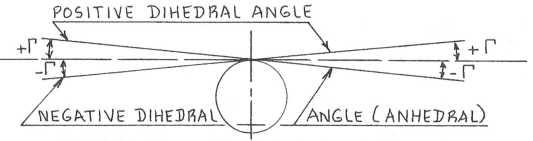

Figure 4.23 Definition of Wing Dihedral Angle

1. Spiral Stability and 2. Dutch Roll Stability:

Positive wing geometric dihedral causes the rolling moment due to sideslip derivative, C_{l_β} to be negative.

This derivative in turn affects both spiral and dutch roll stability:

More negative C_{l_β} means more spiral stability

but also less dutch roll stability! Reference 37 contains detailed discussions of these effects.

Airplanes must have a certain minimum amount of negative rolling moment due to sideslip: dihedral effect. This is needed to prevent excessive spiral instability. Too much dihedral effect tends to lower dutch roll damping.

High wing airplanes have inherent dihedral effect due to wing position while low wing airplanes tend to be deficient in inherent dihedral effect. For this reason low wing airplanes tend to have considerably greater geometric dihedral than high wing airplanes.

Swept wing airplanes tend to have too much dihedral effect due to sweep. This can be offset in high wing airplanes by giving the wing negative dihedral (anhedral). The BAe 146 of Figure 3.21d (Part II) is an example.

3. Ground and water clearance: Airplane wings, nacelles and/or propellers must have a minimum amount of ground and water clearance.

Chapter 9 in Part II presents the minimum ground clearance criteria for land based airplanes. For water based airplanes Reference 14 gives some indications of waterspray directions. Propellers and inlets should not be located in such areas.

In some airplanes positive or negative wing dihedral at the root is used for entirely different reasons. On the F4U Corsair fighter of WWII (See Figure 4.24) the reasons were: 1. low interference drag and 2. Propeller clearance with a reasonably short landing gear.

On the Piaggio 166-DL3 (See Figure 4.25) the reasons are: 1) low interference drag and 2) retain the wing of an earlier Piaggio amphibian which needed the large positive root dihedral to keep the propellers away from water spray during take-off.

In airplanes with highly elastic wings (B52, B747) the elastic deformation of the wing in flight creates extra geometric dihedral. This must be accounted for in the design of such airplanes.

Table 4.11 summarizes the effect of wing dihedral.

Tables 6.1 through 6.12 in Part II provide numerical data on wing dihedral angles used in a wide range of airplanes.

4.1.12 Wing Incidence on the Fuselage: How Much?

The following factors affect the decision of wing incidence angle relative to the fuselage:

1. Cruise drag   2. Floor attitude in cruise

3. Landing gear configuration

Figure 4.26 gives a definition of wing incidence angle, i_w.

1. Cruise drag: If i_w is made too large, the fuselage will 'cruise' nose down thereby probably increasing drag. The opposite is true as well.

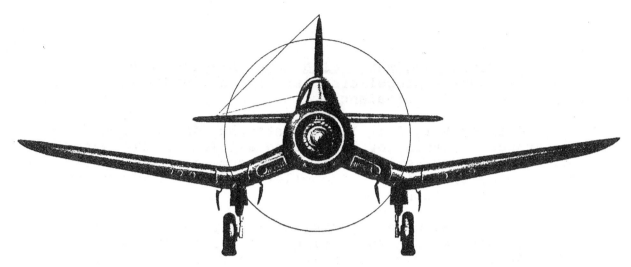

Figure 4.24 Vought F4U Corsair: Example of a Wing with Root Anhedral and Tip Dihedral

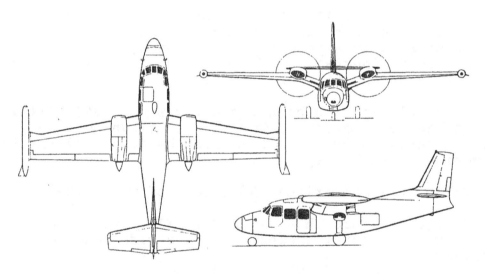

Figure 4.25 Piaggio P166-DL3: Example of a Wing with Root Dihedral and Tip Anhedral

Table 4.11 Summary of the Effect of Wing Dihedral Angle
===

Item	Effect of Dihedral Angle	
	Positive	Negative
Spiral stability	Increased	Decreased
Dutch roll stability	Decreased	Increased
Ground clearance of wing, nacelle, propeller or landing gear	Increased	Decreased

2. Floor attitude in cruise: It is also evident from Figure 4.26 that the 'floor attitude' in cruise is influenced by the choice of i_w.

If the floor attitude in cruise differs too much from horizontal it will be difficult for people to walk. Pushing beverage service carts through the aisles may also become a problem. In cruise, floor levels should not be greater than +/- 2 degrees.

The following equation may be used to determine a preliminary value for i_w:

$$i_w = (C_{L_{cr}})/C_{L_\alpha} + (d\alpha_{o_L}/d\varepsilon_t)\varepsilon_t + \alpha_{o_{L_r}} \qquad (4.11)$$

The cruise lift coefficient, $C_{L_{cr}}$ is assumed to be the lift coefficient for which the floor is supposed to be level.

The derivative $d\alpha_{o_L}/d\varepsilon_t$ may be taken as 0.4 for straight tapered wings with linear twist distribution.

Wing incidence angle can be made variable if both low cruise drag and good forward visibility in the approach mandate it. Examples are the Vought F-8 and the Republic XF-91.

In canard airplanes the relative selection of canard and wing incidence (after airfoils have been decided upon) is of great importance to the stall characteristics of the wing. Wing bending moment distribution due to canard vortex interference is another important consideration in selecting wing incidence.

3. Landing gear configuration: If a tandem landing gear is required (for example: because of a requirement for an uninterrupted bomb bay) the wing must have enough incidence so that the airplane can lift off without rotation. The Boeing B52 of Figure 4.27 is an example of this.

Table 4.12 summarizes the effect of wing incidence angle.

Tables 6.1 through 6.12 contain numerical data in wing incidence angles used on a wide range of airplanes.

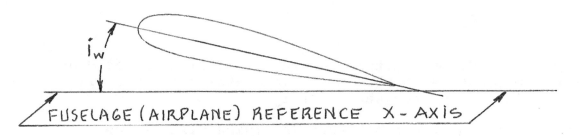

Figure 4.26 Definition of Wing Incidence Angle

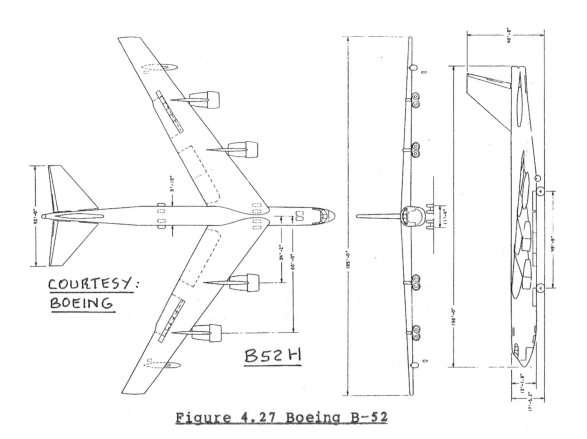

Figure 4.27 Boeing B-52

Table 4.12 Summary of the Effect of Wing Incidence Angle
===

Item	Large i_w	Small i_w
Cruise drag	High	Low
Cockpit visibility	Good	Watch out
Landing attitude in terms of nose gear hitting runway first	Watch out	No problem

4.1.13 Variable Camber (MAW = Mission Adaptive Wing)?

Birds have the ability to vary wing camber in flight in a more or less continuous manner. The advantage of this is the ability to adjust maximum lift and drag almost at will.

Wings with leading and trailing edge devices come close to this albeit for low speed flight only: the flap-placard speed inhibits operation at higher speed. Recent flight tests on an F111 with a Boeing designed variable camber wing (MAW) indicate that fighter maneuvering ability can be significantly enhanced by such a feature. The price is: weight and complexity (cost).

Figure 4.28 illustrates the MAW concept for the wing trailing edge as used in the F111 test program. The leading edge employs a similar system.

Another example of in-flight variable camber is found in the maneuvering flaps used in the Grumman X-29 (Figure 3.46, Part II).

4.1.14 Leading Edge Strakes (Lexes)

By using highly swept relatively sharp leading edge strakes (also called lexes) significant increases in lift at moderate to high angles of attack can be obtained.

The F18 and SR71 are examples of such strakes: see Figures 4.29 and 4.30. Figure 4.31 shows the additional vortex lift which can be obtained from very highly swept lifting surfaces. Lexes in fact are very highly swept lifting surfaces. Reference 36 provides data and theoretical discussions of vortex lift and vortex bursting.

A major design problem is to determine what happens to the strake/body generated vortices as they proceed downstream. If they envelop the vertical tail while still intact, they can enhance directional stability, a synergistic effect. If however, they burst before getting to the vertical tail, they can induce significant structural oscillations and lead to early fatigue problems.

The separation and bursting behavior of vortices shed from slender fuselages and/or strakes can also lead to the so-called wing-rock phenomenon. A modern theory for predicting this phenomenon is given in Reference 38.

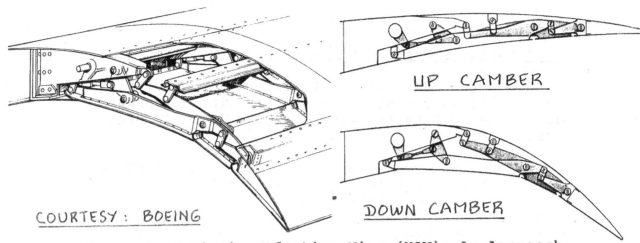

Figure 4.28 Mission-Adaptive-Wing (MAW): An Approach to Continuous Variable Camber

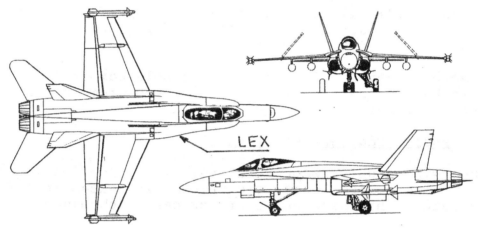

Figure 4.29 F-18: Example of Leading Edge Strakes

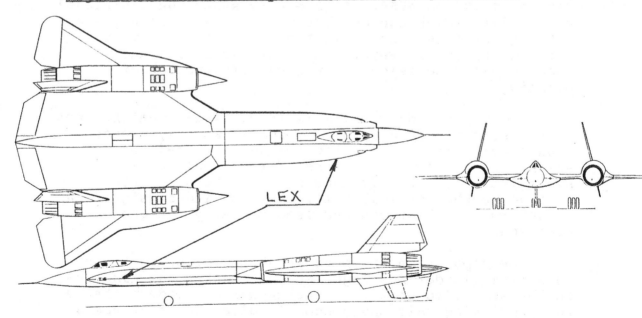

Figure 4.30 YF12A: Example of Leading Edge Strakes

Part III	Chapter 4	Page 200

4.1.15 Planform Tailoring: Why and How?

Many airplanes end up with significant planform irregularities: broken and/or curved leading edges, large fences, snags and leading edge droop and/or extensions. Examples of such planform tailoring are shown in Figures 4.32 through 4.35.

Some reasons for using planform tailoring are:

1. Stall behavior   2. Spin and/or stall entry/recovery behavior   3. Pitching moment behavior at high Mach
4. Aileron buzz   5. Aeroelastic behavior

The reasons for inboard leading edge and for trailing edge extensions were already discussed in Sub-section 4.1.9.

To improve the stall behavior of a wing (i.e. delay stall to a higher angle of attack and/or make the ensuing recovery more gentle) leading edge extensions and/or droop may be used. See Figure 4.32.

Leading edge snags and stall fences have been used in many airplanes as 'aerodynamic afterthoughts' or 'fixes' of problems ranging from high Mach pitching moment to low speed stall behavior. Figures 4.33 and Figure 4.23 (Part II) show examples.

Aileron buzz can occur if the wing sections at the aileron stations develop shocks close to the aileron hingeline. If the aileron is cable controlled (eleastic element) the aileron can develop a severe vibration which is known as aileron buzz. Such problems can be relieved by leading edge extensions (lower t/c and thus delayed shock formation) and sometimes by a row of vortex generators (Learjet 25).

At high subsonic speeds transport and bomber wings tend to develop significant aeroelastic behavior: aft swept wings 'unload' themselves outboard and forward swept wings do just the opposite. These effects can be lessened by assuring that the loci of aerodynamic centers and the elastic axis coincide. If that is achieved, aeroelastic deformations are minimized. The Handley Page Victor (Fig.4.34) uses such a (aero-isoclinic) wing. A remarkable aspect of this wing is that it was developed without the help of digital computers!!

Observe the so-called Küchemann bodies mounted midspan at the trailing edge of the wing. These bodies

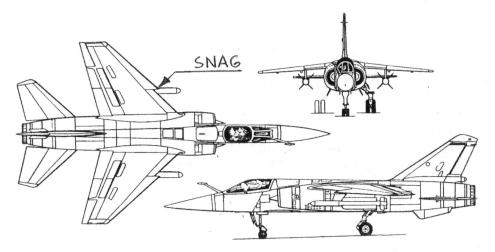

Figure 4.33 Example of Leading Edge Snags: Mirage F1-C

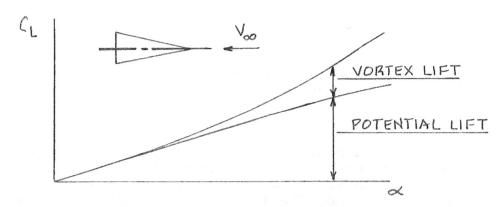

Figure 4.31 Vortex Lift Compared with Potential Lift

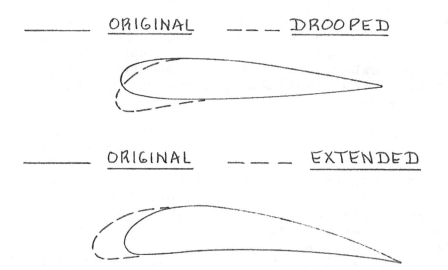

Figure 4.32 Leading Edge Droop and Leading Edge Extension

delay critical Mach number through local area ruling. Several Soviet airplanes also employ these Küchemann bodies to retract the main landing gear.

Figure 4.35 shows another example of significant leading edge tailoring in a high subsonic jet bomber, the AVRoe Vulcan.

Several supersonic airplanes employ leading edge tailoring to achieve the proper balance between acceptable subsonic and supersonic performance. The Concorde of Figure 3.35d (Part II) is an interesting example.

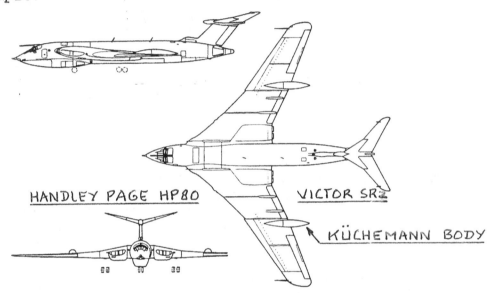

Figure 4.34 Example of Planform Tailoring: HP-Victor

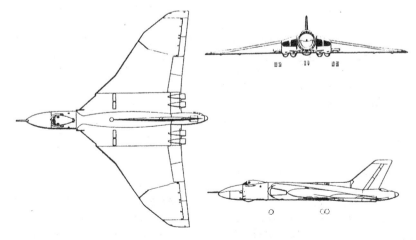

Figure 4.35 Example of Planform Tailoring: AVRoe Vulcan

4.1.16 Area Ruling: When is it Required?

Area ruling is required in those Mach ranges where the drag due to compressibility becomes unacceptably high. Figure 4.36 shows some early experimental results obtained by Whitcomb (Ref.39): with area ruling a large reduction in wave drag can be obtained.

Area ruling has also been succesfully applied in the local sense as follows:

1. Tiptank area ruling (F5)
2. Wing/nacelle area ruling (737 and GP180)

For further discussions of the area rule concept the reader should consult Part VI and Refs 13, 29 and 36.

An interesting observation is that a swept forward wing 'integrates' much smoother into a wing/fuselage combination than a swept aft wing. Figure 4.37 illustrates this point.

4.1.17 Wing Span: When is it Too Large?

For a given wing area, increasing span means increased aspect ratio. The effect of aspect ratio was discussed in Sub-sections 4.1.1 and 4.1.6. Non-aerodynamic constraints may prevent the use of wing spans beyond some specific value. Examples of such constraints are:

1. <u>Aircraft carrier space limitations:</u> This may result in the need to 'fold' wings or to limit the span.

2. <u>Hangar width limitations:</u> If building larger hangars is not justified by aerodynamic efficiencies from increased span, this may be an important constraint.

3. <u>Gate space limitations:</u> If the investment in a change in the terminal/gate infra-structure is too high, this may lead to span limitations. In such a case winglets may be used instead of physical span increases of a wing. The Boeing 747-400 is an example case! The possibility of wing-tip folding may also have to be considered at some point in the future.

4. <u>Cartwheeling:</u> Very large spans can lead to 'cartwheeling' accidents while making S-turns close to the ground. An example of an airplane with a very large wing span was the Hurel-Dubois 321 shown in Figure 4.38. The wing span of this airplane was 148.5 ft with an aspect ratio of 20.2!

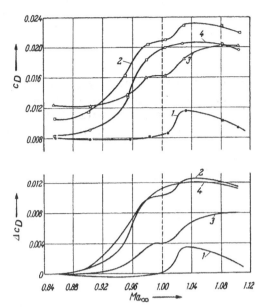
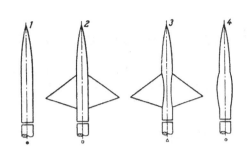

SOURCE: REF. 39

Figure 4.36 Effect of Area Ruling on Compressibility Drag

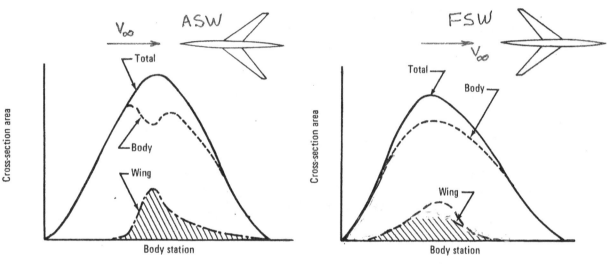

Figure 4.37 Comparison of Fuselage Area Ruling for Aft and Forward Swept Wings

Figure 4.38 Example of a Very Large Span Wing: HD-321

Part III Chapter 4 Page 205

4.1.18 Aerodynamic Coupling

Aerodynamic coupling is the intentionally adding of a closely coupled lifting surface forward of the wing. Figure 4.39 illustrates the vortex coupling which can occur between a wing and a closely coupled canard. This principle was used first on the SAAB 37 Viggen fighter airplane. Figure 4.40 shows the potential effect of this type of coupling on airplane lift-curve characteristics.

4.1.19 Flaps: What Size and Which Type?

The following factors affect the decision of wing flap size and type:

1. High lift requirements
2. Trim considerations
3. Drag considerations
4. Cost, complexity and maintenance.

1. High lift requirements: Figure 4.41 reviews typical values of wing maximum lift coefficients achievable with different types of flaps for an unswept wing with A=6.

Figure 4.42 shows values of maximum lift coefficients achievable with swept wings.

The reader is reminded that required values of overall airplane (trimmed) maximum lift coefficients were obtained from the performance sizing calculations described in Part I. Class I methods for sizing the required flaps were presented in Chapter 7 of Part II. Part VI contains methods for 'fine-tuning' the required flaps.

2. Trim considerations: Flaps cause significant changes in pitching moment. There are two sources for these pitching moments:

1. The pitching moments induced by flap camber.

2. Pitching moments induced by downwash changes on a horizontal tail and/or by upwash on a canard.

To 'trim' out these flap induced pitching moments considerable down loads may be required on a horizontal tail. Similarly, considerable uploads may be required on a canard. The resulting 'trimmed' maximum lift coefficient is usually less than the maximum lift coefficient generated by the flapped wing in a

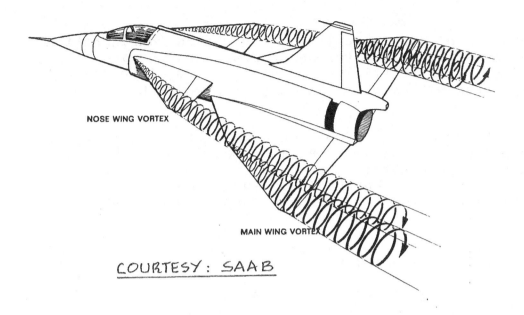

Figure 4.39 Aerodynamic Coupling Between a Wing and a Canard on the SAAB 37 Viggen

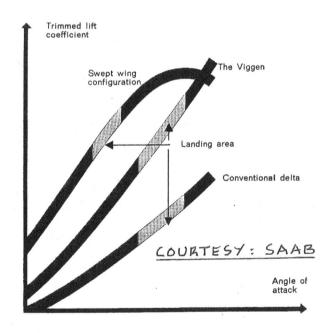

Figure 4.40 Effect of Canard/Wing Coupling on Lift

conventional configuration. In a canard configuration the potential exists for an increase in 'trimmed' maximumt lift coefficient. Interference of the canard tip vortex system with the wing may alter this conclusion!

The reader should be aware that from a performance viewpoint, the only maximum lift coefficient which is of practical interest is the 'trimmed' maximum lift coefficient. Conventional configurations are normally critical at forward c.g. while canard configurations tend to be critical at aft c.g.

For detailed information on lift, drag and pitching moments associated with flaps, see Refs 15, 40 and 41.

Reference 37 contains detailed discussions of airplane trim diagrams. Part VI contains methods for constructing trim diagrams for new designs.

3. Drag considerations: Flap deployment always results in an increase in drag. An important design consideration in the selection of a flap system is the flaps-down lift-to-drag ratio in an engine out climb. Part I addressed this problem by presenting a rapid method for determining thrust/power to weight ratios needed to satisfy engine out climb requirements. The drag estimates used in those sizing methods were very preliminary in nature (Class I). More accurate methods for estimation of drag due to flaps are given in Part VI.

4. Cost, complexity and maintenance: As a general rule of thumb, the higher the trimmed value of maximum lift coefficient with the flaps down, the greater the complexity, the maintenance requirements and therefore the cost. By careful attention to the detail mechanical design of the flaps and the associated systems, cost can be kept within acceptable limits. The Boeing 727 still is an outstanding example of a highly complex yet reliable and maintainable flap system.

Section 4.2 presents examples of a number of flap mechanizations.

4.1.20 Lateral Controls: Type, Size and Location?

The following types of wing mounted lateral control device are prevalent today:

1. Ailerons   2. Spoilers

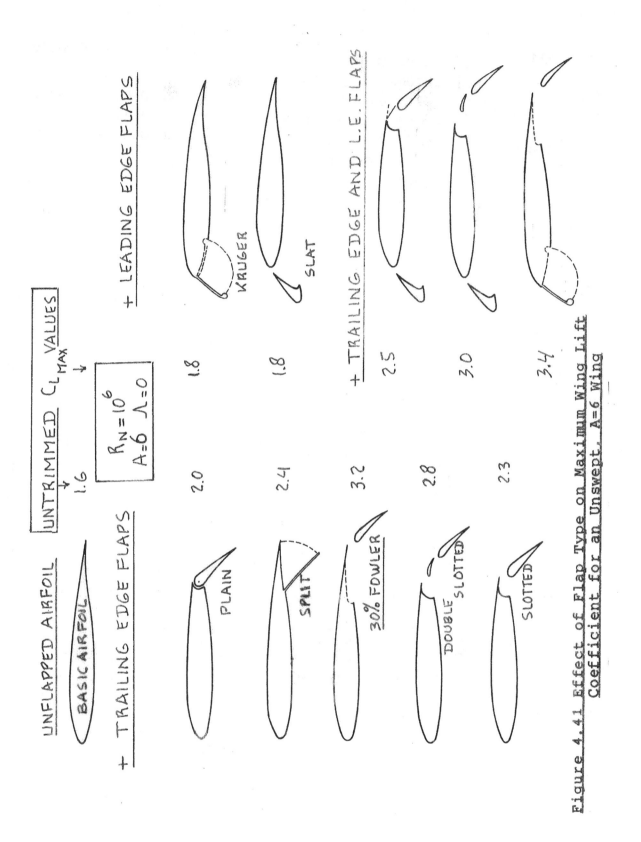

Figure 4.41 Effect of Flap Type on Maximum Wing Lift Coefficient for an Unswept, A=6 Wing

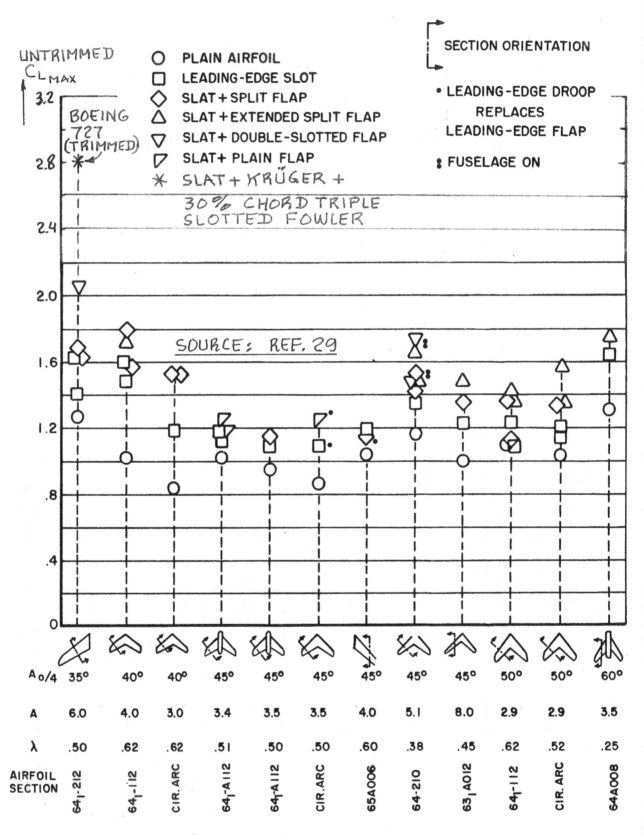

Figure 4.42 Effect of Aspect Ratio, Sweep Angle and Flap Type on Wing Maximum Lift Coefficient

Most airplanes use wing mounted ailerons and/or spoilers for lateral control. The reason is to take advantage of the largest possible moment arm. Fighter airplanes and some military trainers use differential stabilizers to enhance their lateral control effectiveness.

Tables 8.1 through 8.12 in Part II contain data for aileron and spoiler sizes and locations employed on a wide range of airplanes. More detailed methods for sizing of lateral controls may be found in Part VI and Part VII.

1. **Ailerons:** Ailerons are 'plain' flaps mounted close to the wing tips for lateral control. They loose effectiveness at high angles of attack. They create a negative yawing moment, called adverse yaw. The magnitude of adverse yaw may be decreased by the application of 'differential' aileron controls (See Figure 4.43) or by the use of so-called Frise ailerons (See Figure 4.44).

When ailerons are used on swept aft wing airplanes they loose effectiveness at high dynamic pressures because of a phenomenon called 'aileron reversal. This aeroelastic phenomenon can be so severe that outboard ailerons must be 'locked-in-place' in some airplanes : Boeing 707, 727, 747.

At very high sweep angles ailerons loose effectiveness also because of the fact that the outboard flow over the top of the wing tends to become parallel to the aileron hinge line.

The English Electric Lightning fighter employed an interesting tip mounted aileron which neatly overcame the aerodynamic and aeroelastic degradations in effectiveness. Figure 4.45 shows the Lightning configuration.

2. **Spoilers:** Spoilers literally 'spoil' the lift over the part of the surface immediately behind the spoiler. Figure 4.46 illustrates the effect. Spoilers are extremely effective with flaps down. Most high speed airplanes use spoilers for roll control at high speed and a combination of ailerons and spoilers at low speed.

Spoilers generate a positive yawing moment: proverse yaw. Proverse yaw is to be preferred over adverse yaw. However, too much proverse yaw is also disliked by pilots.

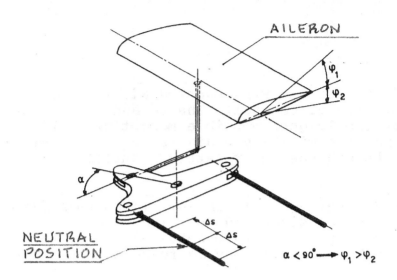

Figure 4.43 Example of Differential Aileron Control

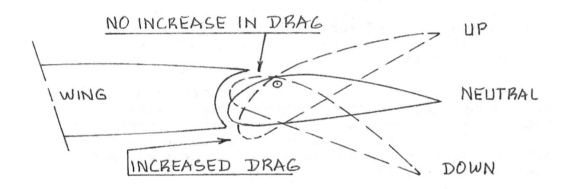

Figure 4.44 Example of Frise Ailerons

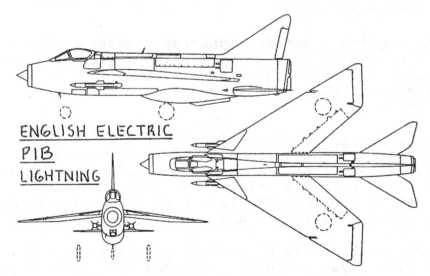

Figure 4.45 Example of Tip Mounted Ailerons

Because spoilers separate the airflow they also cause drag. For this reason it is not a good idea to use spoilers alone when flying on autopilot through turbulent air: drag is continually being 'integrated'. A better solution is to use a small inboard aileron (inboard to avoid aeroelastic effects) for moderate autopilot inputs. For large autopilot inputs a spoiler can be 'picked-up' after using say 15 degrees of the inboard aileron. This type of 'nonlinear' gearing of lateral controls is widely used on transports. Nearly all Boeing transports use this type of lateral control system.

Section 4.2 presents examples of wing mounted lateral control mechanizations.

Part VI contains methods for estimating lateral control effectiveness.

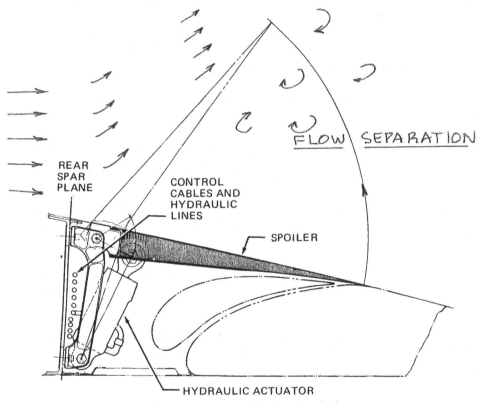

Figure 4.46 Example of Spoiler Operation

4.1.21 Review of Wing Drag Contributions

The wing is responsible for generating most of the lift on an airplane: 90 to 95 percent. The wing is also responsible for generating a large amount of drag: 20 - 40 percent of the total drag of an airplane. It is obviously desirable to design the wing so that it provides the highest possible value of lift-to-drag ratio in those flight conditions where aerodynamic efficiency is important. Such flight conditions are:

*cruise   *loiter   *climb   *engine-out climb

The reader is referred to references 12, 13 and 29 for excellent discussions on the aerodynamic design of wings. A summary of important design considerations is given in the following.

Wings, in addition to generating lift, are responsible for generating the following types of drag:

1. Friction drag
2. Induced drag
3. Compressibility drag
4. Interference drag
5. Profile drag

Detailed procedures for predicting wing drag are presented in Part VI.

1. Friction drag: Friction drag is directly related to wetted area and to the type of boundary layer which is present in a given flight condition.

Wing wetted area is itself related to wing planform area (wing reference area), and to airfoil thickness: Equation (12.1) in Part II shows this.

For a given wetted area the friction drag depends on how much of the boundary layer is laminar and how much is turbulent.

Figure 4.47 shows the major drag reduction which can be obtained with laminar flow as compared to turbulent flow.

Until fairly recently it was generally believed that except for gliders and certain smoothly finished homebuilts, most airplanes were 'turbulent' airplanes. In predicting drag of most airplanes the boundary layer was assumed to be fully turbulent.

References 42 and 43 show that very significant

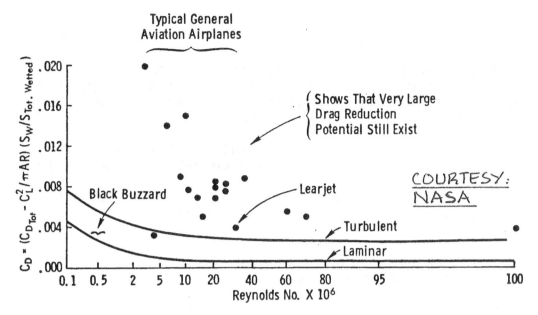

Figure 4.47 Example of Drag Reduction Potential With Laminar Flow

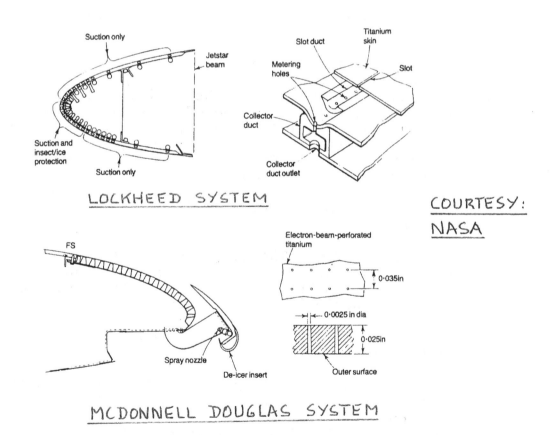

Figure 4.48 Examples of Hybrid Laminar Flow Control

laminar flow runs are indeed feasible at typical operational combinations of Mach Number and Reynold's Number for airplanes ranging from general aviation airplanes to large jet transports.

Reference 44 shows the significant performance advantages which can accrue due to natural laminar flow.

Typical questions which arise in wing design with regard to the possibility of achieving natural laminar flow are:

1. What roughness and waviness criteria need to be used in the wing manufacturing process?

2. What combinations of sweep angle, Mach number and Reynold's Number lead to stable laminar flow?

3. What about insect contamination?

References 45 and 46 provide some answers to these questions.

If natural laminar flow cannot be achieved in a stable, reliable manner, the next approach is that of 'hybrid' laminar flow. Here, the leading edge is artificially kept laminar by boundary layer suction. This then allows longer 'natural' laminar flow runs over the airfoil. Figure 4.48 shows two design approaches to 'hybrid' laminar flow. Reference 47 contains a discussion of these approaches.

Methods for estimating wing friction drag are provided in Part VI.

2. Induced drag: The amount of induced drag generated by the wing is dependent on the amount of lift being generated, on the aspect ratio of the wing and on the so-called Oswald efficiency factor.

Methods for estimating the induced drag of wings are presented in Part VI.

3. Compressibility drag: How much compressibility drag is generated by a wing depends on the Mach Number, the sweep angle, the thickness ratio and the blending of the wing into the fuselage. In chapter 6 of Part II a simple method was provided to select wing sweep and wing thickness so that the wing operates below the drag rise Mach Number in subsonic flow. The compressibility drag in supersonic flow (also referred to as wave drag)

depends also on sweep angle, on thickness ratio on lift coefficient and on Mach number.

Part VI contains methods for estimating the wing contribution to compressibility drag. Area ruling of the wing fuselage intersection is nearly always required to keep the compressibility drag within acceptable bounds.

<u>4. Interference drag:</u> Interference drag is a poorly understood component of drag. It is well known, that interference drag depends greatly on the 'fairing' of a wing into a fuselage. The same can be said for the interference drag created by other bodies installed under or on a wing. Examples of such bodies are: nacelles, tanks, radar pods and typical military stores. In some instances 'local' arear ruling is required to bring the interference drag down. The reader should consult Reference 15 for more information on interference drag.

<u>5. Profile drag:</u> Profile drag in wings comes about only after flow separation has occurred. Properly designed wings do not exhibit significant areas of flow separation in most flight regimes. An exception is formed by vehicles which spend a significant amount of time maneuvering close to the maximum lift coefficient. In those cases profile drag becomes important. Use of experimental data is recommended whenever drag due to flow separation needs to be estimated.

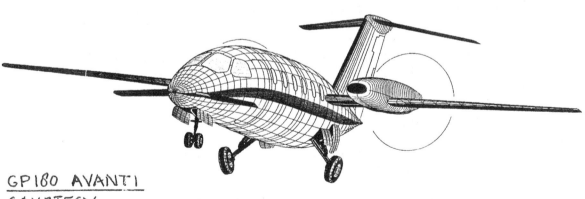

GP180 AVANTI
COURTESY:
GATES LEARJET / PIAGGIO

4.2 STRUCTURAL DESIGN CONSIDERATIONS AND EXAMPLES OF STRUCTURAL LAYOUT DESIGN OF WINGS

The purpose of this section is to provide some guidelines and examples of how the structural layout design and the integration of the wing structure into the airplane can be accomplished.

The material in this section is organized as follows:

4.2.1 Typical Spar, Rib and Stiffener Spacings

4.2.2 Examples of wing structural arrangements

4.2.3 Examples of wing/fuselage integration

4.2.4 Examples of wing cross section design

4.2.5 Examples of lateral control mechanizations

4.2.6 Examples of high lift device mechanizations

4.2.7 Examples of wing skin gages

4.2.8 Maintenance and Access Requirements

Chapter 7 contains a procedure for preparing the overall structural arrangement of an airplane. For additional insight into airplane structural arrangements the reader should consult Chapter 8.

4.2.1 Typical Spar, Rib and Stiffener Spacings

The actual structural arrangement of spars, ribs and skin stiffeners depends very much on the type of airplane being designed and the loads to which it will be subjected. The reader should refer to Refs. 11, 19 and 23 for detailed information on the types of loads to which airplane structures are subjected. Reference 20 contains detailed methods for the preliminary structural analysis of wing structures.

Figure 4.49 defines the locations of major structural components for wings.

Wing Spar Locations: Most airplane wings use a so-called torque-box (wing-box) as the main load carrying component. The torque box should be located to take maximum advantage of the structural height available within the airfoil contours. This will save weight. The

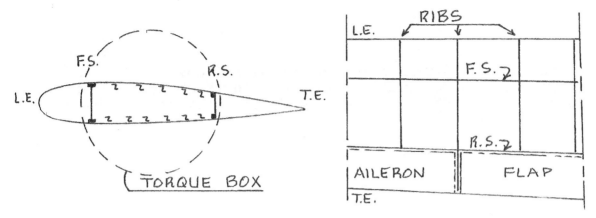

Figure 4.49 Definition of Major Structural Wing Components and Their Location

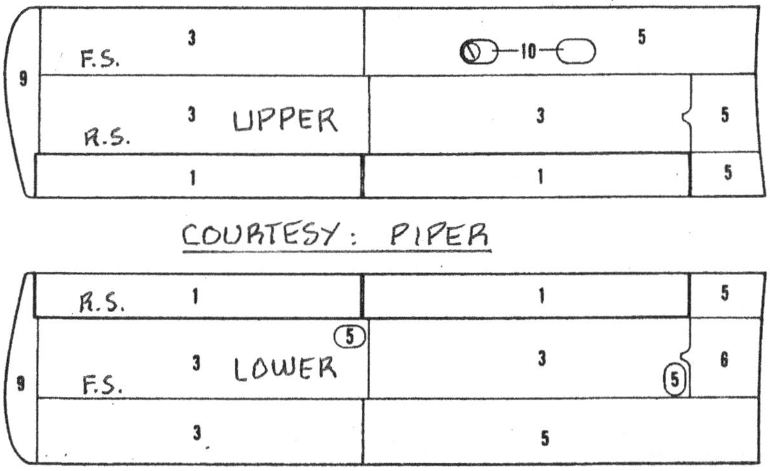

Figure 4.50 Wing Structural Arrangement Piper PA-38-112 Tomahawk

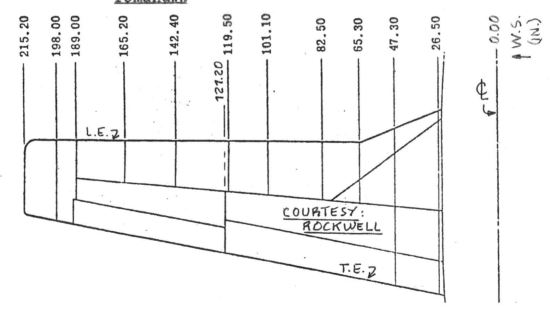

Figure 4.51 Wing Structural Arrangement Rockwell 112B

Part III   Chapter 4   Page 219

torque box is normally closed off by a front spar (F.S.), a rear spar (R.S.) and an upper and lower skin. The spar locations are often constrained by requirements for high lift devices.

Typical spar locations are:

Front spar: 15-30 percent chord

Rear spar: 65-75 percent chord

Multiple spar construction is often applied in the case of fighter wings. The F16 wing has 11 spars, 5 ribs and a machine-tapered skin without stiffeners.

<u>Wing Rib Locations:</u> To help stabilize torque box skins and to serve as attachment points for leading edge skins, trailing edge skins and/or flaps, ailerons and spoilers, wing ribs are used. Typical rib spacings are:

Light Airplanes: 36 inches

Transports: 24 inches

Fighters and Trainers: rib spacings vary widely.

Ribs are always required at those spanwise locations where stores are attached to a wing. Examples of such ribs are shown in Section 4.3.

<u>Wing Stiffener Spacings:</u> These vary widely and depend on the relative stiffness of the wing skin. For example spacings see sub-section 4.2.2.

4.2.2 Examples of Wing Structural Arrangements

A procedure for arriving at the overall structural arrangement for any new airplane is given in Chapter 7. Figures 4.50 through 4.59 present examples of wing structural arrangements for the following airplanes:

Fig. 4.50 Piper Tomahawk   Fig. 4.51 Rockwell 112B

Fig. 4.52 Short Skyvan   Fig. 4.53 McDD DC9-30

Fig. 4.54 McDD DC-10   Fig. 4.55 Boeing 767

Fig. 4.56 Aerospatiale   Fig. 4.57 Douglas A4
         Corvette

Fig. 4.58 Piaggio P166   Fig. 4.59 Canadair CL-215

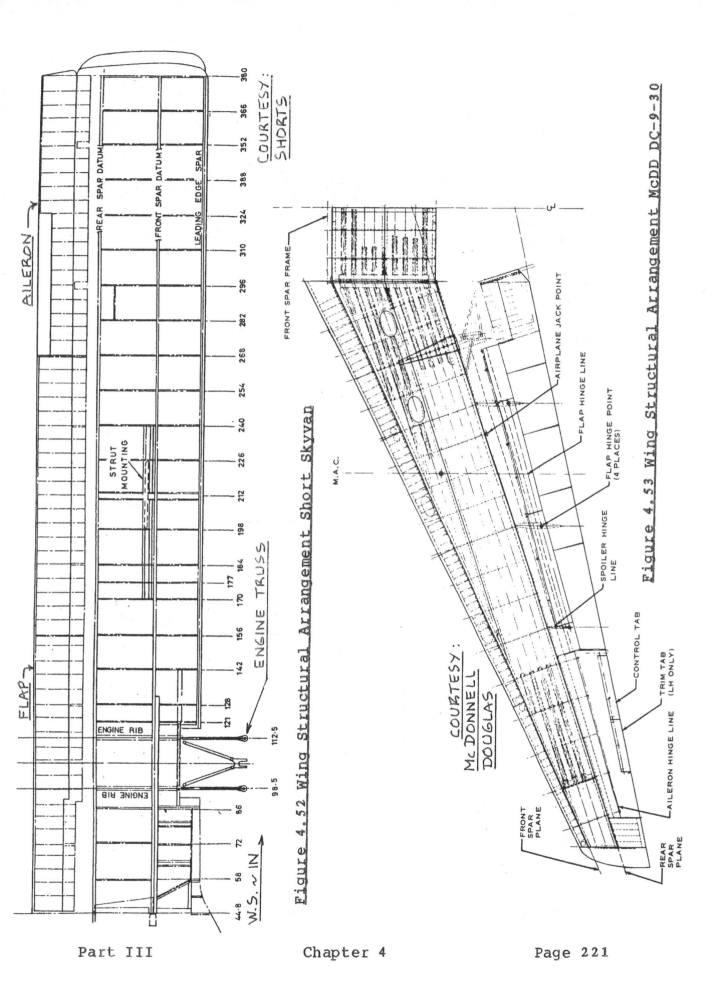

Figure 4.52 Wing Structural Arrangement Short Skyvan

Figure 4.53 Wing Structural Arrangement MCDD DC-9-30

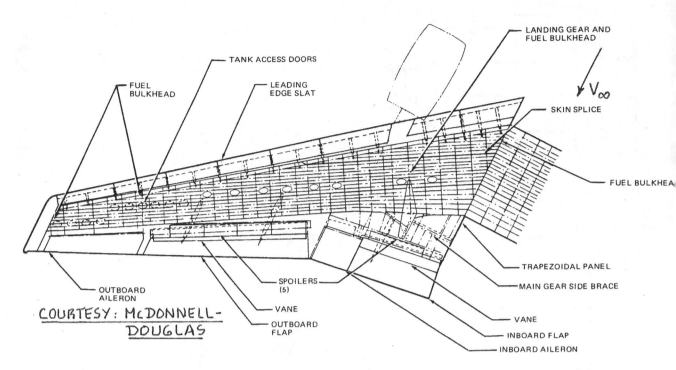

Figure 4.54 Wing Structural Arrangement McDD DC-10

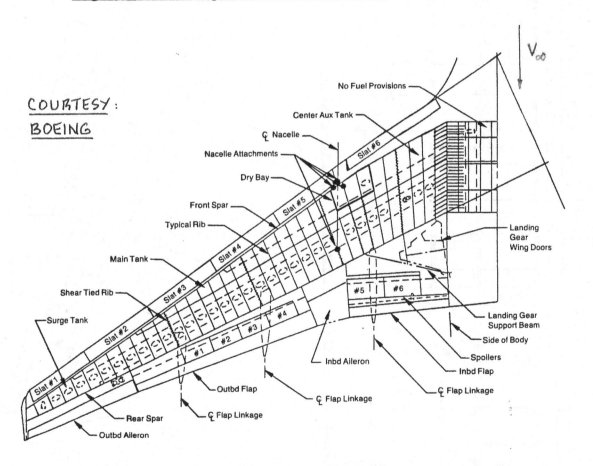

Figure 4.55 Wing Structural Arrangement Boeing 767

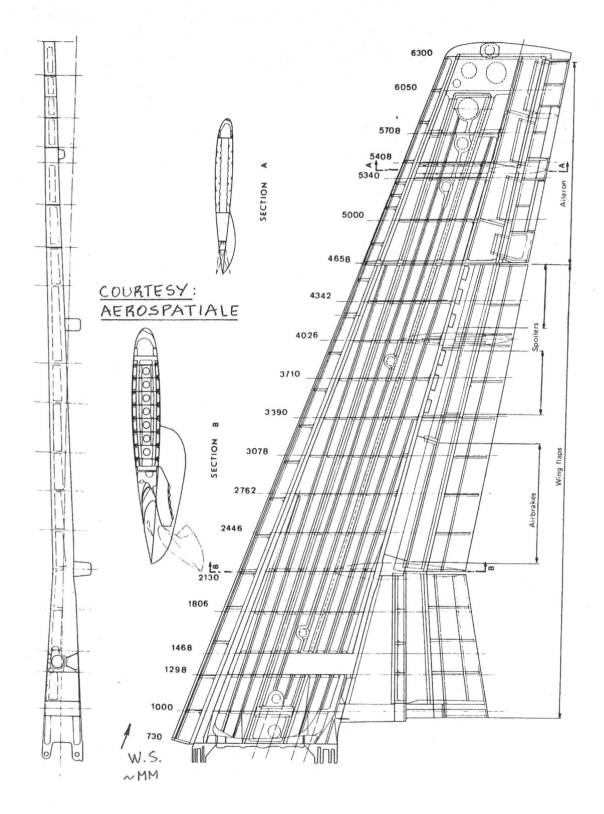

Figure 4.56 Wing Structural Arrangement Aerospatiale Corvette

Part III   Chapter 4   Page 223

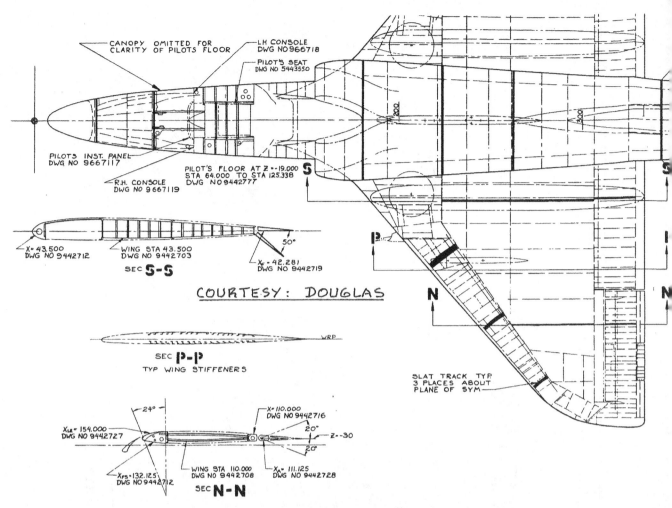

Figure 4.57 Wing Structural Arrangement Douglas A4D-2N

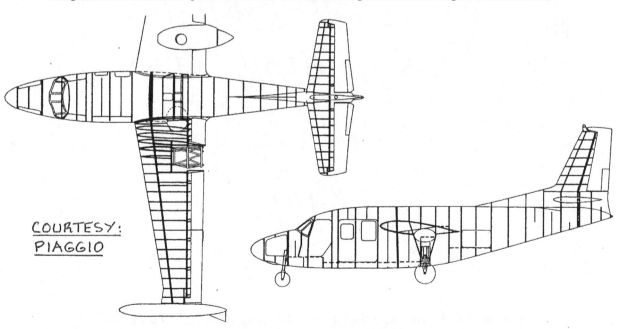

Figure 4.58 Overall Structural Arrangement Piaggio P166-DL3

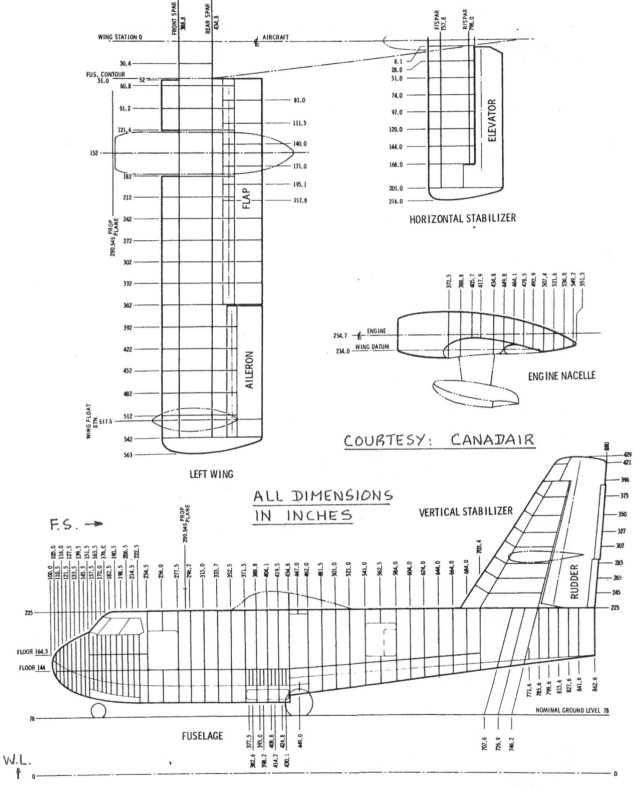

Figure 4.59 Overall Structural Arrangement Canadair CL-215

Note that in most of these structural arrangements the wing stations of major structural elements are defined. A complete structural arrangement must contain that information!

The structural arrangements for the P166 and the CL 215 are given in their entirety.

4.2.3 Examples of Wing/Fuselage Integration

Wings are normally joined to a fuselage in one of the following manners:

1. Wings are 'bolted' to a fuselage 'carry-through' section.

Figures 4.60 and 4.61 are examples of this type of wing/fuselage joint. Note that in Fig. 4.61 the attachments bolts are in tension/compression. In most other wing fuselage joints the attachment bolts are in shear.

Note that in the wing/fuselage attachment of Fig.4.60 the front spar attachment takes care of all bending loads while the rear spar attachment can only transmit a normal force (pin-joint!).

The Corvette wings (Fig. 4.56) are joined to the fuselage through a total of four large bolts in shear.

2. Wings, including the carry-through section are joined to the fuselage by a large number of 'small' bolts.

Figures 4.57 and 4.62 are examples of this type of arrangement

4.2.4 Examples of Wing Cross Section Design

Figures 4.63 and 4.64 provide examples of typical wing cross section designs for the DC9-30 and the DC-10 respectively. Figure 4.57 shows the wing cross sections in a typical fighter.

Figures 4.65 through 4.67 privide details of typical wing box construction methods. Conventional riveted, bonded honeycomb and composite constructions are shown respectively.

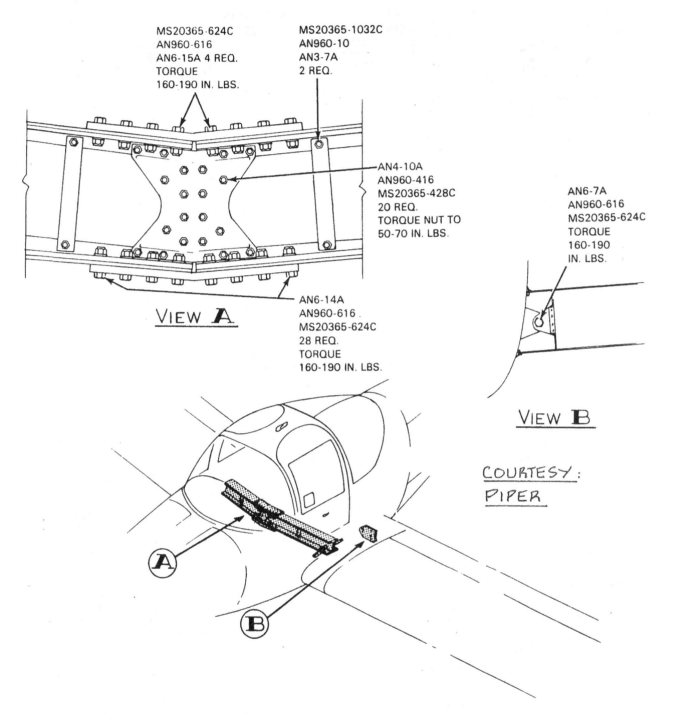

Figure 4.60 Wing/Fuselage Attachment Piper PA-38-112 Tomahawk

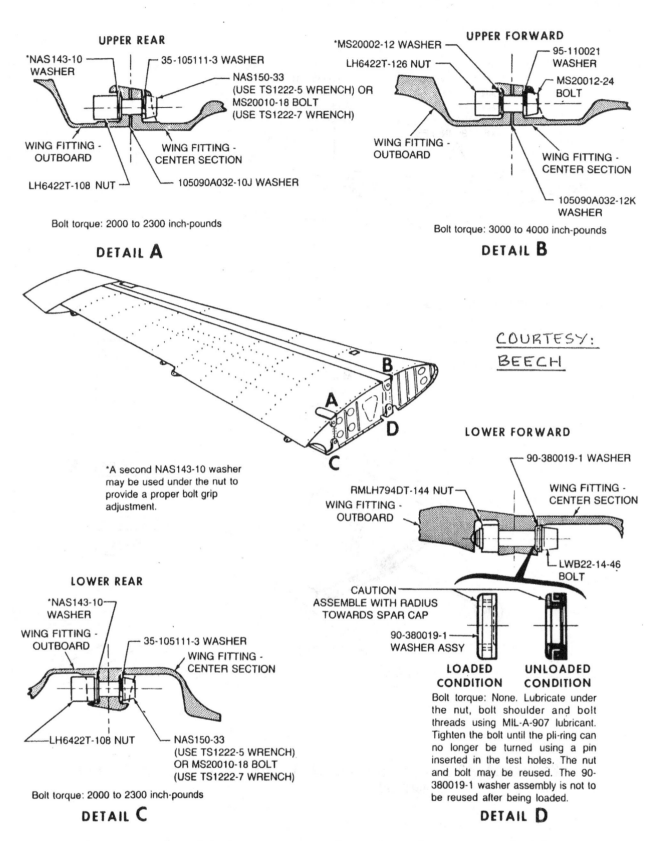

Figure 4.61 Wing/Center Section Attachment Beech King Air F90

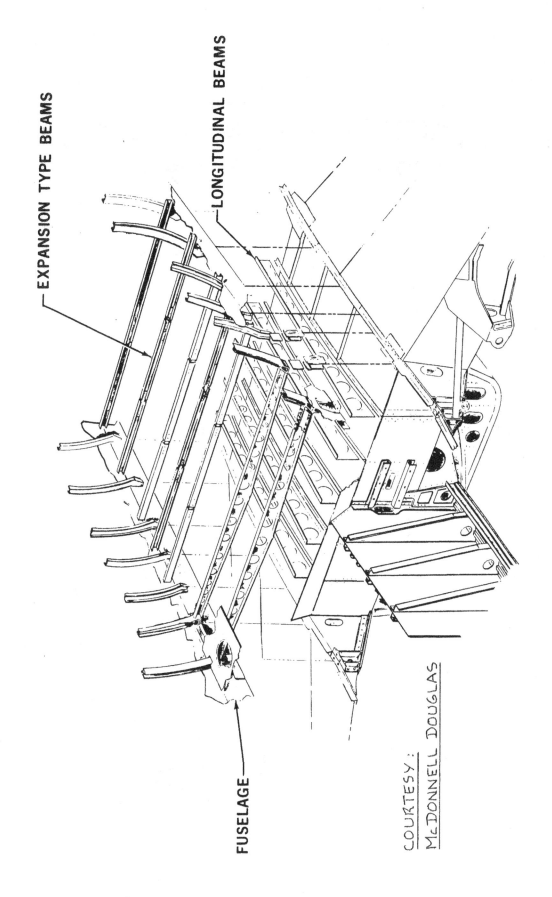

Figure 4.62 Wing/Fuselage Attachment McDD DC-10

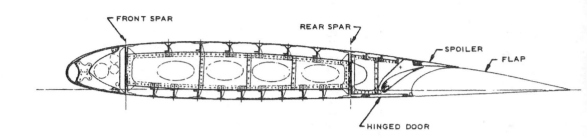

Figure 4.63 Wing Cross Section Design McDD DC-9-30

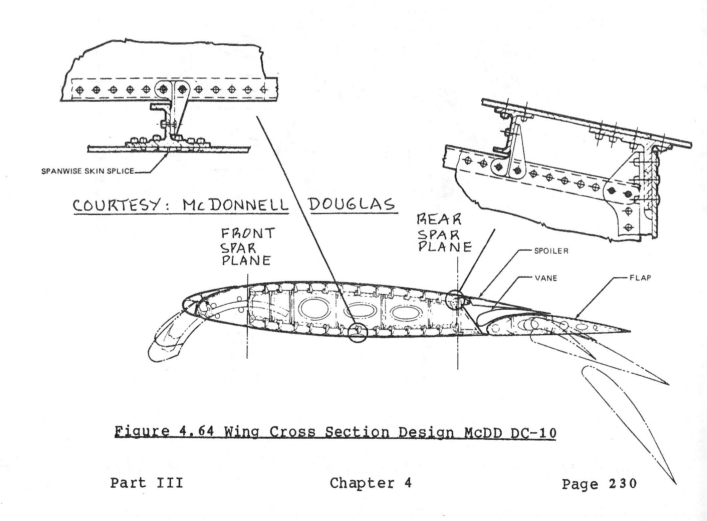

Figure 4.64 Wing Cross Section Design McDD DC-10

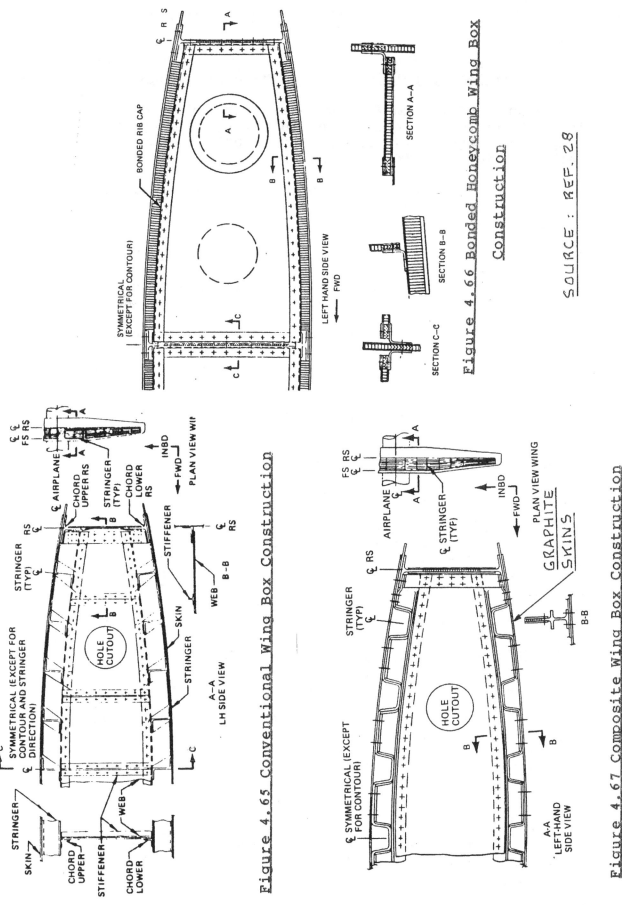

Figure 4.65 Conventional Wing Box Construction

Figure 4.66 Bonded Honeycomb Wing Box Construction

SOURCE: REF. 28

Figure 4.67 Composite Wing Box Construction

4.2.5 Examples of Lateral Control Mechanizations

Figures 4.68 and 4.69 show examples of outboard and inboard aileron mechanizations in a large transport airplane. Figure 4.70 shows a roll control spoiler mechanization also for a transport airplane.

Figure 4.56 and 4.57 show the aileron mechanizations used in a business jet and in a fighter airplane respectively.

4.2.6 Examples of High Lift Device Mechanizations

Figures 4.71 through 4.77 show examples of high lift device mechanizations in a large transport airplane.

Double slotted flap mechanizations are shown in Figures 4.71 through 4.74 while the corresponding leading edge slat mechanizations are shown in Figures 4.75 and 4.76.

A single slotted flap design is seen in Fig. 4.77.

A plain type flap for a fighter airplane is shown in Figure 4.57, cross section NN.

4.2.7 Examples of Wing Skin Gages

Figure 4.78 and 4.79 show examples of typical wing skin gages and materials used in light twin turboprop business airplanes. The actual skin gages used depend very much on the V-n diagram for which the airplane is designed. The reader is referred to Ref. 30 for descriptions of typical structural materials and construction methods used in a wide variety of airplanes.

4.2.8 Maintenance and Access Requirements

Wing structures must be inspectable. That implies inspection covers of a size sufficiently large that inspection and if necessary maintenance functions can be carried out. Figures 4.53 - 4.56, 4.78 and 4.79 give examples of different types of inspection holes. Note that large numbers of these are required.

In many airplanes the leading and trailing edges of the wing are used to house high lift devices, lateral controls and their associated systems. All of these require maintenance and therefore access. The comments made about accessibility in Sub-section 3.3.6 also apply to wing access.

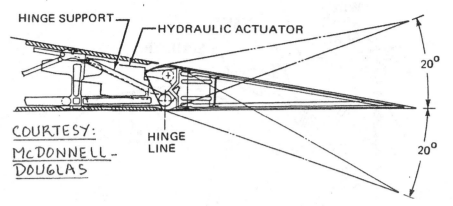

Figure 4.68 Inboard Aileron Installation McDD DC-10

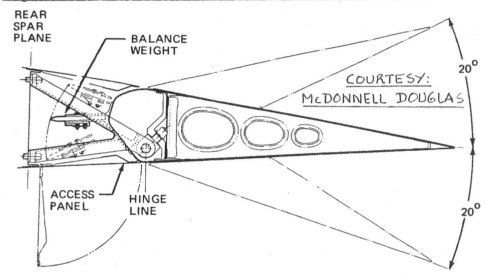

Figure 4.69 Outboard Aileron Installation McDD DC 10

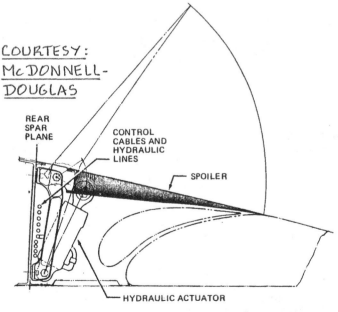

Figure 4.70 Outboard Spoiler Installation McDD DC-10

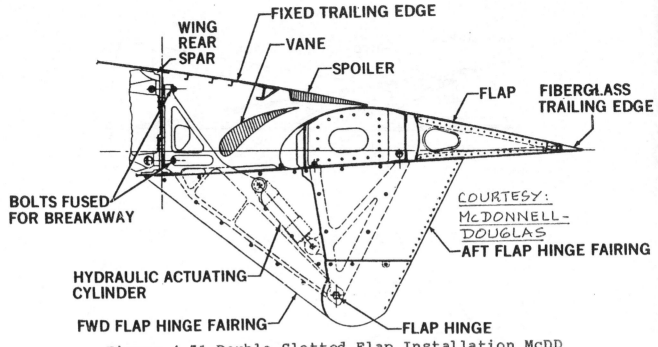

Figure 4.71 Double Slotted Flap Installation McDD DC-9-30

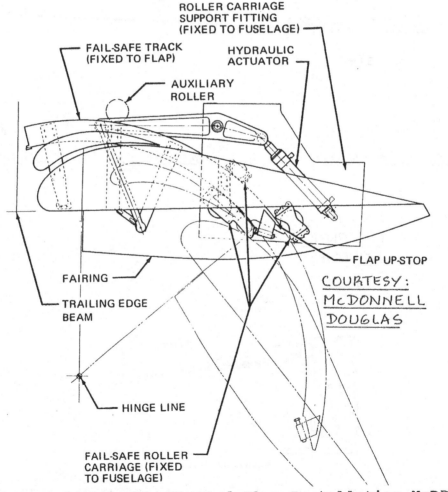

Figure 4.72 Double Slotted Flap Installation McDD DC-10: Inboard Flap, Inboard Support

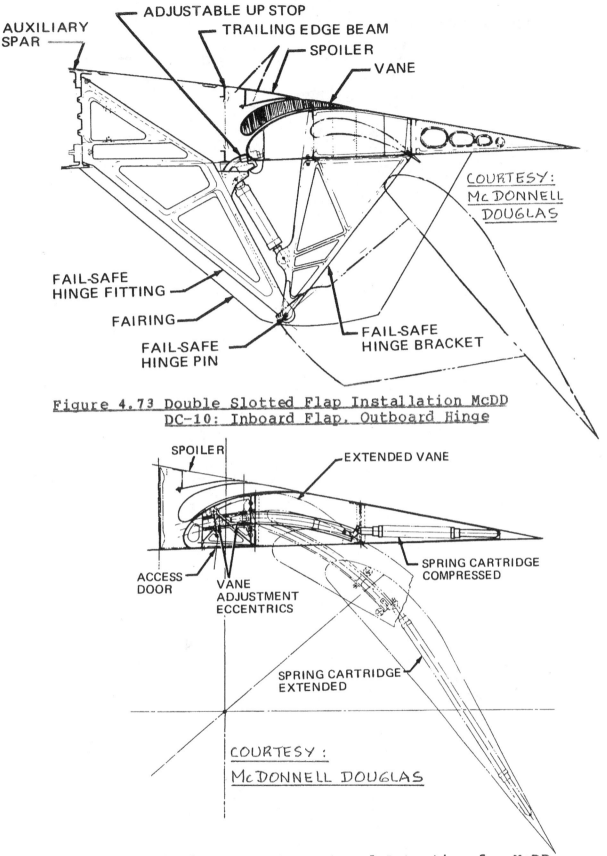

Figure 4.73 Double Slotted Flap Installation McDD DC-10: Inboard Flap, Outboard Hinge

Figure 4.74 Flap Vane Support and Actuation for McDD DC-10: Double Slotted Flaps

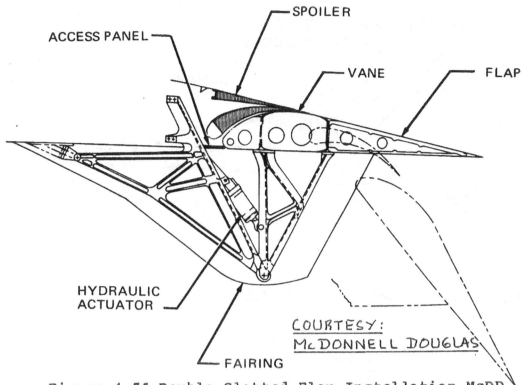

Figure 4.75 Double Slotted Flap Installation McDD DC-10: Outboard Flap Drive Hinge

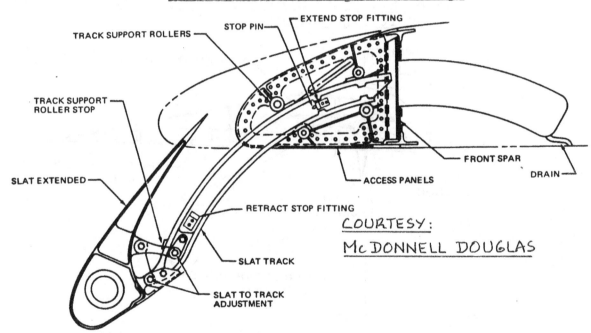

Figure 4.76 Leading Edge Slat Installation McDD DC-10

Figure 4.77 Single Slotted Flap Installation FR T-46

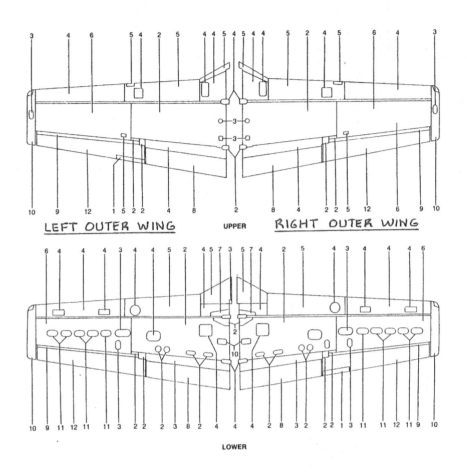

NUMBER	MATERIAL	THICKNESS IN INCHES
1.	2024-T3	.016
2.	2024-T3	.020
3.	2024-T3	.025
4.	2024-T3	.032
5.	2024-T3	.040
6.	2024-T3	.090
7.	2024-T42	.064
8.	6061-T6	.020
9.	6061-T6	.025
10.	6061-T6	.032
11.	6061-T6	.250
12.	AZ-31B-0 (Magnesium Alloy)	.020

COURTESY: BEECH

Figure 4.78 Typical Wing Skin Gages and Materials in a Beech King Air F90: Outer Wings

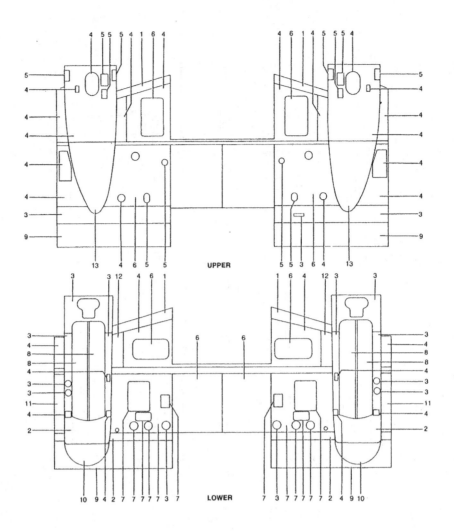

NUMBER	MATERIAL	THICKNESS IN INCHES
1.	2024-T3 (Four layers of .012 material are bonded together with Epibond 104 adhesive)	.012
2.	2024-T3	.020
3.	2024-T3	.025
4.	2024-T3	.032
5.	2024-T3	.040
6.	2024-T3	.050
7.	2024-T3	.063
8.	2024-T42 (Bonded Honeycomb)	.016
9.	2024-T42	.020
10.	2024-T42	.025
11.	2024-T42	.032
12.	2024-T42	.040
13.	6061-T62	.040

COURTESY: BEECH

Figure 4.79 Typical Wing Skin Gages and Materials in a Beech King Air F90: Center Section

4.3 MILITARY DESIGN CONSIDERATIONS

In many military airplanes there is a need for carrying external stores, for folding wings and for incorporating variable wing sweep.

Figure 4.80 shows an example of a typical wing 'hard point' and the method used to attach stores via a so-called weapons pylon. Note the major strengthening that is used at such a 'hard point'.

Particularly in the case of aircraft carrier based airplanes the need for wing folding (and sometimes vertical tail folding) often exists. Figures 4.81 through 4.84 show examples of folding applications.

A typical design problem associated with such folds is the way any controls are routed through such folding joints. Figure 4.85 shows an example of a wing fold joint. The controls are routed through this joint with a method shown in Figure 4.86.

Figure 4.87 shows a typical variable sweep wing pivot construction method. The reader is reminded of the fact that all wing loads must be passed through that one pivot. It is evident that this will incur a major structural weight penalty. In part V this penalty is assessed as 17.5 percent of the wing weight! Figure 4.87 also shows an application of a 'swiveling' wing store. In a variable sweep airplane this can become an additional complication.

4.4 DETAILED OVERALL STRUCTURAL ARRANGEMENTS

Because wing structural design cannot always be easily separated from the structural layout design of the entire airplane a number of overall structural layouts are presented to conclude this chapter on wing design. Figures 4.88 through 4.92 show structural cutaways for the following airplanes:

Figure 4.88 Cessna 441 Conquest
Figure 4.89 Cessna Citation II
Figure 4.90 Caproni Vizzola C22J
Figure 4.91 Fairchild Republic A-10
Figure 4.92 Douglas A4D-2N

For additional examples of wing structural arrangements the reader is referred to the airplane cutaway drawings in Chapter 9.

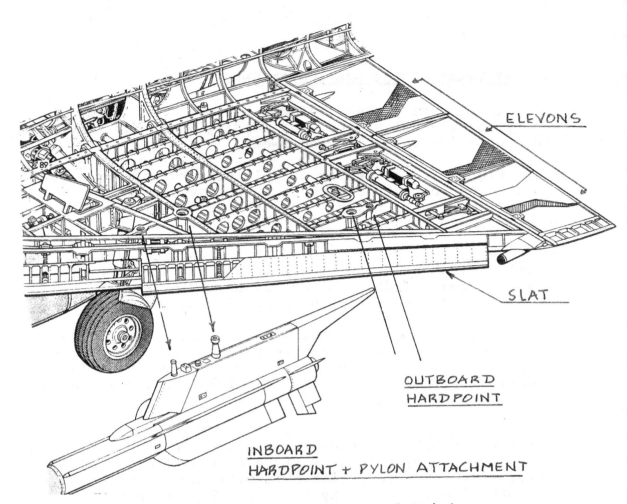

Figure 4.80 Example of a Wing Hard Point

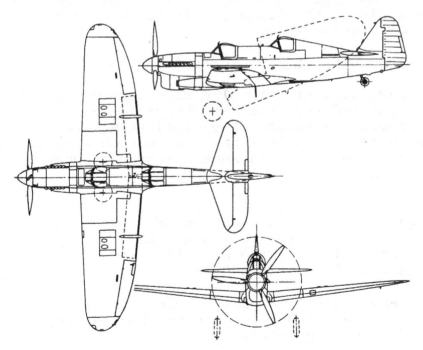

Figure 4.81 Wing Folding Fairey Firefly T1

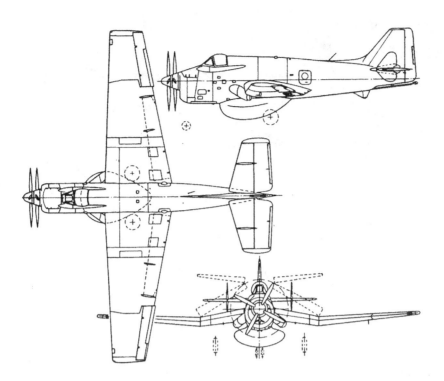

Figure 4.82 Wing Folding Fairey Gannett AEW3

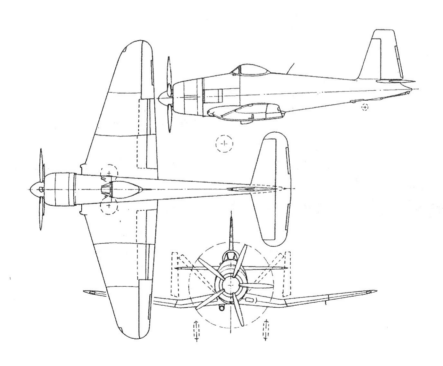

Figure 4.83 Wing Folding Blackburn B-48

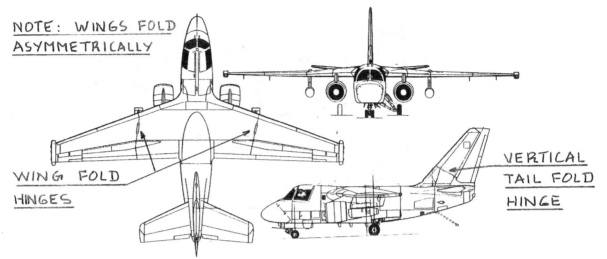

Figure 4.84 Wing and Vertical Tail Folding Lockheed S3-A

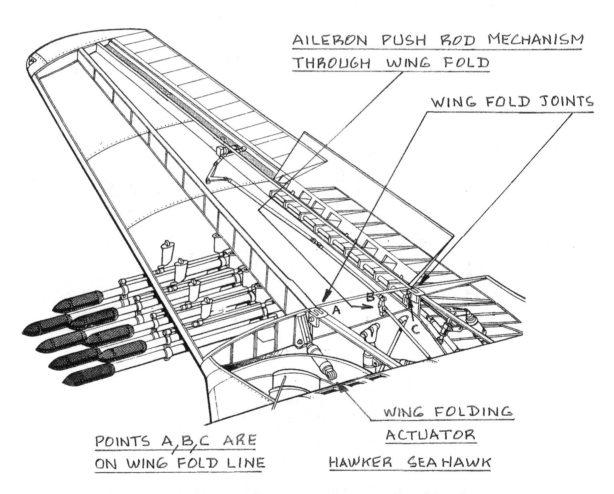

Figure 4.85 Example Wing Folding Mechanization

Part III Chapter 4 Page 242

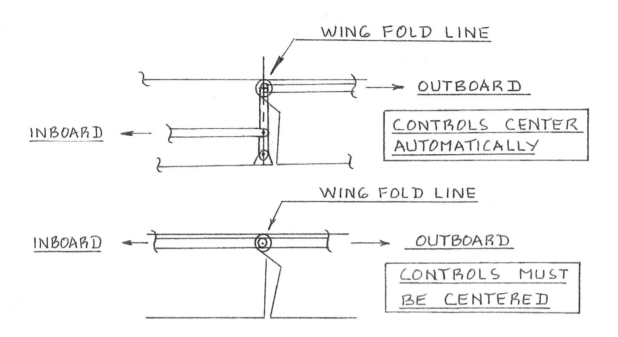

Figure 4.86 Controls Routing Through Wing Fold

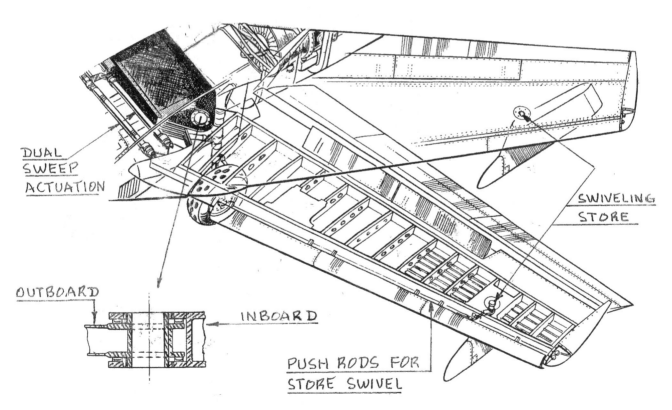

Figure 4.87 Example of Wing Pivot Construction and Swiveling Store Application

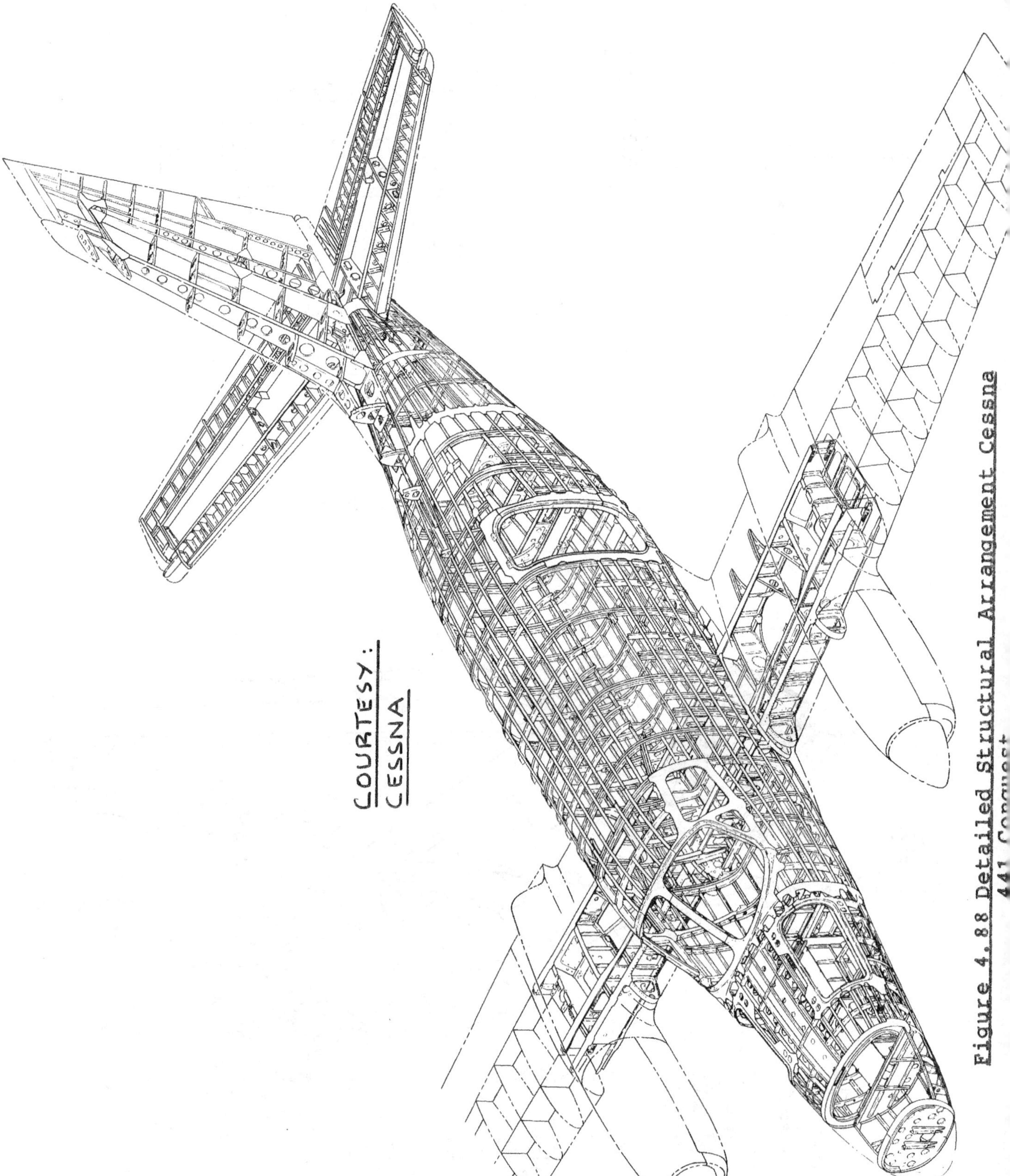

COURTESY: CESSNA

Figure 4.88 Detailed Structural Arrangement Cessna 441 Conquest

Part III Chapter 4 Page 244

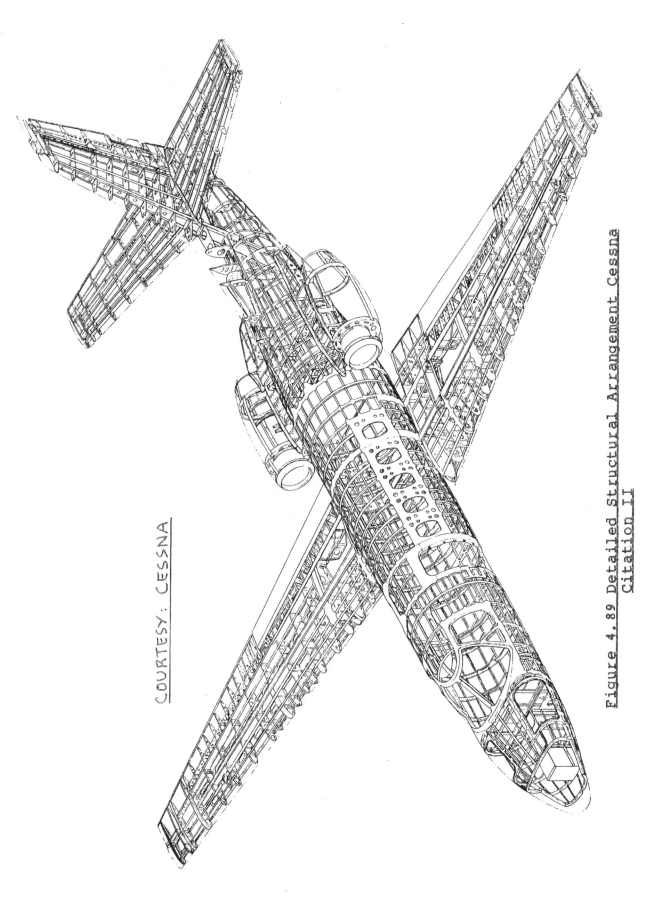

Figure 4.89 Detailed Structural Arrangement Cessna Citation II

COURTESY: CESSNA

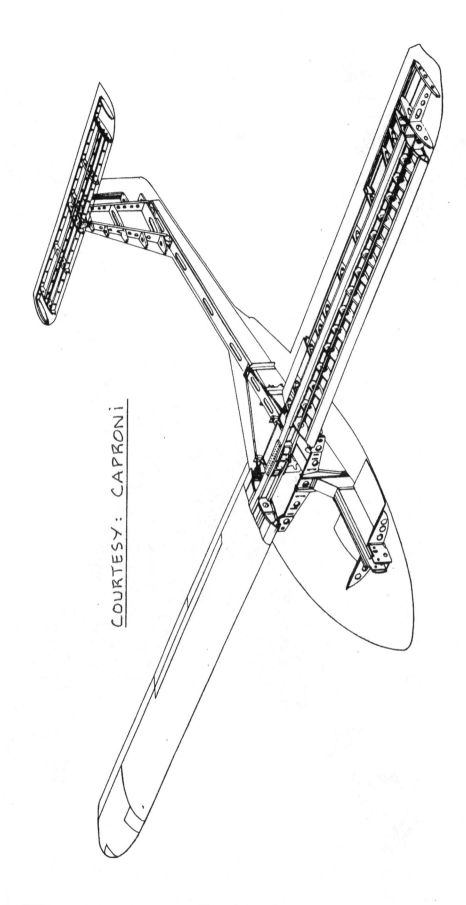

Figure 4.90 Detailed Structural Arrangement Caproni Vizzola C22J

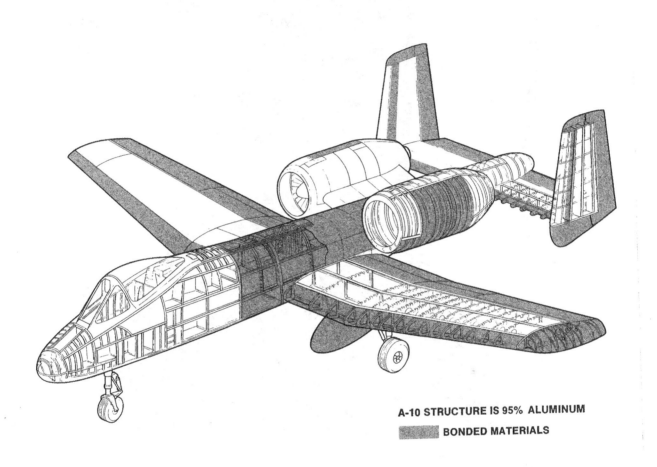

Figure 4.91 Detailed Structural Arrangement Fairchild Republic A-10

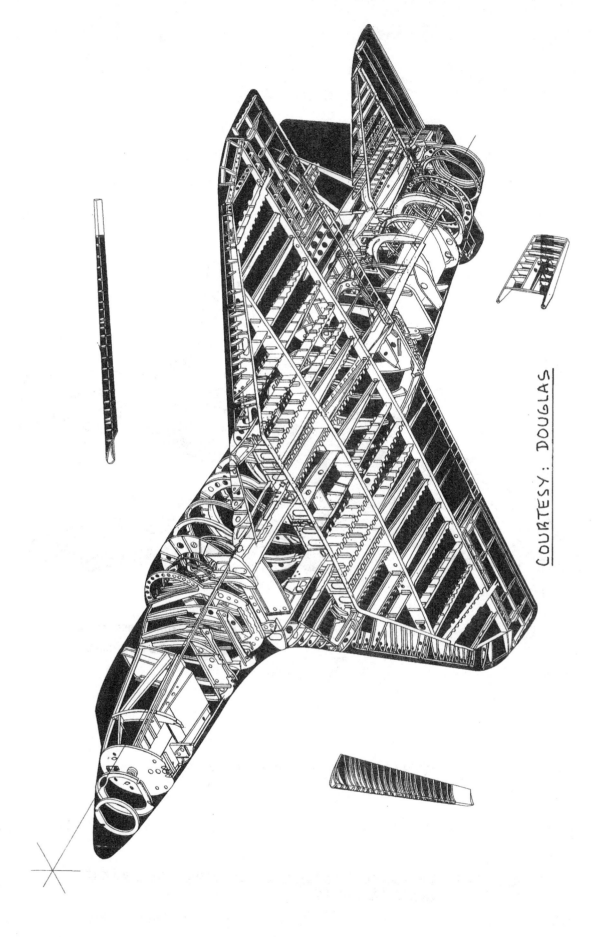

Figure 4.92 Detailed Structural Arrangement Douglas A4D-2N

5. EMPENNAGE LAYOUT DESIGN

The purpose of this chapter is to provide design considerations, design data and design examples for the layout design of the empennage.

A step-by-step procedure for arriving at a satisfactory Class I preliminary empennage layout is presented in Chapter 8 of Part II. That procedure is meant to be used together with p.d. sequence I of Chapter 2 in Part II. During the next phase of empennage design (Class II in p.d. sequence II as outlined in Chapter 2 of Part II) it is recommended to use the same procedure, but now augmented with the empennage design considerations presented in this chapter.

The reader should also review the large number of empennage/wing configurations presented in Chapter 3 of Part II. It is always useful to determine what has been done by various manufacturers.

Section 5.1 presents a general discussion of empennage design aspects: aerodynamic as well as operational.

Section 5.2 contains a discussion of empennage design integration considerations. Guidelines for structural design of the empennage are also given. Furthermore the structural integration of the empennage into the entire airplane configuration is discussed with examples.

5.1 EMPENNAGE CONFIGURATION: AERODYNAMIC AND OPERATIONAL DESIGN CONSIDERATIONS

An overview of airplane configurations including some discussions of empennage configurations is presented in Chapter 3 of Part II. A step-by-step procedure for arriving at a satisfactory preliminary empennage layout is contained in Chapter 8 of Part II.

The purpose of this section is to present additional design information relative to the choice of the empennage configuration.

References 12, 13, 14 and 37 should be consulted for further information on empennage design.

The following empennage configuration aspects will be discussed:

    5.1.1 Conventional (Tails aft), canard or three-surface?
    5.1.2 Additional empennage configuration choices.
    5.1.3 Empennage and control surface size: stability, control and handling considerations.
    5.1.4 Stall and spin considerations.
    5.1.5 Empennage planform design.
    5.1.6 Empennage airfoil design or selection.

Finally, a review of empennage drag contributions is presented:

    5.1.7 Review of empennage drag contributions.

5.1.1 Conventional (Tails Aft), Canard or Three-surface?

The choice of conventional, canard and/or three-surface empennage configurations is strongly coupled with the overall airplane configuration design philosophy:

Important aspects to be considered are:

1. Achievable trimmed lift-to-drag ratio

2. Achievable trimmed maximum lift coefficient (flaps up)

3. Achievable trimmed maximum lift coefficient in landing and/or in take-off (flaps down)
   NOTE: This must be done AT THE CRITICAL C.G. LOCATION FOR THE CONFIGURATION BEING STUDIED

4. Distribution of major airplanes masses (examples are: engines, fuel, payload) and their relation to the weight and balance problem

5. Structural design synergism

6. Good looks

__1. Achievable trimmed lift-to-drag ratio:__ Ref.48 shows that three-surface configurations can achieve higher trimmed cruise lift-to-drag ratios than either conventional or canard configurations. It is shown that three-surface configurations can achieve this at ANY location of the c.g.!

However, Ref.48 also observes, that these conclusions are based on the so-called Prandtl/Munk requirement of elliptical lift distributions and that these conclusions may be invalid if actual spanload distributions are accounted for.

In Ref.49 it is shown that if actual spanload distributions are accounted for, the conventional configuration has higher trimmed L/D value.
The Ref.48 data are valid for one c.g. location only and the method does not account for the effect of the propulsive installation on the lift distribution. Particularly for propeller driven airplanes this effect is known to be important.

Conclusion: no concensus exists with respect to the question which type of configuration yields the highest trimmed L/D in cruise.

2. Achievable trimmed maximum lift coefficient flaps up: No systematic studies have been carried out to determine which configuration can yield the maximum trimmed lift coefficient in a flaps up condition or with the use of maneuvering flaps. Unpublished work done by Grumman and by Rockwell seems to indicate the fact that in fighters the canard canard configuration has a definite edge in this regard.

3. Achievable trimmed maximum lift coefficient in landing and/or in take-off (flaps down): No systematic studies have been carried out to determine which configuration can yield the maximum trimmed lift coefficient in take-off and/or in landing (flaps down) conditions.

If canard/wing interference from canard tip vortices can be minimized the canard configuration and the three-surface configuration have (in principle) the higher trimmed maximum lift capability. The reason is quite simply the additive lift due to the canard when compared to the down lift from a conventional tail.

NOTE: This argument will be invalid for inherently unstable airplanes: in those cases the tail lifts up!

The reader should consult Part VII for detailed discussions of trim diagrams for stable and for unstable configurations.

4. Distribution of major airplane masses and their relation to the weight and balance problem: With the engines located forward (Cessna 172, Beech King Air, Boeing 737 and 747) the overall weight and balance of the airplane virtually dictates a conventional empennage

configuration. To minimize wetted area it is always desirable to have as little empennage area as possible.

To achieve the latter, the empennage surfaces must be placed in locations which maximize the product of the lift-curve-slope and the moment arm of each empennage surface! Figure 5.1 illustrates the principles involved here.

With the engines located aft (Rutan Varieze, Beech Starship I, Piaggio P 180, Boeing 727, Il-62) a conventional empennage arrangement becomes awkwardly large at some point. The Varieze wing/engine/fuselage combination has so much longitudinal stability that the only practical solution is a pure canard configuration. The 727 and the Il-62 use a very highly swept vertical tail to gain enough moment arm to still 'get away' with a conventional empennage arrangement.

5. Structural design synergism: In some cases it is possible by a unique combination of structural components to achieve a particularly favorable ratio of empty weight to take-off weight. The GP180 is an example: the wing torque box, the rear pressure bulkhead and the main landing gear are essentially attached to a common fuselage structure. Couple this with the favorable wing/fuselage intersection (mid wing) and a significant synergism has been achieved. What makes airplane configuration design decision making so difficult is that this type of 'synergism' needs to be weighed against other (sometimes negative) aspects of a particular layout. To make such decisions based on 'hard' data requires a lot of 'up-front' engineering effort. Time for such an effort is not always available.

Another example is the design case history of the Boeing 727. Ref.50 shows that in the early 727 design studies the landing gear was retracted into Küchemann bodies (See Fig.4.34 for an example), there was no acceptable spot for the APU and the rear cargo door was simply too small. By deciding to retract the gear into the fuselage (via a yehudi in the wing) which resulted in the need for a local enlargement of the fuselage cross section, sufficient room was created for the APU as well as for a cargo door of acceptable size.

These examples should indicate to the reader the importance of synergistic design thinking.

6. Good looks: This aspect of airplane design should not be considered as trivial. The 'good looks' question is obviously a very subjective one: de gustibus

non disputandum! An airplane designer should always consider the aesthetics of his creations. An example of this is the swept vertical tail on Cessna single engine airplanes. Those sweep angles are incorporated not for high Mach reasons but only for 'good looks'.

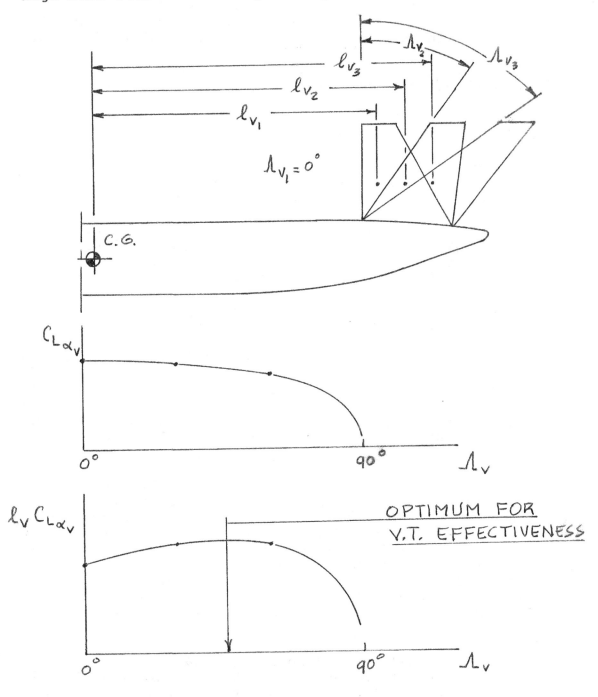

Figure 5.1 Illustration of Optimization of the Aerodynamic Effectiveness of an Empennage Surface

5.1.2 Additional Empennage Configuration Choices

Once the fundamental questions regarding the overall empennage configuration have been settled, a number of detailed configuration choices remain:

1. V-tail
2. T-tail
3. Single vertical tail
4. Multiple vertical tail
5. Twin boom tail
6. Vertical tail on wing

1. V-tail: Examples of V-tail (Butterfly tail) applications are found in the following airplanes:

1. Beech Twin Quad (Fig.5.2)
2. Beech Bonanza (Fig.5.3)
3. Fouga Magister (Fig.5.4)
4. Heinkel He211 (Fig.5.5)

Potential advantages of the V-tail configuration are small savings in wetted area and in weight when compared with a conventional empennage arrangement.

A so-called 'mixer' unit is needed to achieve uncoupled longitudinal and directional control. Figure 5.6 shows an example of such a 'mixer' in an airplane with a mechanical flight control system.

2. T-tail: Examples of T-tail applications are found in Figures 3.4, 3.8, 3.17, 3.21 and 3.32 in Part II.

From a viewpoint of vertical tail effectiveness per unit area, the 'best' locations for a horizontal tail in relation to the vertical tail is either the T-tail or the low tail configuration. The horizontal tail acts like an 'end-plate' which increases the lift-curve-slope of the vertical tail. Part VI contains a detailed procedure which accounts for this 'end-plate' effect.

From a structural weight viewpoint the low horizontal tail location is to be preferred over the T-tail location. This statement is not necessarily true in cases such as the 727, the DC9 and the BAC 111. In those airplanes (because of the aft engine mounts on the fuselage) it would be very difficult to find an acceptable 'low' position for the horizontal tail. Also, by sweeping the vertical tail aft, moment arm is gained for the T-tail mounted horizontal tail. This saves horizontal tail area.

Most T-tail mounted horizontal tails are given a slight amount of anhedral. Flutter calculations for the T-tail arrangement usually show that doing this saves weight.

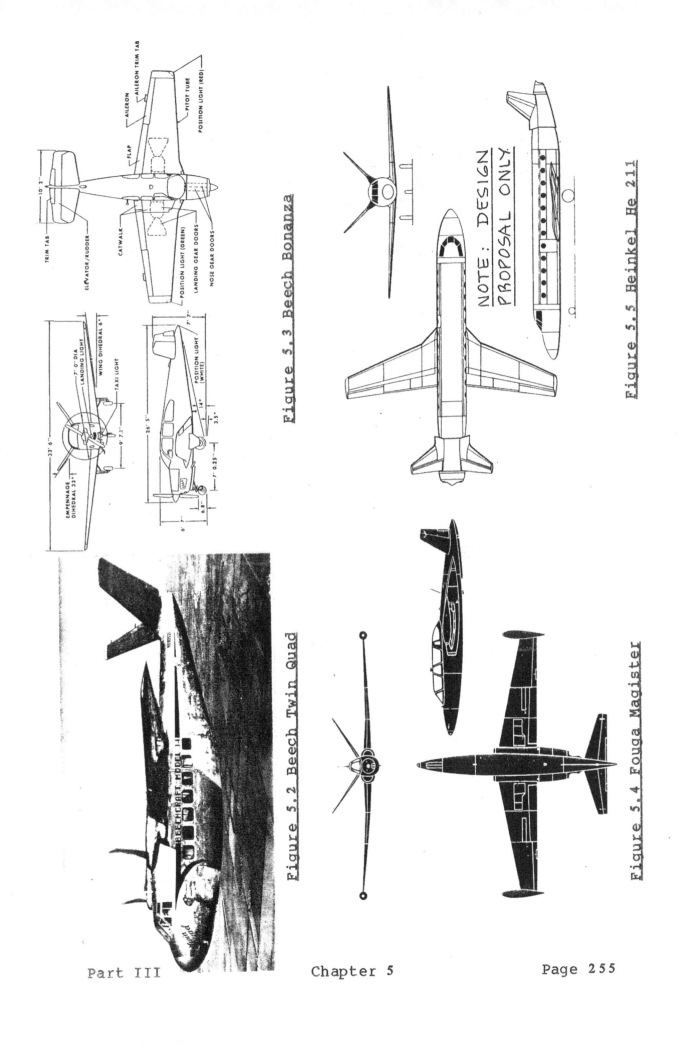

Figure 5.3 Beech Bonanza

Figure 5.5 Heinkel He 211

Figure 5.2 Beech Twin Quad

Figure 5.4 Fouga Magister

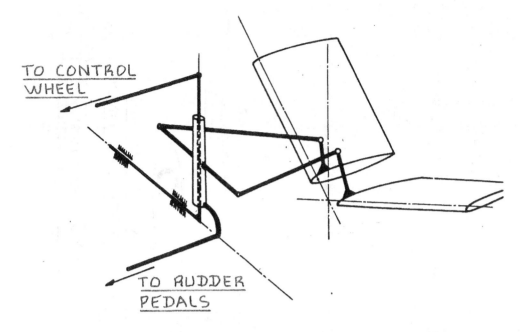

Figure 5.6 Control Mixer for a V-tail

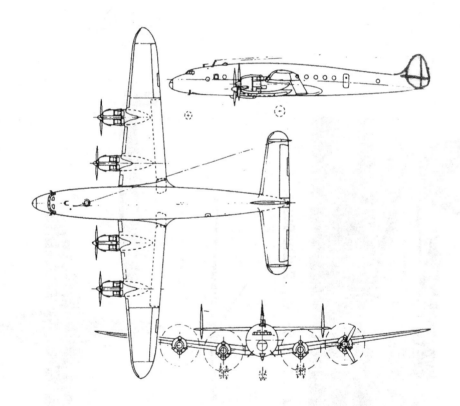

Figure 5.7 Lockheed Constellation

T-tail configurations often are prone to the so-called 'deep-stall trim-point problem'. This problem is discussed in more detail in sub-section 5.1.4.

3. Single vertical tail: Most airplanes have the single vertical tail configuration. Chapter 3 of Part II contains examples of single vertical tail applications.

From a weight viewpoint as well as from several aerodynamic viewpoints the single vertical tail is the most effective one for civil airplanes in the low subsonic speed range. The reason is that a single vertical tail can be built with a higher aspect ratio and therefore be more effective per unit area.

Potential disadvantages of large single vertical tails are:

1. Large rolling moment due to rudder deflection: to 'fix' this requires an aileron/rudder interconnect system.

2. Problems with servicing and hangering because of excessive height.

For supersonic airplanes a single vertical tail is usually not a good design solution. The reason is the fact that the vertical tail normally is located in an expansion wave behind the wing. The dynamic pressure at the vertical tail tends to be less than the free stream dynamic pressure. For such airplanes a vertical tail located below the wing would be best above Mach 1.

Solutions to this problem are: folding vertical fins (Mig-27 Flogger), fixed ventral fins and/or multiple vertical tails. Many fighters in Chapter 3, Part II use some of these solutions.

The XB-70A of Fig.3.34b (Part II) employs folding wingtips to enhance directional stability in supersonic flight and at the same time reduce the aft shift of aerodynamic center which is a problem in nearly all supersonic airplanes.

If there are good reasons not to use either a low horizontal tail or a T-tail, a so-called 'cruciform' arrangement may be used. Examples of 'cruciform' arrangements are found in Part II as:

1. Dassault Falcon 50 (Fig.3.13d)
2. BAE HS 125 (Fig.3.15b)
3. Cessna Crusader (Fig.3.8c)
4. Lockheed Jetstar (Fig.3.13a)

A disadvantage of the cruciform arrangement is the lower vertical tail effectiveness due to lack of 'end-plate' effect. In some airplanes it is possible to use the cruciform arrangement to place the horizontal tail outside the slipstream to prevent tail buffeting (fatigue). This can be an important consideration in propeller driven airplanes with high power-to-weight ratios.

An interesting aspect of the Lockheed Jetstar is that the entire vertical pivots so that its leading edge sweep angle is altered. This is done to achieve longitudinal trim!

4. <u>Multiple vertical tails:</u> Examples of multiple vertical tail applications are:

1. Lockheed Constellation (Fig.5.7)
2. Grumman E2C Hawkeye (Fig.3.30c in Part II)
3. Antonov AN-22 Antheus (Fig.3.28d in Part II)
4. Fairchild Republic T-46A (Fig.3.23b in Part II)
5. Grumman F-14A Tomcat (Fig.3.25a in Part II)
6. F.R. A-10A Thunderbolt II (Fig.3.27c in Part II)
7. McDD F-15C Eagle (Fig.3.27b in Part II)

Potential advantages of multiple vertical tails are:

1. low rolling moment due to rudder deflection
2. redundancy to combat damage (A-10A!!)
3. in propeller driven airplanes: slipstream augmentation of one of the vertical tails in an engine out situation

5. <u>Twin boom tails:</u> Examples of twin boom tail applications are:

1. Cessna Skymaster (Fig.3.9c in Part II)
2. Fairchild Packet II (Fig.3.41 this part)
3. Armstrong Whitworth Argosy (Fig.3.2 this part)
4. WSK Mielec M-15 (Fig.3.11c in Part II)
5. Eris (Fig.13.3 in Part II)

Twin boom configurations are usually heavier that layouts with a conventional fuselage. One reason is that the fuselage of a twin boom airplane tends to have a low fineness ratio which increases drag which in turn results in higher required installed power and thus weight.

In cargo transports the high wing twin boom layout can be advantageous from a loading and unloading viewpoint.

If the booms can also be used to carry fuel (this makes sense only for extremely long range airplanes) the twin boom design can in fact come out lighter. An example of such a case is the Rutan Voyager 'around-the-world' airplane.

<u>6. Vertical tail on wing</u>: The Beech Starship I and the XB-70A are examples of this approach (Figures 3.42 and 3.34b, Part II respectively).

There have not been many instances where the vertical tail was mounted on the fuselage, forward of the c.g. This approach would reduce directional stability! The AFTI F16 and an experimental German modified F4 both of which used their 'vertical canards' to reduce directional stability and to allow pure sideforce control, are notable exceptions.

5.1.3 Empennage Size: Stability, Control and Handling Considerations

A step-by-step Class I method for empennage and control surface sizing is presented in Chapter 8 of Part II. In Class II empennage sizing methods much more detailed analyses of the stability and control characteristics of an airplane are needed. A detailed review of airplane stability and control theory and applications may be found in Ref.37. Part VII contains Class II methods for determining the required stability, control and handling characteristics of airplanes. This sub-section will highlight some of the most important aspects of empennage design as it relates to stability, control and handling requirements.

<u>WARNING:</u> Whether or not an airplane is structurally rigid or elastic can make a great deal of difference to stability and control behavior. The reader should not automatically assume that all airplanes are reasonably rigid. Most jet transports and even fighters in high 'g' flight conditions should be considered as 'elastic' airplanes. Ref.37 contains methods for analyzing the effects of aeroelasticity on stability and control.

5.1.3.1 Longitudinal considerations

From a viewpoint of longitudinal stability, control and handling, the horizontal empennage surfaces must satisfy the following requirements:

1. Longitudinal stability requirements
2. Longitudinal control requirements
3. Longitudinal stick (or wheel) force requirements

<u>1. Longitudinal stability requirements:</u> Longitudinal stability at forward and at aft c.g. must be consistent with requirements for static, dynamic as well as maneuvering stability. Stability requirements (inherent or de-facto) determine primarily the size of the horizontal empennage surfaces, once the disposition (i.e. moment arms) has been decided.

The stability requirements of Ref.51 should be used in conjunction with the methods of Part VII to determine the adequacy of any proposed empennage configuration and size.

<u>2. Longitudinal control requirements:</u> The following longitudinal control requirements must be considered:

2a. Control power for trim at forward and aft c.g. must be consistent with the operational flight envelope and weight of the airplane.

2b. Control power for take-off rotation at forward c.g. (for tricycle gear airplanes) and at aft c.g. (for taildraggers) must be consistent with the operational flight envelope and weight of the airplane. If control power for take-off rotation is not sufficient a consequence is that the take-off fieldlength is much larger than predicted.

2c. Control power for maneuvering (in calm and in gusty air) must be consistent with the operational flight envelope, c.g. location and weight of the airplane. This requirement is particularly important in the case of inherently unstable (highly augmented) airplanes.

2d. Control power must be sufficient to allow for any requirements for artificial static and/or dynamic stability.

Control power requirements determine both the size and the maximum lift of the horizontal empennage surfaces. Ref.51 should be used in conjunction with the methods of Part VII to assure that empennage and control surface sizes are sufficient.

   <u>3. Longitudinal stick (or wheel) force requirements:</u> These requirements cover such concepts as: stick-force speed gradients, trim speed and return-to-trim-speed, stick-force per 'g' and maximum incremental stick force needed to cope with changes in airplane configuration such as: power or thrust changes, flap changes and/or failures.

   References 11 and 51 should be consulted for details. Ref.37 and Part VII contain detailed methods for analyzing stick force requirements.

<u>5.1.3.2 Lateral-Directional considerations</u>

   From a viewpoint of lateral-directional stability, control and handling the vertical empennage surfaces must satisfy the following requirements:

1. Lateral-directional stability requirements
2. Lateral-directional control requirements
3. Lateral-directional stick (or wheel) and rudder pedal force requirements.

   <u>1. Lateral-directional stability requirements:</u> The following lateral-directional stability requirements must be considered:

   1a. Lateral stability must be consistent with requirements for static and dynamic stability at all c.g. locations and the operational flight envelope of the airplane. This requirement is usually dominated by the inherent lateral stability designed into the wing. It is shown in Ref.37 that wing dihedral angle, wing sweep angle and wing location on the fuselage dominate the magnitude and the sign of the stability derivative C_{l_β}.

   Since the wing is designed primarily by performance and operational considerations, the empennage is often used to 'fine-tune' the lateral stability of airplanes. Examples of this fine-tuning are: F-4 and AV8B.

   1b. Directional stability must be consistent with requirements for static and dynamic stability at all c.g. locations and the operational flight envelope of the airplane. This requirement often dictates the size of the vertical tail.

2. Lateral-directional control requirements: The following lateral-directional control requirements must be considered:

2a. Lateral control power must be sufficient to meet the time-to-bank and response requirements of Refs 11 and 51.

2b. Directional control power must be sufficient to handle the most critical engine-out situation (V_{mc}).

2c. Directional control power must be sufficient to allow for cross-wind landing conditions

2d. Directional control power must be sufficient to allow for maneuvering.

2e. Directional control power must be sufficient to provide for any requirements for artificial static and/or dynamic stability.

These control power requirements determine the size, type and location of ailerons, spoilers, differential stabilizers and rudders. Requirement 2b) also may determine the maximum lift capability demanded of the vertical tail with full rudder deflection. References 11 and 51 should be used in conjunction with the methods of Part VII to assure the correct size of the lateral-directional control surfaces.

3. Lateral-directional stick (or wheel) and rudder pedal force requirements: It takes a definite physical effort by a pilot to move the primary cockpit controls. It is obvious that the cockpit control forces needed by the pilot to satisfy the control power requirements outlined under 2) must be within the capabilities of the pilot. References 11 and 51 specify the allowable control force limits for temporary and for prolonged situations. Ref.37 and Part VII contain methods for the analysis of pilot control force requirements.

5.1.4 Stall and Spin Characteristics

For satisfactory stall and spin characteristics it is necessary that sufficient control power and stability can be maintained up to levels of angle of attack and sideslip consistent with the operational requirements for the airplane.

The following characteristics will be discussed:

1. Stable and unstable pitch breaks
2. The stall scenario
3. The deep stall trim problem
4. Pitch-up in high speed airplanes
5. Spin departure
6. Spin Recovery

<u>1. Stable and unstable pitch breaks:</u> The C_m-C_L behavior of the airplane at forward and at aft c.g. is of major importance here as well as the associated behavior of the C_L-α curve.

Figure 5.8a illustrates the C_L-C_m behavior for airplanes with a stable pitch break. It is noted that stable pitch breaks are required for FAR23 airplanes.

Figure 5.8b shows the C_L-C_m character for airplanes with an unstable pitch break. Unstable pitch breaks are acceptable for FAR25 and for military airplanes. However, depending on the dynamic behavior of the airplane following pilot induced stall entry (or gust induced stall entry) the airplane may have to be outfitted with stick-shakers and/or stick-pushers. In the latter case, the value for $C_{L_{max}}$ which may be used in the certification of the airplane is <u>not</u> the 'aerodynamic' $C_{L_{max}}$ but instead is a value somewhere between 'stick-shaker C_L' and 'stick-pusher C_L' (See Fig.5.8b). This can lead to significant performance penalties, particularly in field length.

<u>2. The stall scenario:</u> Two types of configurations will be discussed: conventional (aft tail) configurations and canard configurations.

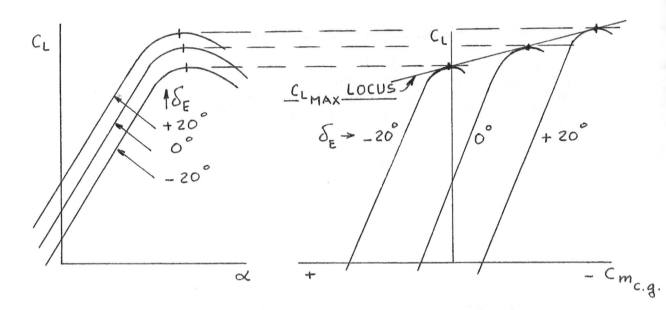

Figure 5.8a Stable Pitch Break Behavior

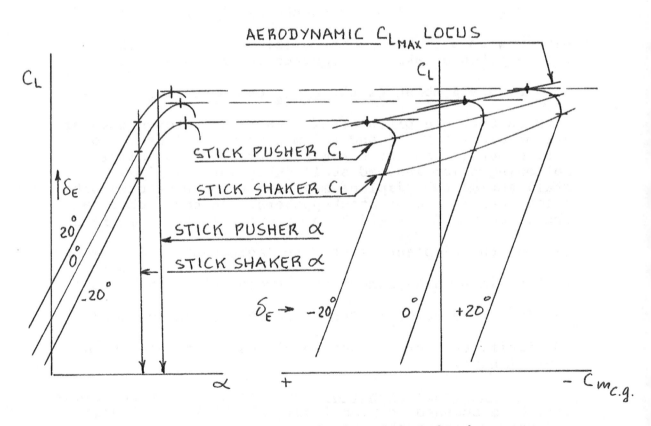

Figure 5.8b Unstable Pitch Break Behavior

2a) Conventional (aft tail) configurations:

The high angle of attack pitching moment behavior of a conventional airplane depends on the behavior of the wing-body and on that of the horizontal tail.

The stall scenario for a conventional airplanes is as follows:

Once the wing stalls along the inboard trailing edge, the downwash from that part of the wing over the horizontal tail disappears. This is seen by the horizontal tail as a positive increase in angle of attack. This creates a nose down pitching moment which is perceived by the pilot as a 'stable' pitch break.

<u>However:</u> the breakdown of the flow over the inboard part of the wing can change the wing contribution to airplane pitching moment. If the wing contribution to pitching moment becomes 'larger positive' than the tail contribution is 'negative', an unstable pitchbreak results.

Figure 5.9 shows the classical pitch stability boundary for wings in terms of aspect ratio and sweep angle. Note that wings with very high aspect ratios will have unstable pitchbreaks even at low sweep angles. The tail must be designed to overcome this if a net stable break for the entire airplane is required.

Figure 5.10 identifies four regions for the location of the horizontal tail: A, B, C and D.

Figure 5.11 shows the net result of adding the horizontal tail in each of these regions:

- in Region A: the wing wake does not affect the tail. This is normally the best place for a horizontal tail. With flaps down, the flap wake may alter this!

- in Region B: same as A at low speeds. However, at high subsonic speeds, particularly when maneuvering, there could be a problem.

- in Region C: the horizontal tail will enter the wing wake only when the latter is unstable.

- in Region D: a reversal of the C_m-C_L curve will usually occur leading to the 'deep-stall' trim-point phenomenon which is discussed under 3.

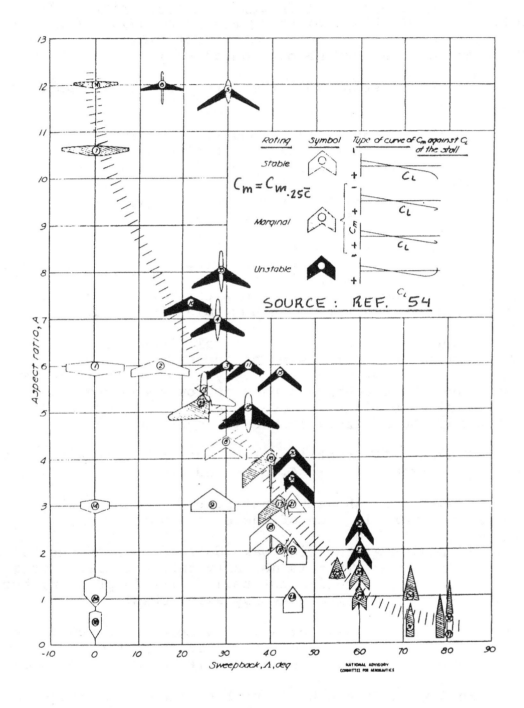

Figure 5.9 Wing Alone Pitch Break Stability Boundary

Part III Chapter 5 Page 266

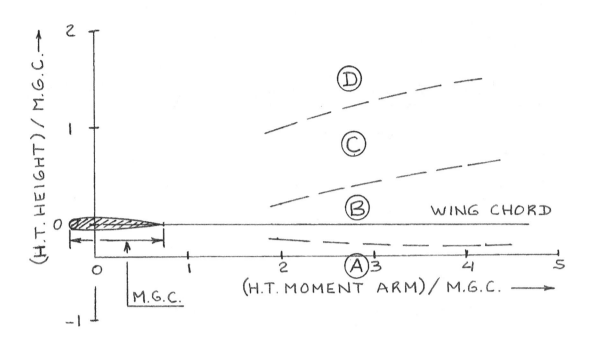

Figure 5.10 Four Regions for Horizontal Tail Location in Relation to Pitch Break Stability

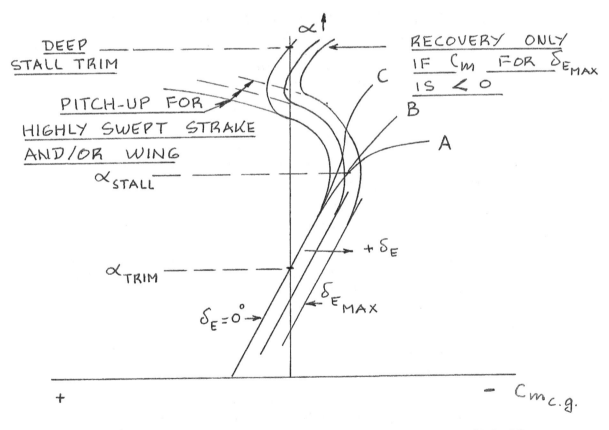

Figure 5.11 Pitch Break Behavior of Airplanes with the Horizontal Tail Located in the Regions of Figure 5.10

2b) Canard configurations: These must be designed so that the canard stalls before the wing.

The stall scenario for a canard airplane should be as follows:

When the canard stalls, a nose-down pitching moment is automatically generated: the canard lift drops and the canard downwash at the wing disappears which has the effect to slightly increase wing lift. Both factors contribute to the nose-down pitching moment.

The following design decisions must be made:

i) What type airfoil to use in the canard surface.
ii) What incidence angle should the canard have.

i) The selection of canard airfoil affects the value of $C_{L_{max}}$ which can be generated by the canard. If the canard airfoil is a laminar flow airfoil its sensitivity to transition to turbulent flow (for example due to flying into rain) must be carefully considered if major trim changes are to be avoided. Whether any such trim changes are acceptable depends on available control power as well as on the required stick (or wheel) force to compensate for these trim changes.

The reader should be aware of the fact that these considerations apply to ALL flap configurations which may be operationally encountered!

ii) The canard incidence angle determines when canard stall will occur relative to wing stall.

Sofar it has been assumed that the canard incidence angle is 'fixed'. If the canard incidence is variable it must be recognized that this makes canard stall a function of the trim state (i.e. c.g. location). This can be undesirable.

In canard equipped fighters, the canard incidence is used as the primary longitudinal control. In that case the flight control system must be able to detect situations of impending stall and prevent the pilot from entering flight conditions from which recovery may be questionable. This can be done by control signal limiting.

WARNINGS: 1.) The canard configuration shown in Figure 3.43 in Part II, has unacceptable stall recovery characteristics, despite the fact that it has a stable pitch break: Read page 83 in Part II!!

2. Figure 5.12 illustrates the potential interference effect of a canard on wing spanwise lift distribution. This effect can have an adverse effect on wing stall: in an extreme case it could induce wing tip stall. The reader should also be aware of the fact that this type of interference may result in induced drag penalties (deviation from elliptical spanloading) as well as in root-bending-moment penalties. This type of canard-wing interference can be reduced by adding camber to the wing as suggested in Figure 5.12. The Beech Starship I incorporates such camber in its wing.

3. The deep stall trim problem: Airplanes with T-tail empennage configurations and the wing relatively aft on the fuselage usually have the horizontal tail in the region marked D in Figure 5.10. Figure 5.11 shows the 'deep-stall' trim point which then may occur. When an airplane is 'parked' in that flight condition recovery is possible only with sufficient longitudinal control power and even then after considerable loss of altitude.

To obtain the required level of control power for deep stall recovery a large span horizontal tail is needed: this keeps the outboard part of the tail out of the wing/nacelle wake.

Most swept wing T-tail airplanes are equipped with stick-shakers and stick-pushers to prevent pilots from entering the deep stall trim point.

4. Pitch-up in high speed airplanes: In high speed airplanes the wing inboard section is usually given a very high sweep angle (or a highly swept strake). In such cases the center of pressure of the wing will move forward rapidly at high angles of attack: the trailing edge will begin to separate while the leading edge start to develop additional vortex lift! The resulting unstable pitchbreak is impossible to overcome by reasonably sized horizontal tails. Nearly all airplanes with variable swept wings have this problem. Figure 5.11 also illustrates this problem. In such airplanes the flight control system must be automated to prevent the pilot from inadvertently entering the strong, uncontrollable pitch-up region.

To maintain controllability in spins (not required

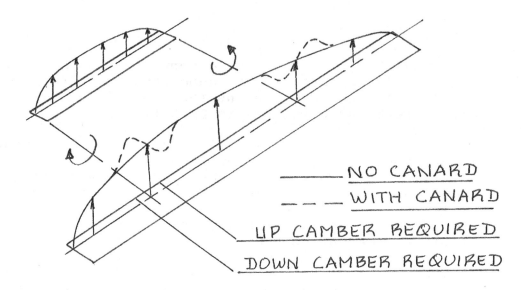

Figure 5.12 Effect of Canard Downwash on the Wing Span Lift Distribution

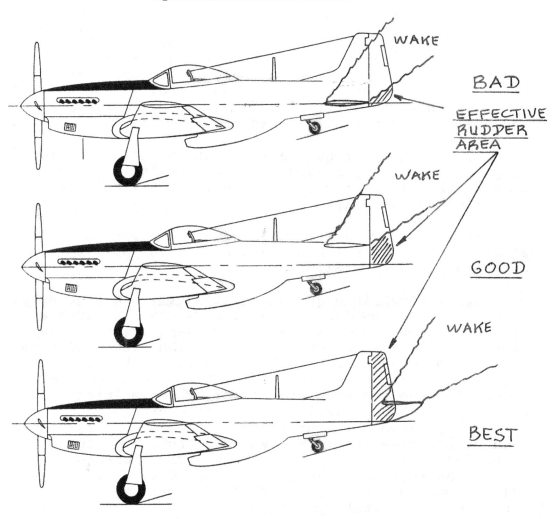

Figure 5.13 Empennage Configurations as Related to Stall/Spin Characteristics

for FAR25 airplanes and for comparable military airplanes) it is necessary to prevent separated wakes from interfering with control surfaces. Figure 5.13 illustrates desirable empennage configurations from a stall and spin viewpoint.

5. Spin Departures: Once an airplane is placed in an angle of attack where stall occurs, it is essential that the airplane be 'spin-resistant' to prevent a so-called spin departure. Ref.52 shows that to prevent inherent spin departures the following parameter must be positive:

$$C_{n_{\beta_{dyn}}} = \{C_{n_{\beta_B}} - (I_{zz_B}/I_{xx_B})C_{l_{\beta_B}} \tan\alpha\}\cos\alpha > 0 \qquad (5.1)$$

Note: the stability derivatives C_{n_β} and C_{l_β} in Eqn.(5.1) are in the body axis system!

It is normally necessary to run a windtunnel test to determine how this parameter varies with angle of attack.

6. Spin recovery: Once the airplane is in a spin it is required in FAR 23.221 (Ref.11) that recovery can be effected by control procedures which do not require exceptional pilot skills. The placement of the longitudinal and directional flight controls is of great importance in determining whether an airplane can meet this requirement.

Another way to make an airplane spin resistant is to design the wing so that auto-rotation is delayed to much higher angles of attack. Once a wing stalls the roll damping derivative normally reverses sign. It is this fact which 'drives' the spinning motion in most airplanes. By careful planform and airfoil design it is possible to delay roll damping reversal to much higher angles of attack. Ref.53 provides a useful overview of recent NASA research results in this area. Ref.14 contains an excellent discussion of airplane spin characteristics.

NOTE: there is no requirement that commercial transports be recoverable from a spin. These airplanes are not supposed to be operated near flight conditions where spin could be possible.

For fighter airplanes a requirement for recovery from extreme angles of attack (up to 90 degrees!) is that the planform centroid is behind the most aft c.g. This requirement can be readily verified from a threeview drawing and a weight and balance calculation.

5.1.5 Empennage Planform Design

The discussions in Chapter 4 on the subject of wing planform design also apply to empennage planform design. Chapter 8 in Part II contains tabulated information on planform design parameters used in a large number of airplanes.

For reasons of structural weight it is usually not feasible to use high aspect ratios in empennage surfaces. The reader should not deviate too far from the aspect ratios of Tables 8.1-8.12 in Part II!

In high speed airplanes the empennage should be designed so that its critical Mach numbers are above that of the wing.

Historically the subjective judgement of 'good or bad looks' has had a significant influence on the planform design of empennage surfaces. In many airplanes the planform design of the vertical tail reflects almost a company 'trademark'!

5.1.6 Empennage Airfoil Design or Selection

Most horizontal and vertical tails use symmetrical airfoils. The reason is that these surfaces must be able to provide 'lift' in either direction.

If the horizontal tail is found to be critical in either 'download to trim at forward c.g., flaps down' or in take-off rotation an inversely cambered airfoil may be useful to keep down the size of the horizontal tail. In the take-off rotation case, the down lift capability must be determined in the presence of ground effect!

Tables 8.1-8.12 show which airfoil types are used in the empennage surfaces of twelve types of airplanes.

For canard surfaces the problem of airfoil selection and/or design is far less straightforward. The designer should make a list of the critical lift requirements placed on any canard design. The effect of Reynolds number, Mach number and surface condition in icing conditions and rain conditions must be carefully assessed before selecting a canard airfoil.

5.1.7 Review of Empennage Drag Contributions

The empennage generates the same types of drag as the wing:

1. Friction drag
2. Induced drag
3. Compressibility drag
4. Interference drag
5. Profile drag

The total drag generated by the empennage varies from 10 - 20 percent of the total drag of an airplane, depending on the size and disposition.

Detailed procedures for predicting empennage drag are contained in Part VI.

The discussion of friction drag, compressibility drag, interference drag and profile drag for the wing in Chapter 4 applies almost verbatum to the empennage. For that reason it will not be repeated.

The induced drag generated by the empennage depends on the amount of lift which the empennage must provide.

Keep in mind that the induced drag due to individual empennage surfaces is independent of the direction in which the lift is generated: induced drag is proportional to the square of the lift coefficient.

The conditions under which a conventional tail and/or a canard lift up or down to provide moment trim for an airplane are delineated in Part VII. The prevailing situation is as follows:

*Conventional tails usually provide down lift to trim.
*Canards usually provide up lift to trim.

Note: although a canard does normally lift 'up', this can interfere with the wing in an unfavorable manner. Figure 5.12 illustrates the potential downwash effect of a lifting canard on the wing. This effect needs to be accounted for in determining the actual aerodynamic performance of the wing. By providing down/up camber in the wing as indicated in Fig.5.12 this problem can be eliminated for one flight condition only! The cruise flight condition normally would be the one for which this is done.

Canards can have a favorable interference effect on a wing due to the interaction of canard vortices with wing vortices. Figure 4.39 illustrates this for the Viggen fighter. Recent examples of this type of configuration are shown in Figures 5.14 and 5.15.

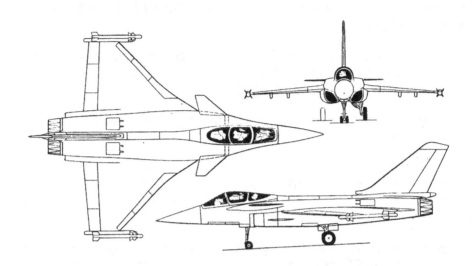

Figure 5.14 Dassault Breguet Rafale

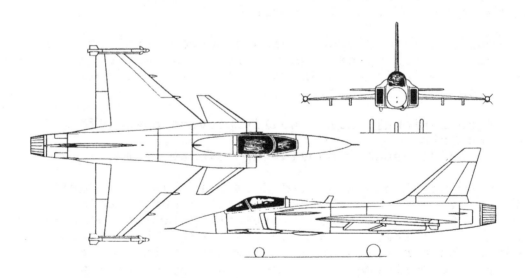

Figure 5.15 SAAB 2110 Gripen

5.2 STRUCTURAL AND INTEGRATION DESIGN CONSIDERATIONS FOR THE EMPENNAGE

The purpose of this section is to provide guidelines and examples of how the structural layout design and the integration of the empennage structure into the airplane can be accomplished.

The decision about empennage size and disposition (overall empennage configuration) has been made as a result of the work described in Section 5.1.

The material in this section is organized as:

5.2.1 Typical spar, rib and stiffener spacings.

5.2.2 Examples of empennage structural arrangements

5.2.3 Examples of fuselage/empennage integration and/or vertical/horizontal tail integration

5.2.4 Examples of empennage cross section design

5.2.5 Examples of longitudinal control mechanizations

5.2.6 Examples of directional control mechanizations

5.2.7 Examples of empennage skin gages

5.2.8 Maintenance and access requirements

A procedure for preparing the overall structural arrangement of an airplane is given in Chapter 7. For further insight into empennage structural arrangements the reader should consult Chapter 8.

5.2.1 Typical Spar, Rib and Stiffener Spacings

The actual arrangement of empennage spars, ribs and stiffeners depends very much on the type of airplane being designed and the loads to which it will be subjected. The reader should refer to Refs. 11, 19 and 23 for detailed information on the types of loads to which empennange structures are subjected. Reference 20 contains detailed methods for the preliminary structural analysis of empennage structures.

Figure 5.16 defines the locations of major structural components for empennage surfaces. Tail booms

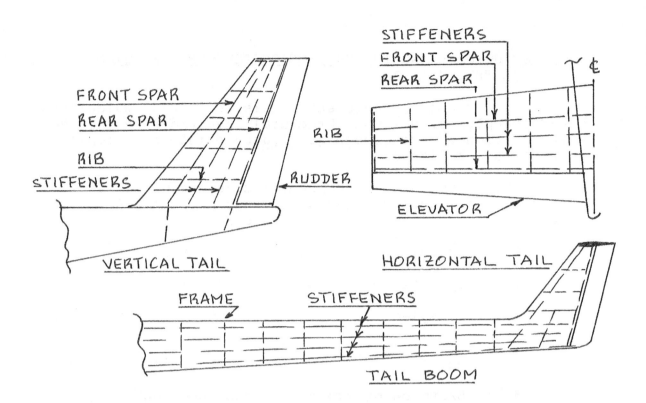

Figure 5.16 Definition of Empennage Spar, Rib and Stiffener Locations

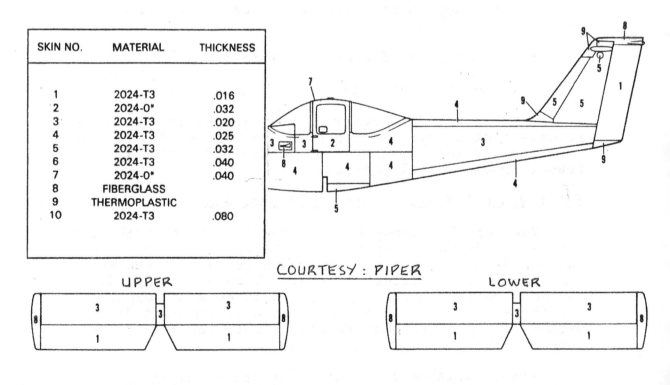

Figure 5.17 Empennage Structural Arrangement Piper PA-38-112 Tomahawk

in the case of twin boom configurations are included in this definition.

Empennage Spar Locations: Most airplanes use a so-called torque-box to carry the main loads on empennage surfaces. The torque-box should be located to take maximum advantage of the airfoil thickness distribution.: this will save weight. A torque-box is normally closed off by a front spar (F.S.), a rear spar (R.S.) and an upper and lower skin. The rear spar location is often constrained by control surface size requirements. The front spar location is not normally so constrained in the case of empennage surfaces.

Typical spar locations are:

Front spar: 15-25 percent chord

Rear spar: 70-75 percent chord

Multiple spar construction is often used in the case of fighter empennage surfaces.

Empennage Rib Locations: To help stabilize the torque-box of the empennage and to serve as anchors for control surface attachment brackets, ribs are used. Typical empennage rib spacings are:

Light airplanes: 15 - 30 inches

Transports: 24 inches

Fighters and Trainers: rib spacings vary widely.

Remember: wherever 'point loads' are expected a rib will be required. Examples of point loads are: control surface loads, taillet loads (Beech Model 1900, see Figure 3.16b), intersections with other surfaces.

Empennage Stiffener Spacings: These vary widely depending on the airplane type. The relative empennage skin stiffness determines the number and type of stiffeners needed: See Sub-section 5.2.2 for examples.

5.2.2 Examples of Empennage Structural Arrangements

A procedure for arriving at the overall structural arrangement for any new airplane is presented in Chapter 8.

Figures 5.17 through 5.24 present examples of

empennage structural arrangements for the following airplanes:

    Fig. 5.17 Piper Tomahawk   Fig. 5.18 Rockwell 112B

    Fig. 5.19 Short Skyvan   Fig. 5.20 McDD DC9-30

    Fig. 5.21 McDD DC-10   Fig. 5.22 Boeing 767

    Fig. 5.23 Aerospatiale   Fig. 5.24 Douglas A4
             Corvette

Note that in several of these structural arrangement drawings the fuselage stations, buttock line locations and waterline locations are indicated. A complete structural arrangement must contain that information!

Additional examples for the structural arrangement of the empennage may be found in Figures 4.58, 4.59 and in Section 4.4.

5.2.3 Examples of Fuselage/Empennage Integration and/or Vertical/Horizontal Tail Integration

The fuselage frames or bulkheads where empennage surfaces are attached must be able to carry the empennage loads into the fuselage, usually the fuselage skin. This requires proper location of empennage spars in relation to their fuselage attachment frames and vice versa. Figure 5.25 illustrates two types of vertical-tail/fuselage-frame attachment methods. Figures 4.88 and 4.89 provide more specific examples.

The fuselage structural depth at the location of empennage attachment should not be too shallow: bending and torsion moments exerted on the fuselage by the empennage can be very high! This is the reason for the upper fuselage bulges in most military cargo transports: the aft loading ramps cut so severely into the fuselage structure as to leave little depth for empennage attachment. This depth is then created by bulging the fuselage upward at the point of vertical tail attachment. Examples of this type of fuselage 'bulging' are seen in Figures 3.28c, 3.29b and 3.30a in Part II.

Figures 5.20 and 5.22 provide examples of the structural empennage integration for a transport and for a business jet.

Figures 5.26 and 5.27 show examples of horizontal/vertical tail integration in the case of small T-tail airplanes. Figure 5.27 also gives a good idea about the component breakdown from a manufacturing viewpoint.

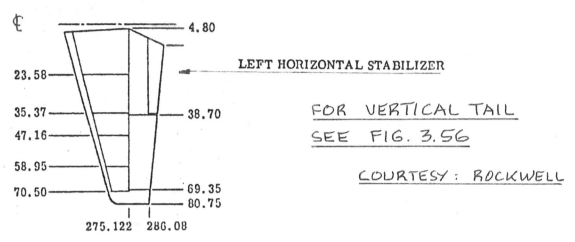

Figure 5.18 Empennage Structural Arrangement Rockwell 112B

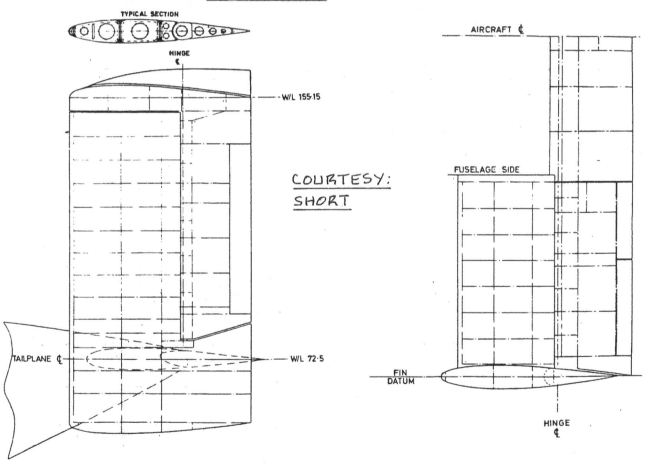

Figure 5.19 Empennage Structural Arrangement Short Skyvan

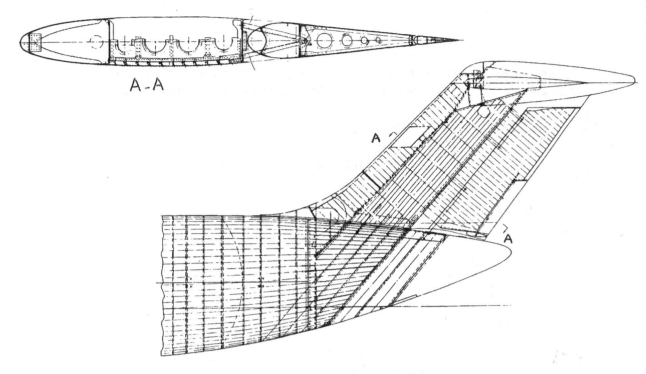

Figure 5.20a Vertical Tail Structural Arrangement
McDonnell Douglas DC9-30

COURTESY:
McDONNELL DOUGLAS

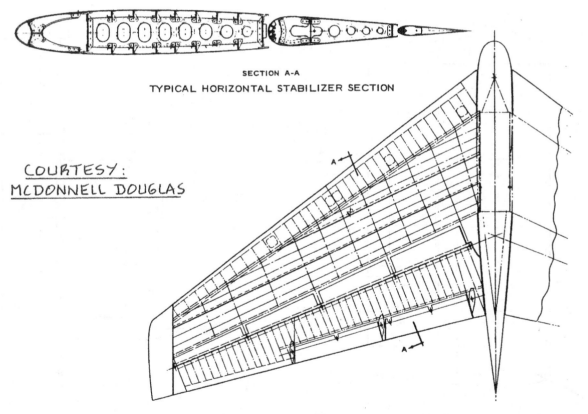

Figure 5.20b Horizontal Tail Structural Arrangement
McDonnell Douglas DC9-30

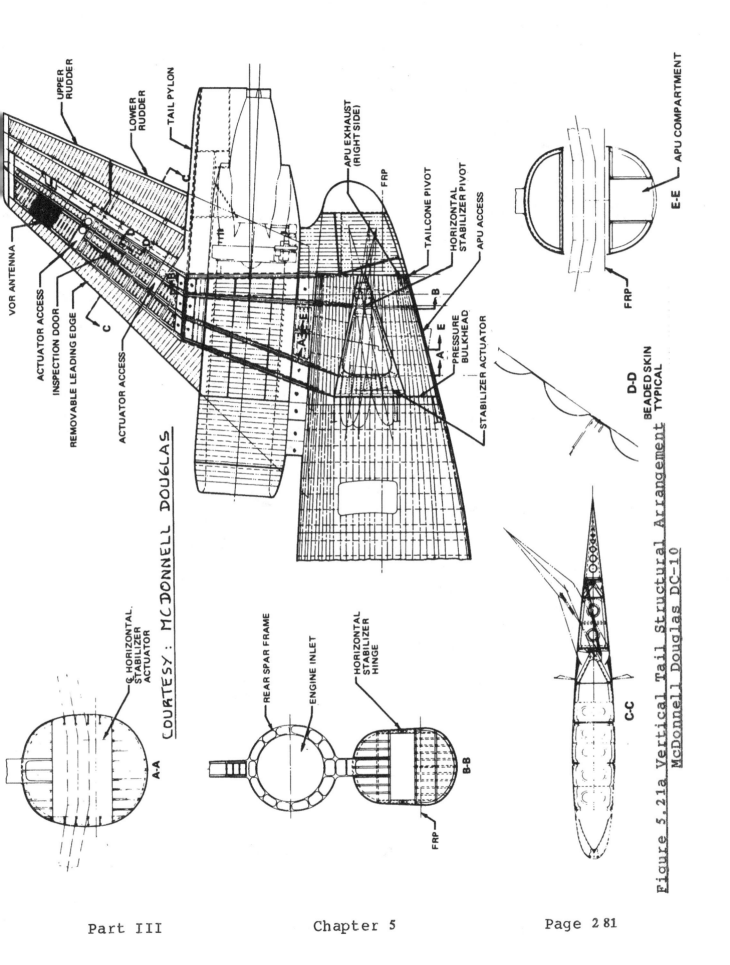

Figure 5.21a Vertical Tail Structural Arrangement TYPICAL
McDonnell Douglas DC-10

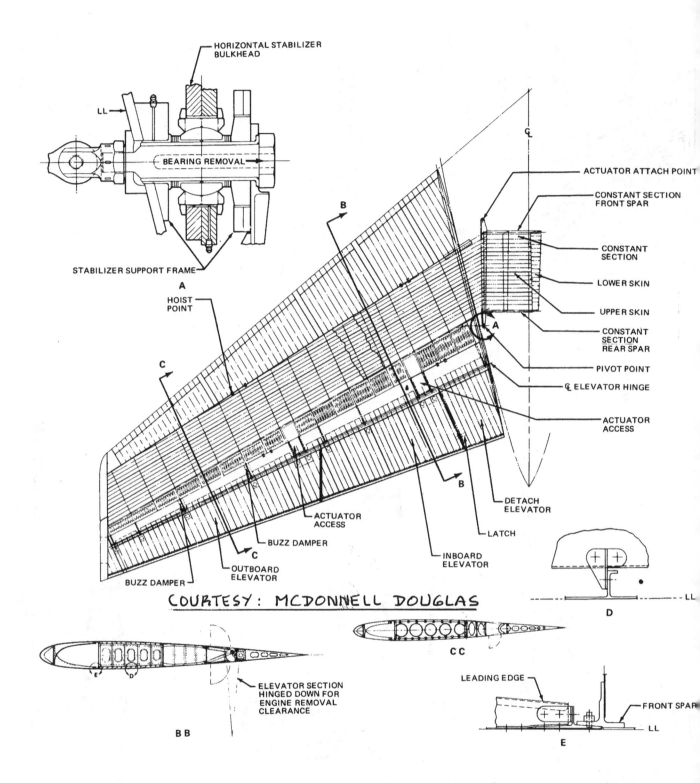

Figure 5.21b Horizontal Tail Structural Arrangement
McDonnell Douglas DC-10

Part III   Chapter 5   Page 282

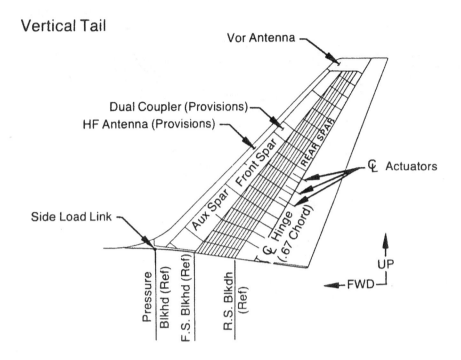

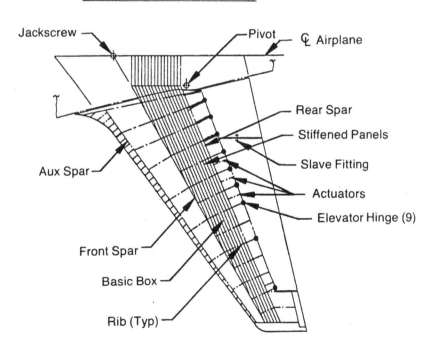

Figure 5.22 Empennage Structural Arrangement Boeing 767

Part III　　　　　Chapter 5　　　　　Page 283

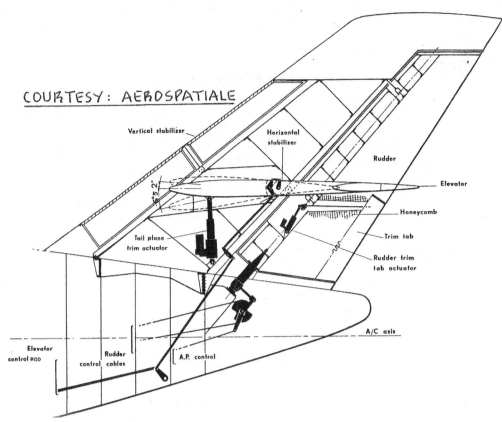

Figure 5.23a Vertical Tail Structural Arrangement
Aerospatiale Corvette

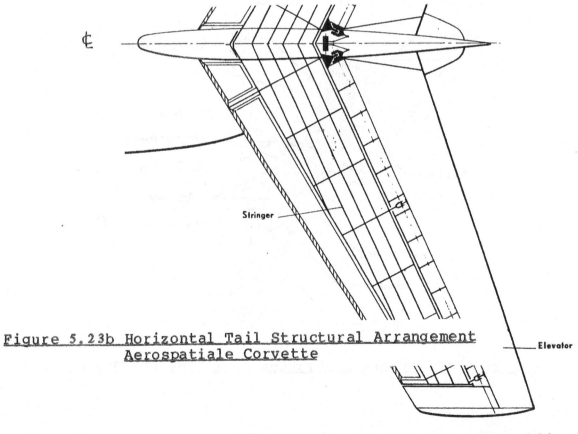

Figure 5.23b Horizontal Tail Structural Arrangement
Aerospatiale Corvette

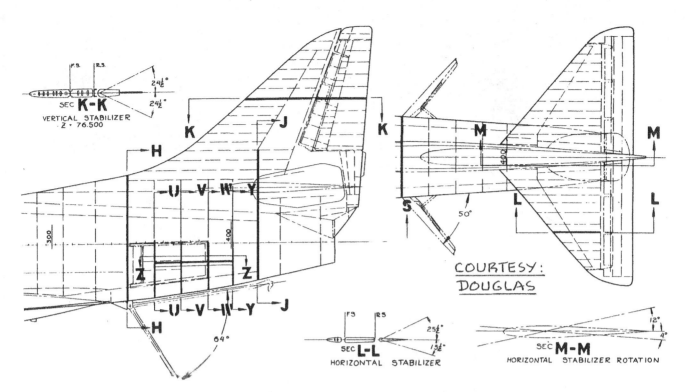

Figure 5.24 Empennage Structural Arrangement Douglas A4

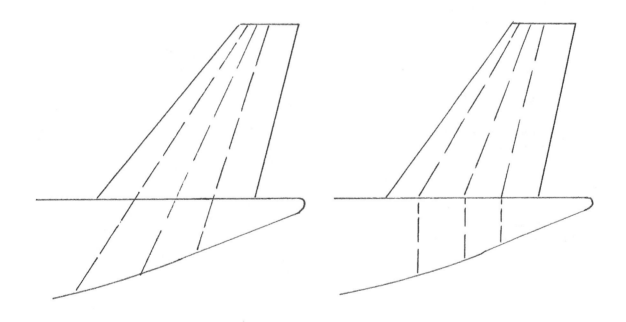

Figure 5.25 Example of Two Methods for Vertical Tail to Fuselage Attachment

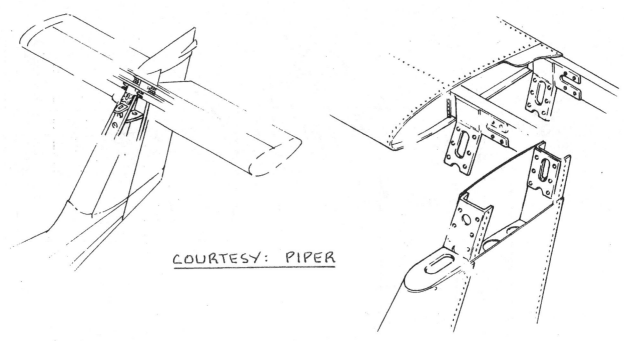

Figure 5.26 T-tail Structure Piper PA-38-112 Tomahawk

1. Dorsal Fin
2. Vertical Stabilizer
3. Fairing
4. Horizontal Stabilizer

5. Elevator
6. Rotating Beacon
7. Fairing
8. Tail Fairing and Tail Light

9. Elevator Control Horn
10. Elevator Tab
11. Rudder
12. Rudder Tab

13. Rudder Torque Tube
14. Rudder Control Horn
15. Rudder Bell Crank Hinge Bracket
16. Rudder Hinge Bracket

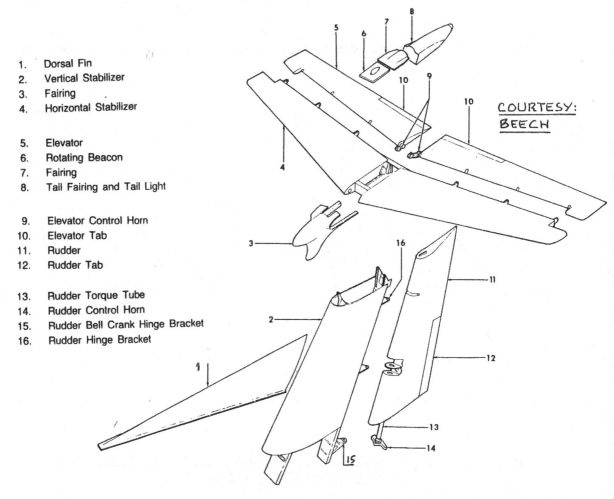

Figure 5.27 T-tail Structure Beech King Air F90

5.2.4 Examples of Empennage Cross Section Design

Figures 5.19, 5.20 and 5.21 present examples of typical cross section designs for vertical and horizontal tails. Observe the similarity with wing cross section designs shown in Chapter 4.

5.2.5 Examples of Longitudinal Control Mechanizations

Figure 5.28 shows the structural mechanization of the elevator control of the Piper Tomahawk, a typical light airplane.

Figures 5.20 and 5.21 provide examples of elevator mechanizations in the DC9 and DC10 respectively.

Examples of elevator control surface designs are seen in Figures 5.28, 5.29 and 5.30.

In most high performance airplanes the entire horizontal tail is used for trim and in most fighters also for primary longitudinal control. In such cases the stabilizer pivots about a fixed point and one or more actuators are used to control the stabilizer. Figures 5.21 and 5.22 show two examples.

A unique solution to obtaining variable incidence stabilizer angles was used by Lockheed on the XF-90 and on the Jetstar. Figure 5.31 shows the XF-90 and identifies the pivot location for the vertical tail: by in effect varying the leading edge sweep angle of the vertical tail, the horizontal stabilizer was given variable incidence!

5.2.6 Examples of Directional Control Mechanizations

This type of rudder design results in much improved rudder effectiveness when compared to a single hinge rudder.

Figure 5.20a shows a rudder plus tab arrangement as used in the DC-9.

Figure 5.27 shows the rudder of the Beech King Air. The rudder is actuated by cables which attach to the control horn labelled no.14 in Fig. 5.27.

5.2.7 Examples of Empennage Skin Gages

Figure 5.17 gives an example of the empennage skin

gages in a very light airplane: Tomahawk. Figure 5.32 shows the same for a Beech King Air.

For fighters and transports the skin gages vary widely depending on the size of the airplane, its performance envelope and on loads.

5.2.8 Maintenance and Access Requirements

The comments made in Chapters 3 and 4 in regard to maintenance and accessibility also apply here. Because of the smaller sizes associated with empennage surfaces there is always a strong temptation to make inspection and access covers too small. This saves weight but is not conducive to safety: any design feature which inhibits accessibility should be considered a detriment to flight safety!

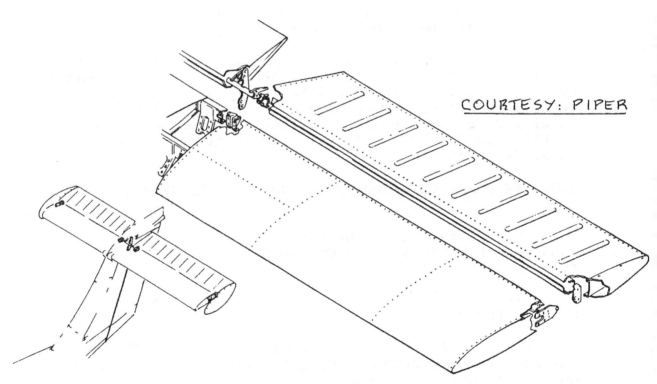

COURTESY: PIPER

Figure 5.28 Elevator Arrangement Piper PA-38 112 Tomahawk

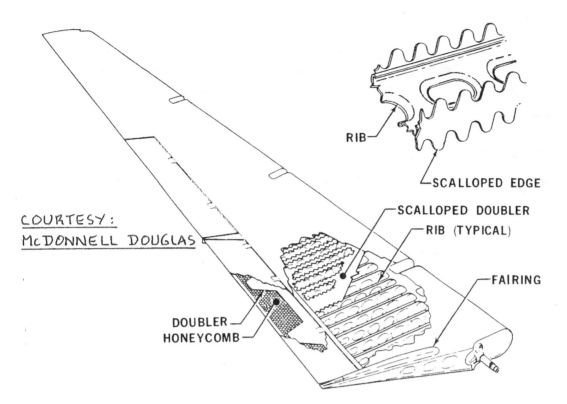

Figure 5.29 Elevator Construction McDD DC9-30

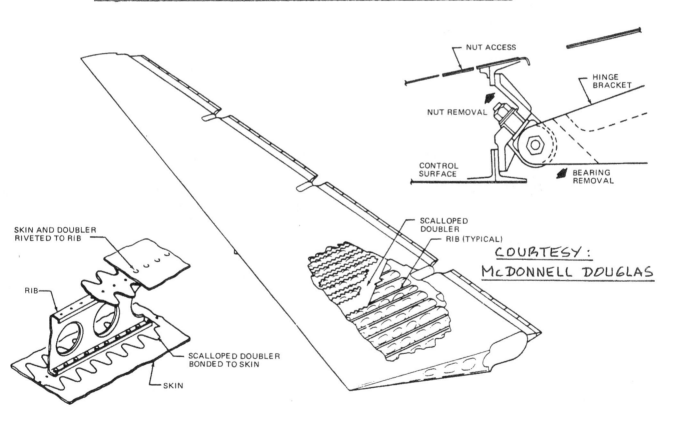

Figure 5.30 Elevator Construction McDD DC-10

Part III Chapter 5 Page 289

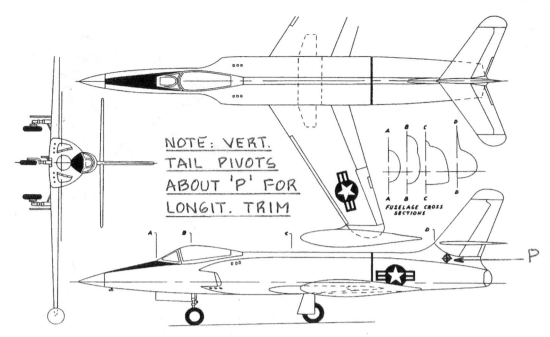

Figure 5.31 Lockheed XF-90

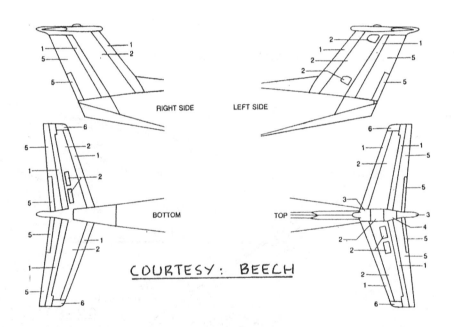

ITEM	MATERIAL	THICKNESS IN INCHES
1	2024T3	.032
2	2024T3	.040
3	Glass Cloth (MIL-C-9084, Type VIII, with MIL-R-7575 Resin)	3 Plies
4	6061T4	.032
5	6061T6	.020
6	Laminated of 2024-T3 and Glass Cloth (MIL-C-9084, Type VIII)	.020 3 Plies

Figure 5.32 Typical Empennage Skin Gages in a Beech King Air F90

6. INTEGRATION OF THE PROPULSION SYSTEM

A Class I procedure for selecting propulsion system type, how many engines to use and where to put these engines was presented as Step 5 of preliminary design sequence 1 as discussed in Chapter 2 of Part II.

The purpose of this chapter is to provide additional information to guide the reader in making these difficult design decisions. The material in this chapter is organized as follows:

6.1 Presentation of engine and propeller data

6.2 Relation between flight envelope and engine type

6.3 Installed thrust/power, inlet and efficiency considerations

6.4 Stability and control considerations (thrust line location and inclination)

6.5 Structural considerations (whirlmode flutter, extension shafts, propeller blade excitations)

6.6 Maintenance and accessibility considerations

6.7 Safety considerations

6.8 Noise considerations

6.9 Example engine and propeller installations

NOTE: Methods for calculating propeller thrust, installed engine thrust/power and other significant engine performance characteristics are presented in Part VI.

6.1 PRESENTATION OF ENGINE AND PROPELLER DATA

In this section tabulated and pictorial data are provided for those engine and propeller characteristics which are important to the preliminary layout designer. The material is organized as follows:

    6.1.1 Propellers
    6.1.2 Piston engines
    6.1.3 Turbo/propeller driven engines
    6.1.4 Turbojet and turbofan engines
    6.1.5 Propfan engines

For more detailed discussions of the fundamental thermodynamic, performance and mechanical aspects of engines the reader should refer to the following references depending on propulsion system type:

For propellers: Refs.12, 14, 55, 56 and 57

For piston engines: Refs.12, 14, 58 and 59

For turbojets and turbofans: Refs.12, 60 and 61

For propfans Refs.62 and 63

6.1.1 Propellers

Methods for computing installed propeller thrust data are presented in Part VI. The following subjects will be addressed:

1. Propeller efficiency
2. Examples of propeller thrust data
3. Examples of existing propellers
4. Propeller weight data

<u>1. Propeller efficiency:</u> Figure 6.1 shows examples of the installed efficiencies which can be expected from propellers in the subsonic speed range.

The following equation defines what is meant by installed propeller efficiency, η_p:

$$THP = \eta_p SHP \qquad (6.1)$$

Installed propeller efficiency depends on such factors as:

* Activity factor, AF
* Airfoil(s)
* Pitch distribution
* Blockage
* Number of blades, n_p
* Tip Mach number
* Single or Counter rotation
* Disk loading/Power loading

Figure 6.1 also shows the improvement in efficiency which can be obtained from counter rotating propellers: the reason is the recovery of swirl losses caused by any individual propeller. The price paid for this improved efficiency is higher installed weight and increased complexity.

Note also in Figure 1, that 'advanced' turboprops do quite well in the high subsonic speed range. The word

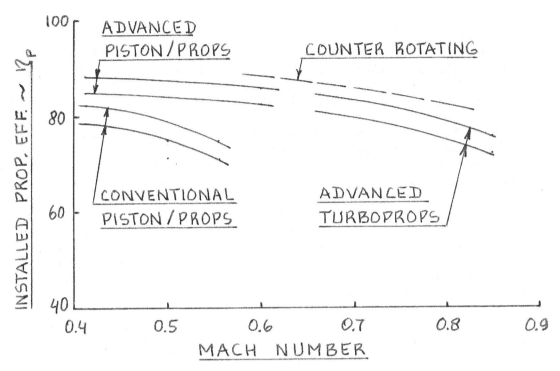

Figure 6.1 Potential Installed Propeller Efficiencies

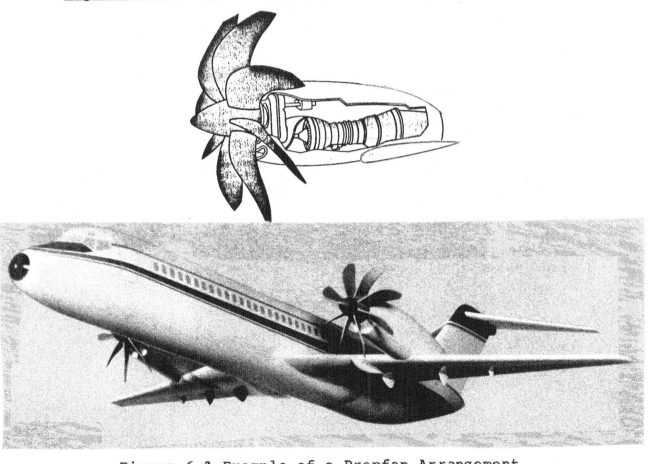

Figure 6.2 Example of a Propfan Arrangement

'advanced' means that thin, supercritical airfoils and swept blades are used. An extreme consequence of this is shown in Figure 6.2 which shows a possible 'propfan' arrangement in a transport type airplane.

Figure 6.3 shows how the number of blades and the diskloading affect ideal propeller efficiency. It is seen that multibladed propellers have a significant performance advantage. The price for this again is increased complexity and increased weight.

2. Examples of propeller thrust data:

Figure 6.4 presents typical free propeller thrust data as a function of speed and propeller rpm and shaft horsepower. Methods for computing both free and installed propeller thrust data are outlined in Part VI.

3. Examples of existing propellers:

Tables 6.1 and 6.2 present detailed design data for current technology as well as for advanced technology propellers.

Figure 6.5 shows an example of a fixed (but ground adjustable) pitch propeller. The pitch angle can be adjusted by inserting 'pitch blocks' with different angles.

Figure 6.6 illustrates the blade geometry used in a conventional variable pitch propeller. The mechanism used to achieve variable pitch capability varies from one manufacturer to another. An example mechanism is shown in Figure 6.7.

Figure 6.8 shows the blade construction of a modern four-bladed propeller. This propeller also is fully reversible, a feature essential in slowing down airplanes on icy surfaces.

4. Propeller weight data:

Methods for estimating propeller weight are presented in Part V. Table 6.3 presents weight data for several propellers. Additional data are found in Tables 6.1 and 6.2.

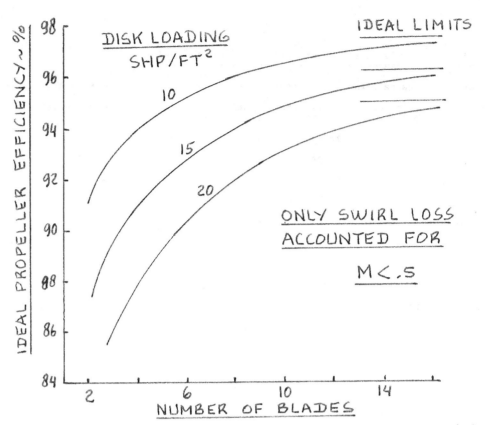

Figure 6.3 Effect of Number of Blades and of Disk Loading on Propeller Efficiency

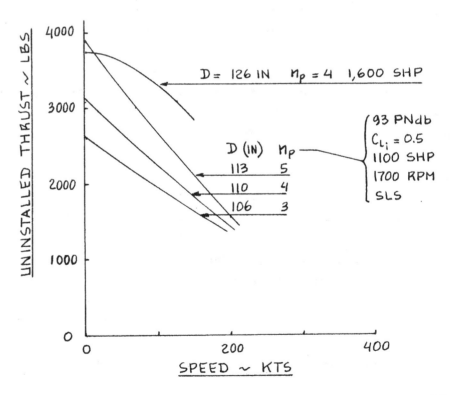

Figure 6.4 Example of the Variation of Propeller Thrust with Flight Speed at Sealevel

Table 6.1 Current Technology Propeller Data (From Ref.64)

Airplane	Cessna 172N	Cessna 210M	Cessna 414A	Cessna 441	Cessna A188B	STAT
Engine	Lycoming O-320-H2AD	Continental IO-520-L	Continentel TSIO-520-NB	Garrett TPE-331-8	Continental IO-520D	PW PT65
Power, hp	160	285	310	635	285	1,270
Prop. RPM	2,700	2,700	2,700	2,000	2,700	1,700
Diameter, in	75	80	76.5	90	80	110
No. of blades	2	3	3	3	3	3
Tipspeed, fps	885	942	901	785	942	816
Cr.speed, kts	120	170	215	295	105	290
Activ. Factor	170	243	267	390	240	360
t/c at .75R	0.085	0.081	0.083	0.065	0.080	0.063
Airfoil	RAF-6	Clark-Y	RAF-6	NACA 16-64	RAF-6	NACA 16-64
Tip sweep deg	0 deg	0 deg	0 deg	1 deg	0 deg	0.5
Proplets	none	none	none	none	none	none
Weight, lbs	36	68	70	117	65	166
Material	Al.	Al.	Al.	Al.	Al.	Al.

Table 6.2 Advanced Technology Propeller Data (From Ref.64)

Airplane	Cessna 172N	Cessna 210M	Cessna 414A	Cessna 441	Cessna A188B	STAT
Engine	Lycoming O-320-H2AD	Continental IO-520-L	Continentel TSIO-520-NB	Garrett TPE-331-8	Continental IO-520D	PW PT65
Power, hp*	145/147	270/272	260/263	515/525	275/278	1075/1080
Prop. RPM*	2665/2388	2653/2399	2699/2429	2051/1822	2653/2399	1818/1627
Diameter, in	83	90	85	100	90	120
No. of blades	2	4	4	4	4	5
Tipspeed, fps*	965/865	1042/942	1001/901	895/795	1042/942	952/852
Cr.speed, kts	120	170	215	295	105	290
Activ. Factor	170	243	267	390	240	360
t/c at .75R	0.060	0.060	0.063	0.040	0.060	0.040
Airfoil	all have advanced NASA propeller airfoils					
Tip sweep	all have 25 degrees tip sweep					
Proplets	all have proplets with height to radius ratio of 0.05					
Weight, lbs	32	55	60	99	48	127
Material	E-glass	Kevlar	Kevlar	Kevlar	Kevlar	Kevlar

* first number meets FAR 36/second number meets FAR 36 minus 5 db

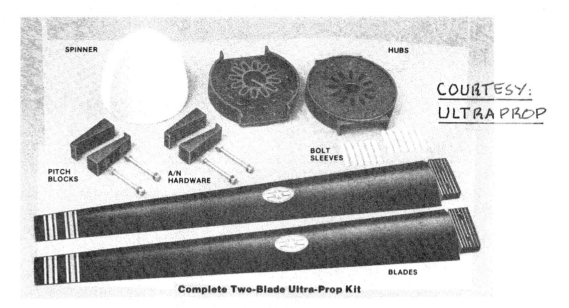

Figure 6.5 Example of a Fixed Pitch Propeller

Station Inches from Centerline	Width-In.		Thickness-In.		*Angle - Degrees				Edge Alignment-In.		Face Alignment-In.	
					New Blade		Reworked Blade					
	Min	Max	Min	Max	Min	Max	Min	Max	Min	Max	Min	Max
9	5.157	5.344	1.761	1.885	6.6	7.6	6.6	7.6	2.355	2.511	0.844	0.934
12	5.966	6.090	1.182	1.232	4.6	5.6	4.6	5.6	2.699	2.761	0.518	0.580
15	6.293	6.417	0.817	0.867	2.2	3.2	2.2	3.2	2.803	2.865	0.319	0.381
18	6.478	6.602	0.675	0.725	0.0	0.0	0.0	0.0	2.886	2.948	0.260	0.322
24	6.598	6.722	0.525	0.575	5.8	6.2	5.9	6.1	2.939	3.001	0.198	0.260
30	6.144	6.206	0.407	0.457	9.9	10.3	10.0	10.2	2.723	2.785	0.149	0.211
36	5.054	5.116	0.298	0.348	12.4	12.8	12.5	12.7	2.237	2.299	0.103	0.165
39	4.349	4.411	0.245	0.295	13.4	13.8	13.5	13.7	1.922	1.984	0.081	0.143
42	3.574	3.636	0.193	0.243	14.2	14.6	14.3	14.5	1.577	1.639	0.060	0.122
45	2.741	2.803	0.100	0.150	14.9	15.3	15.0	15.2	1.205	1.267	0.021	0.083

* For the purpose of measuring blade angles, a plane passing through the thrust face at the 18-inch station is the reference plane. For stations either side of the 18-inch station, the angles are measured on opposite sides of the reference plane.

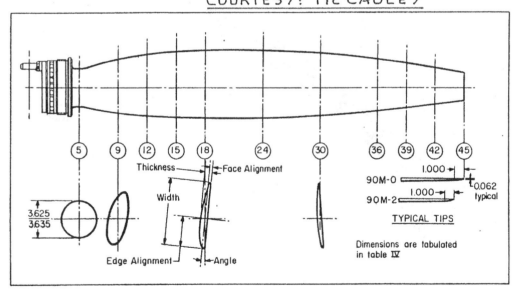

Figure 6.6 Example of Propeller Blade Geometry and Airfoils for a Low Speed Variable Pitch Propeller

Part III　　　　　Chapter 6　　　　　Page 297

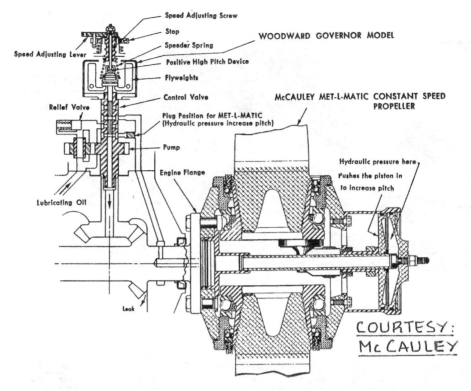

Figure 6.7 Example of a Variable Pitch Mechanism

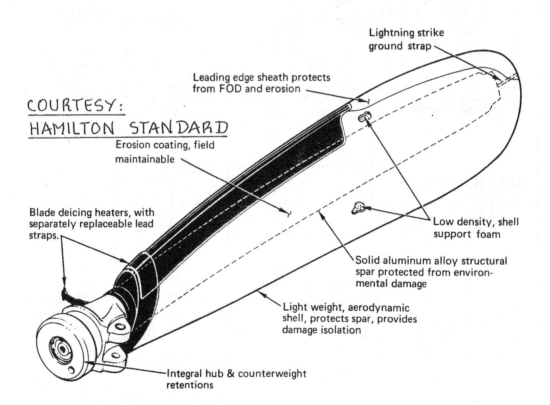

Figure 6.8 Blade Construction of a Modern Four Bladed, Reversible Pitch Propeller

Table 6.3 Propeller Weight Data
================================

Airplane	TOHP	TO RPM	Prop. Diam.	No. of Blades	Weight*	Propeller Control F/V/R**
	hp	rpm	in		lbs	
Cessna 180	225	2,660	84	2	62	F
Cessna 310	240	2,600	80	2	68	V
Beech Baron 95-55	290	2,625	76	2	97	V
Aero Comm. 560	320	2,060	90	2	103	V
DHC-6 Twin Otter	620	2,110	102	3	134	V/R
Beech 99	850	2,000	97	3	131	V/R
Shorts Skyvan	715	2,000	99	4	143	V/R
Beech A-100	680	2,200	90	4	154	V/R
Shorts SD3-30	1,120	1,700	111	5	218	V/R
Hamilton Standard Propfan for 150 Pax Airliner	3,000	1,000	120	8	430	V/R

* Weights exclude spinners. Typical spinner weights 5-12 lbs

** F = Fixed pitch   V = Variable pitch   R = Reversible

Part III   Chapter 6   Page 299

6.1.2 Piston Engines

Airplane piston engines are available in three configurations:

1. Horizontally opposed cylinders
2. Radially arranged cylinders (radials)
3. In-line and/or V- arranged cylinders.

The great majority of modern piston engines are of the horizontally opposed type. Figures 6.9 and 6.10 show an example of a horizontally opposed, aircooled piston engine.

Figure 6.11 shows a 14-cylinder, double row radial piston engine. This engine is no longer in production but hundreds are still available. It is installed in the Canadair CL-215 (Figure 3.31b in Part II) which is still in production.

Smaller radials are still in production in Poland. They are used primarily in agricultural airplanes.

The in-line and/or V- arranged engine configurations were very popular in WWII but have lost favor with most airframers. Small in-lines are still in production in Czechoslovakia. They are used primarily in aerobatic and trainer airplanes.

Reference 30 contains basic design data on piston engines.

Piston engines rapidly loose power with altitude. To counteract this, a so-called 'super-charger' is used. Superchargers use the exhaust gasses from the cylinders to drive a turbine which in turn drives a compressor. The latter compresses the air which is then fed into the cylinders. Super-chargers were in widespread use in WWII in fighter and bomber airplanes. A modern super-charger installation is shown in Figure 6.12.

During preliminary design of airplanes basic data on engine geometry, weight, power output and specific fuel consumption must be available. Ref. 30 provides this information. Tables 6.4 and 6.5 list such preliminary design information for a number of piston engines. The gemetric data used in these tables are defined in Fig.6.13.

Figures 6.14 and 6.15 provide examples of manufacturers engine data for two typical piston engines.

6.1.3 Turbopropeller Engines

A turbopropeller engine is a turboshaft engine which drives a gearbox which in turn drives a propeller. Examples of modern turbopropeller engines are presented in Figures 6.16 and 6.17.

Tables 6.6 and 6.7 present basic design data for a number of turbopropeller engines. Note that propeller weights are <u>not</u> included in these data, <u>but the gearboxes are.</u> More design data on turboshaft engines (no gearbox) and turbopropeller engines (with gearbox but without propeller) are found in Ref.30.

Figures 6.18 and 6.19a present typical manufacturers performance information for turbopropeller engines. Figure 6.19b also provides the installation information needed to integrate this powerplant into an airframe.

6.1.4 Turbojet and Turbofan Engines

The difference between a turbojet and a turbofan engine is primarily in the management of air: in a turbojet all the inlet air passes through the combustion process (BPR = 0), while in a turbofan part of the air is 'by-passed' through a fan which accelerates the bypassed air aft (BPR > 0).

At subsonic speeds turbofans have essentially replaced turbojets in most applications. The reason is the much lower specific fuel consumption associated with the turbofan.

For supersonic operations there is a trend toward replacement of pure turbojets with low by-pass ratio turbofans. What complicates the installation of engines in a supersonic airplane is the requirement for special inlets and frequently the requirement for afterburning (=augmentation).

Figure 6.20 through 6.22 show cutaways of modern turbofan engines for subsonic applications. Performance and geometry data for two of these engines are presented in Figures 6.23 and 6.24. Figure 6.25 presents data for the GE TF39 engine (similar to the PW JT9D engine).

Figures 6.26 through 6.29a show examples of afterburning turbojet and turbofan engines, all for use in supersonic airplanes. Thrust and sfc data for the F100 engine of Fig.6.29a are given in Figure 6.29b.

Tables 6.8 through 6.11 contain basic design data on selected turbojet and turbofan engines. Ref.30 contains information on a large number of turbojets and turbofans.

6.1.5 Propfan Engines

Figures 6.30 and 6.31 show examples of two types of propfan engines. The associated performance predictions for these powerplants are reflected in Figures 6.32 and 6.33 respectively. These engines and their derivatives are expected to become operational in the early 90's. At the time of publication of this text none had been certificated for civil nor for military operation.

NOTE: This note applies to all types of engines which have a gasgenerator as the primary source of power:

Engine weight data for these engines DO NOT include the weight of any associated nacelles, inlets and exhaust pipes and/or nozzles. Manufacturers thrust and sfc data are normally given for 'ideal' inlets (bellmouths) and for 'ideal' exhaust systems. The reason is that engine manufacturers cannot anticipate the different types of engine installations an airframe manufacturer may wish to use. Section 6.9 contains examples of nacelle, inlet and exhaust system configurations.

COURTESY: AVCO-LYCOMING

Figure 6.9 AVCO-Lycoming O-320 Series Horizontally Opposed 4-Cylinder Piston Engine

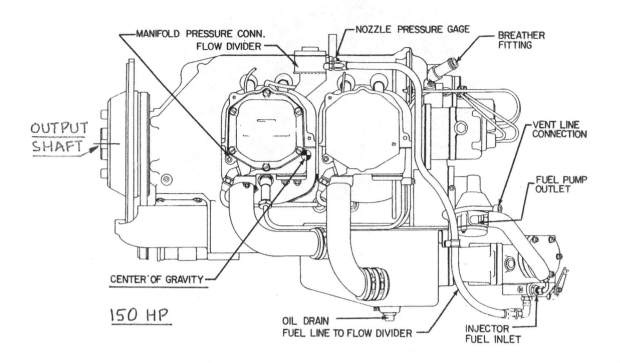

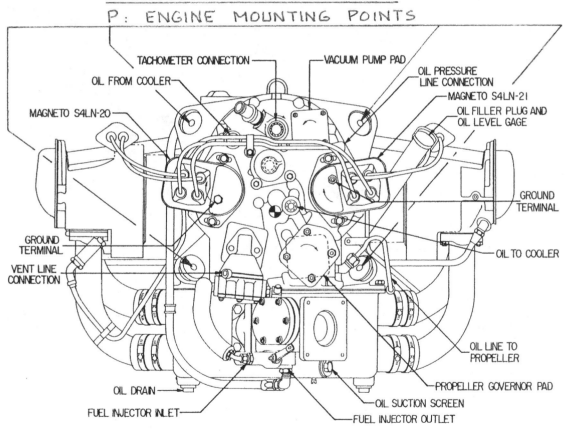

Figure 6.10 Side View and Rear View of AVCO-Lycoming O-320 Series 4-Cylinder Piston Engine

Figure 6.11 Pratt and Whitney R2800 Double Row Radial 14-cylinder Piston Engine

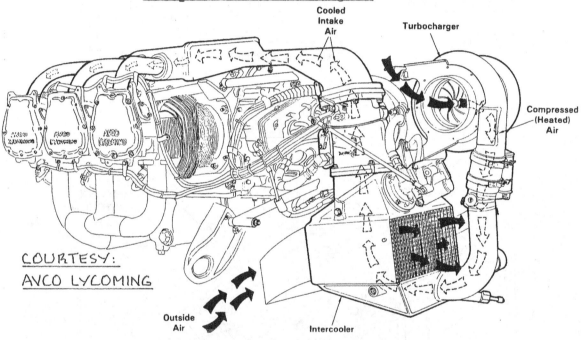

Figure 6.12 Typical Supercharger Installation

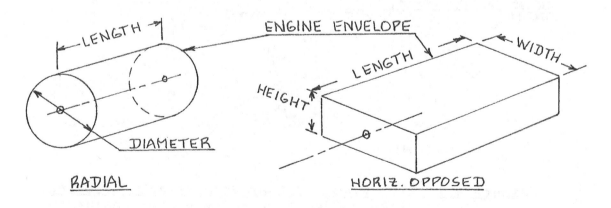

Figure 6.13 Definition of Piston Engine Geometric Data

Part III　　　　　　　　　　Chapter 6　　　　　　　　　　Page 304

Table 6.4 Manufacturer Performance Data for Piston Engines
===

Type	Avco Lycoming				PZL-3S
	IO-320-B1A	IO-360-A1B6D	TIO-540-A1A	TIGO-541-E1A	
Supercharged	no	no	yes	yes	no
Direct drive	yes	yes	yes	geared 0.8	yes
Max. T.O. Power (hp)	160	200	310	425	600
at Prop. RPM/to alt.	2,700/SL	2,700	2,575/15K	2,133/15K	2,200/SL
SFC (lbs/hp/hr)	0.51	0.50	0.565	0.753	0.610
Cruise Power* (hp)	120	150	233	319	415
at RPM	2,350	2,450	2,575	1,833	2,000
SFC (lbs/hp/hr)	0.489	0.481	0.516	0.502	0.510
No. of cylinders	4	4	6	6	7(radial)
Dry weight** (lbs)	287	330	540	700	906
Length (in)	33.6	31.3	51.3	57.6	43.7
Width (in)	32.2	34.3	34.3	34.9	49.9
Height (in)	19.2	19.4	22.7	22.7	diam.
Octane	91/96	100/130	100/130	100/130	91+

*normally at 75 percent rated power, static sealevel.

**includes accessories needed for operation

Table 6.5 Manufacturer Performance Data for Piston Engines
===

Type	Franklin/PZL***			Teledyne-Continental		
	2A-120	4A-235-B2	6A-350-C1	O-200A	TSIO-	IO-520-A
Supercharged	no	no	no	no	no	no
Direct drive	yes	yes	yes	yes	yes	yes
Max. T.O. Power (hp)	60	125	220	100	225	285
at Prop. RPM, SLS	3,200	2,800	2,800	2,750	2,800	2,700
SFC (lbs/hp/hr)	0.53	0.52	0.460	0.60	0.62	0.50
Cruise Power* (hp)	45	94	165	75	169	214
at RPM	2,200	2,080	2,100	2,450	2,550	2,500
SFC (lbs/hp/hr)	0.620	0.440	0.480	0.585	0.52	0.452
No. of cylinders	2	4	6	4	6	6
Dry weight** (lbs)	137	224	333	218	300	471
Length (in)	23.7	30.5	32.1	28.5	35.3	41.4
Width (in)	30.7	31.5	31.6	31.6	33.1	33.6
Height (in)	22.7	25.1	27.5	23.2	23.7	19.8
Octane	100/130	100/130	100/130	80/87	100/130	100/130

*normally at 75 percent rated power **includes accessories needed for operation

***Franklin engines are manufactured in Poland by PZL

PERFORMANCE

Rated Power (Sea Level) . . . 100
Take-Off Power (Sea Level) . . 100
Cruise Power (Sea Level) . . . 75
Power corrected to 29.92 in. Hg.
60° F. Inlet Air Temp.

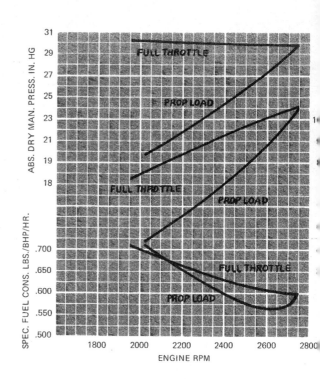

OPERATING DATA

Oil Required . . . Below 40°F: SAE-20
                 Above 40°F: SAE-40
Oil Temperature . . . . . 75°F.–170°F.
Oil Pressure, Idling . . . . .10 p.s.i. Min.
Oil Pressure, Cruising . . . .30-60 p.s.i.
Oil Sump Capacity . . . . 6 quarts
Allowable Temperatures:
   Cylinder Head . . . . . .525°F. Max.
   Cylinder Barrel . . . . .290°F. Max.
   Oil Temperature . . . . . .225°F. Max.

SPECIFICATIONS

Type Certificate Number 252
Number of Cylinders 4
Bore (Inches) 4.06
Stroke (Inches) 3.88
Displacement (cubic inches) 201
Type of Propeller Drive, Flanged . . . . Direct
Compression Ratio 7.0:1

RPM at Rated Power 27
RPM at Take-Off 27
Cruising RPM 25
Recommended Fuel Grade Min. 80
Dry Weight (With Accessories Below Included) . . . 217
Crankshaft Rotation . . . . Clockw
Ignition . . . . D

COURTESY:
TELEDYNE CONTINENTAL

STANDARD EQUIPMENT

ITEM	MAKE	WEIGHT
Spark Plugs	A.C.	1.75
Magnetos	Slick	12.12
Ignition Harness (Shielded)	Continental	3.81

ITEM	MAKE	WEIG
Starter	Delco-Remy	15
Generator	Delco-Remy	10
Oil Cooler	Harrison	4
Tachometer Drive	Continental	0

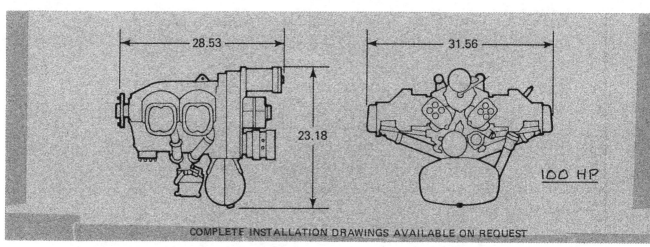

Figure 6.14 Tel.-Continental O-200-A Manufacturer Data

PERFORMANCE

    Rated Power (Sea Level) 225
    Take-Off Power (Sea Level) 225
    Recommended Cruise Power (Sea Level) 172
    Critical Altitude
        Rated Manifold Pressure 20,000
        Cruise Power. 25,000
    Power Corrected to 29.92 In. Hg 60°F. Inlet
    Air Temperature

OPERATING DATA

    Oil Required . . . . Below 40°F: SAE No. 30
                         10W-30 Optional
                         Above 40°F: SAE No. 50
    Oil Temperature . . . . 75°F–240°F
    Oil Pressure, Idling . . . . 10 p.s.i. Min.
    Oil Pressure, Cruising . . . . .30-60 p.s.i.
    Oil Sump Capacity . . . . 10 quarts
    Allowable Temperatures:
        Cylinder Head . . . . 460°F Max.
        Cylinder Barrel . . . . 310°F Max.

SPECIFICATIONS

Type Certificate Number. E9CE
Number of Cylinders 6
Bore (inches). 4.438
Stroke (inches) 3.875
Displacement (cubic inches). 360
Type of Propeller Drive, Flanged. . . . . .Direct
Compression Ratio. 7.5:1
Max. Continuous M.P., to 20,000 Ft.
    Critical Altitude, In. Hg. 37
RPM @ Cruise Power. 2600
RPM @ Rated Power. 2800

Max. Cruise M.P., to 25,500 Ft.
    Critical Altitude, In. Hg. 30
Recommended Fuel, Grade Min. 100/130
Dry Weight (With Accessories Below Included). . . 300.25
Crankshaft Rotation. . . . . Clockwise
Ignition . . . . Dual
Dimensions With Standard Equipment Installed
    Length. 35.34
    Width. 33.11
    Height. 23.74

COURTESY: TELEDYNE CONTINENTAL

STANDARD EQUIPMENT

ITEM	MAKE	WEIGHT
Spark Plugs.	A.C.	3.03
Fuel Injector System.	Continental	8.75
Magnetos	Scintilla	11.00
Ignition Harness (Shielded).	Continental	2.41
Starter	Delco-Remy	16.38
Alternator.	Autolite	10.81
Oil Cooler.	Harrison	4.25

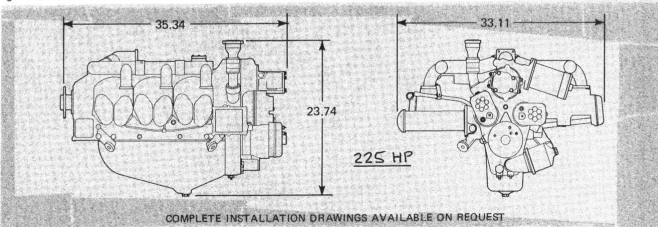

225 HP

COMPLETE INSTALLATION DRAWINGS AVAILABLE ON REQUEST

Figure 6.15 Tel.-Continental TSIO-360-C Manufacturer Data

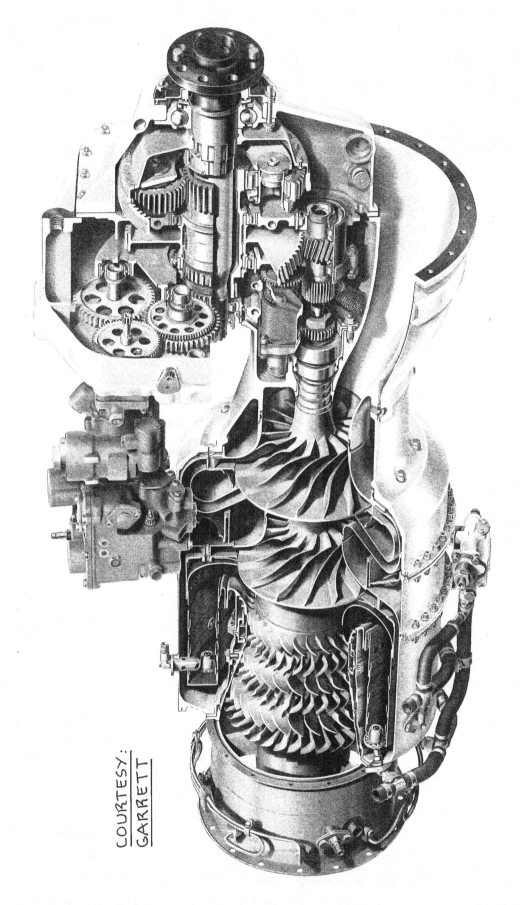

Figure 6.16 Garrett TPE331 Turboprop Engine Cutaway

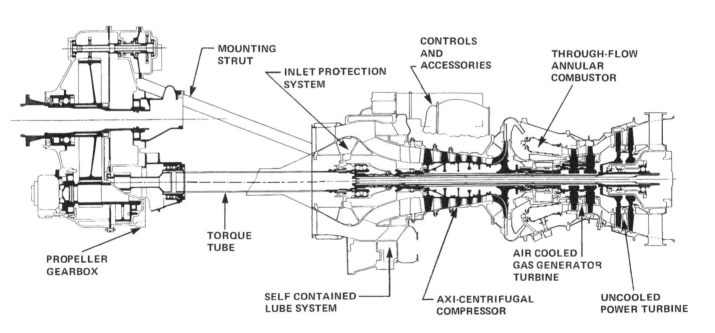

Figure 6.17 General Electric CT7-5 Turboprop Engine Layout and Perspective

Part III   Chapter 6   Page 309

Table 6.6 Manufacturer Performance Data for Turboprop Engines

Type	Pratt and Whitney of Canada Ltd.					
	PT6A-21	PT6A-41	PT6A-65R	PW115	PW120	PW124
For SLS (static shaft horsepower, hp):						
Max. T.O.	550	850	1,173	1,500	1,800	2,150
Max. Cont.	550	850	1,173	1,500	1,700	NA
Max. Cruise	550	850	956	1,500	1,619	2,030
Max. Massflow (lbs/sec)	6.1	8.0	9.4	14.3	14.8	NA
For SLS (specific fuel consumption, lbs/ESHP/hr):						
Max. T.O.	0.630	0.591	0.549	0.529	0.499	0.473
Max. Cont.	0.630	0.591	0.564	0.529	0.506	NA
Max. Cruise	0.649	0.591	0.581	0.529	0.514	NA
SHP/Altitude/speed	316/20K/ 245kts	488/20K/ 245kts	549/20K/ 245kts	861/20K/ 245kts	929/20K/ 245kts	1,165/20K/ 245kts
Rated Propeller RPM	2,200	2,000	1,700	1,300	1,200	1,200
Weight (lbs)	303	380	464	841	921	1,060
Length (in), cold	62	67	74	81	84	84
Max. diam. (in)	19	19	19	25	25	25
Application	Beech C90	Piper Cheyenne III	Shorts 360	EMB-120	DHC-8	BAe-ATP

Table 6.7 Manufacturer Performance Data for Turboprop Engines

Type	General Electric		Rolls R. Dart RDa7 TS1637	Rolls R. Turbomeca AZ-14	Garrett TPE 331-3	T.Lyc. LTP101
	CT7-5A	CT64-820				
For SLS (static shaft horsepower, hp):						
Max. T.O.	1,699	3,133	1,835	800	840	592
Max. Cont.	1,476	2,745	1,835	800	840	592
Max. Cruise	1,417	2,745	1,650	720	770	NA
Max. Massflow (lbs/sec)	10	26.2	23.5	5.5	7.8	4.8
For SLS (specific fuel consumption, lbs/ESHP/hr):						
Max. T.O.	0.456	0.486	0.676	0.521	0.548	0.550
Max. Cont.	0.465	0.505	0.676	0.521	NA	NA
Max. Cruise	0.471	0.505	0.676	0.532	NA	NA
SHP/Altitude/speed	1,655/ 15+K/NA	NA	1,220/ 15K/0kts	NA	NA	NA
Rated Propeller RPM	1,200 (estim.)	1,160	1,400	1,783	1,600	1,924
Weight (lbs)	676	1,145	1,369	454	355	241
Length (in), cold	80.4	110.1	97.6	80.6*	43.5	30.9
Height (in)	31	20.1	37.9	22.9	26.0	18.6
Width (in)	26	diam.	diam.	22.9	21.0	diam.
Application	SF340	G222	F-50	BAe Jetstream	BAe Jetstream	P166-D1-3

*includes propeller!

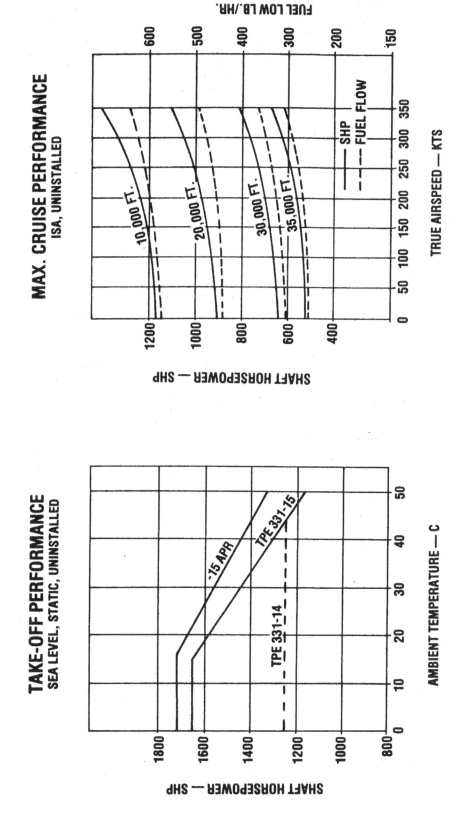

Figure 6.18 Uninstalled Performance Data for Garrett TPE331-14 Turboprop Engine

Part III　　　Chapter 6　　　Page 311

PT6A-41

PERFORMANCE

Guaranteed calibration stand performance at 2000 RPM propeller speed, sea level static output.

Ratings:	ESHP	SHP	Specific Fuel Consumption lb/eshp/hr
Take-off			
Max. Continuous	903	850(1)	.591
Max. Climb			
Max. Cruise	903	850(2)	.591

(1) Available to 106°F
(2) Available to 84°F

COURTESY: PRATT & WHITNEY

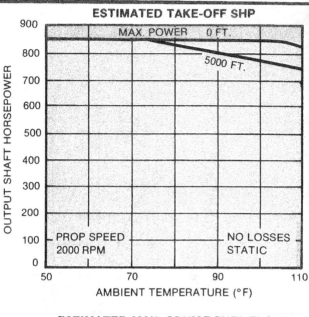

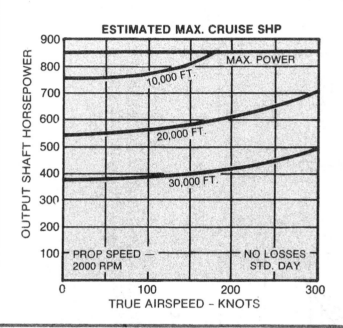

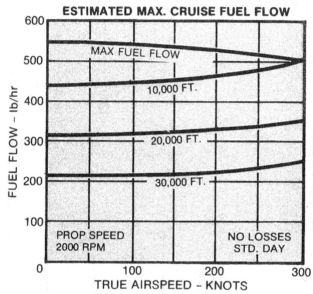

STANDARD EQUIPMENT

- Fuel System — including fuel pump and altitude compensated fuel control unit
- Engine Ignition System — without power source
- Gas Temperature Thermocouples
- Integral Oil Tank
- Torquemeter
- Marinization
- Inlet Screen
- Accessory Drives
  — power section: propeller control unit and tachometer generator
  — gas generator section: starter generator and tachometer generator

ADDITIONAL EQUIPMENT

- Fireseal mounting rings
- Fuel heater with hoses

OPTIONAL EQUIPMENT

- Propeller Control System including propeller control unit
- Accessory Drives
  — gas generator section: three (3) aircraft accessories
  — power section: propeller overspeed governor
- Torque limiter
- Compressor wash ring
- Provision for fuel flowmeter
- Provision for fuel temperature

Figure 6.19a Uninstalled Performance Data for Pratt and Whitney PT6A-41 Turboprop Engine

The PT6A-41 is a free turbine propulsion engine incorporating a multi-stage compressor, single-stage compressor turbine, and independent two-stage power turbine driving the output shaft through integral planetary gearing. A single annular combustion chamber, 14 simplex fuel nozzles and two spark igniter plugs comprise the combustion system. Engine accessories are conveniently grouped on the rear of the engine. The free turbine design permits selection of propeller speed from a wide RPM band to suit aircraft operation.

Engine dry weight (with standard equipment); 380 lbs.

Output shaft speed (max.): 2000 RPM.

COURTESY: PRATT & WHITNEY

Sold under current edition of Production Specification No.723.

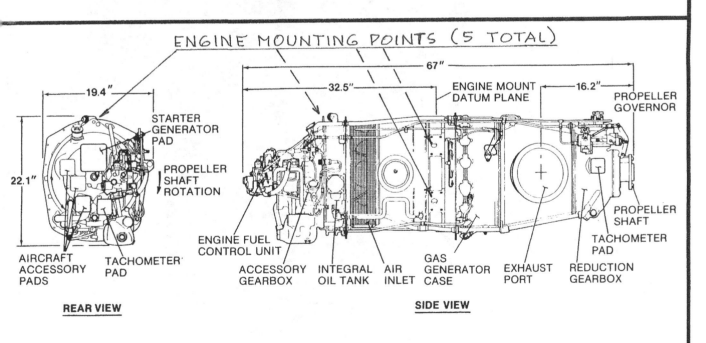

Figure 6.19b Installation Geometry for Pratt and Whitney PT6A-41 Turboprop Engine

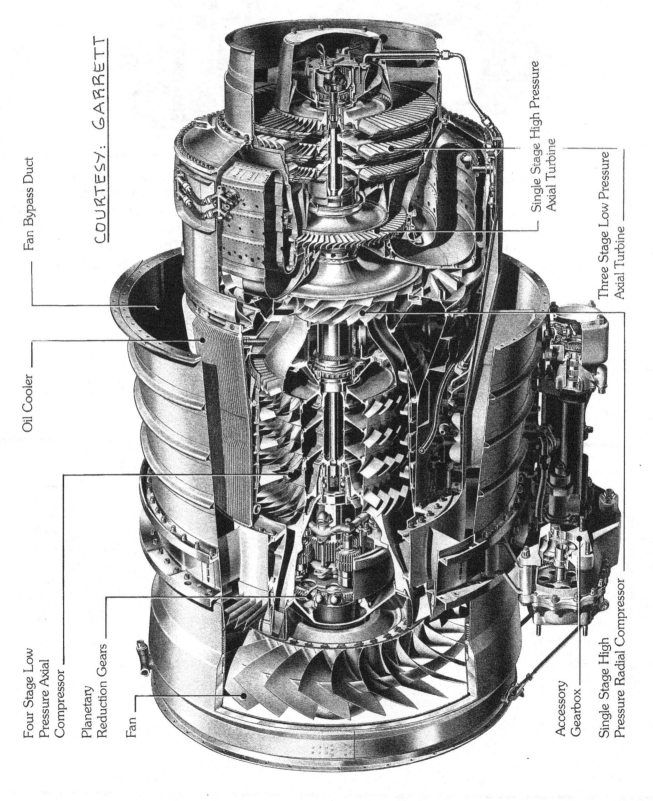

Figure 6.20 Cutaway of Garrett TFE731 Turbofan

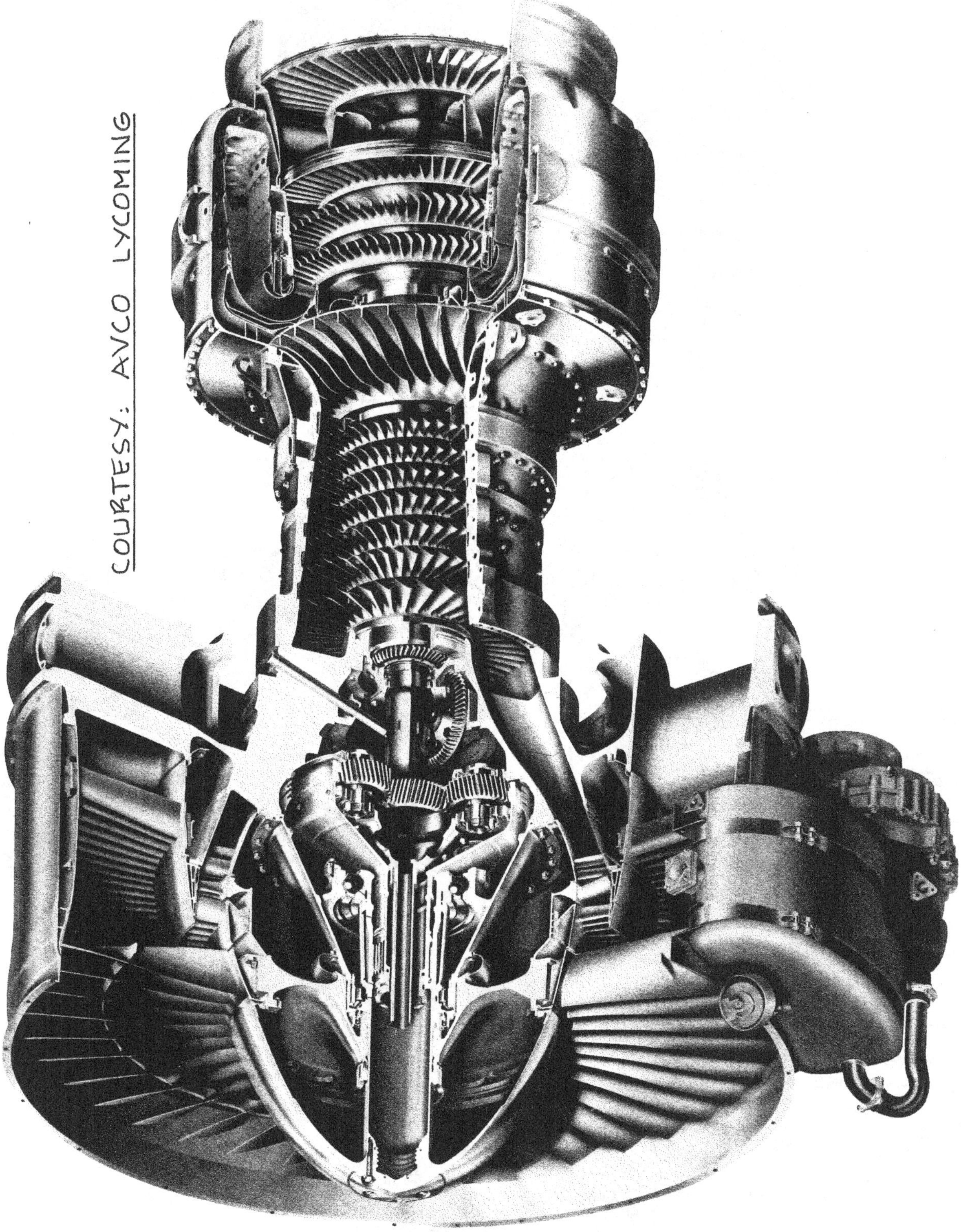

COURTESY: AVCO LYCOMING

Figure 6.21 Cutaway of AVCO Lycoming ALF-502R-5 Turbofan

Part III Chapter 6 Page 315

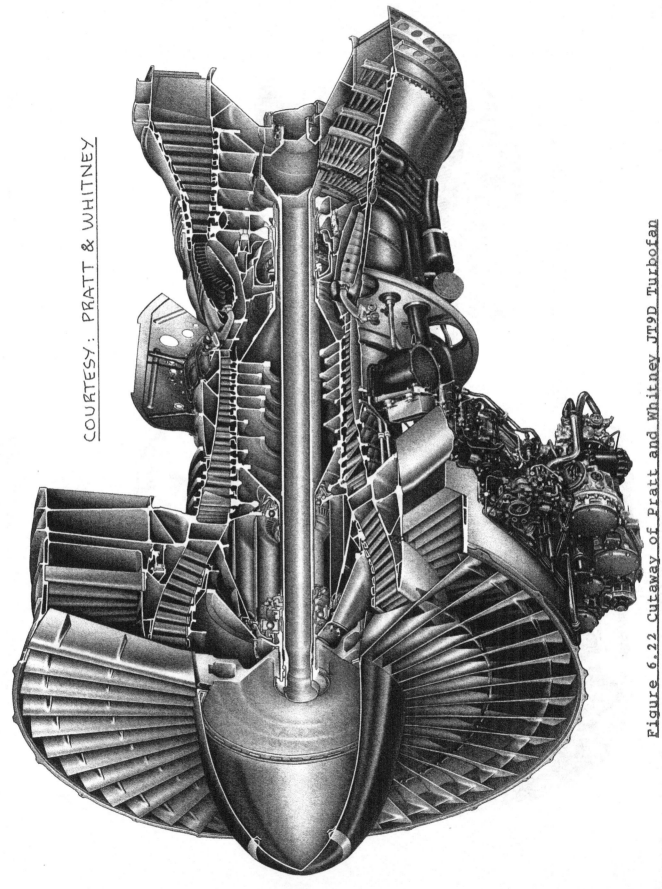

Figure 6.22 Cutaway of Pratt and Whitney JT9D Turbofan

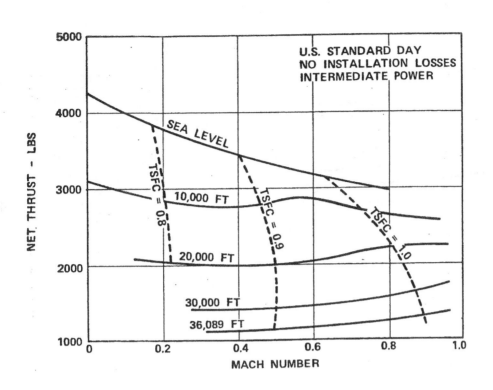

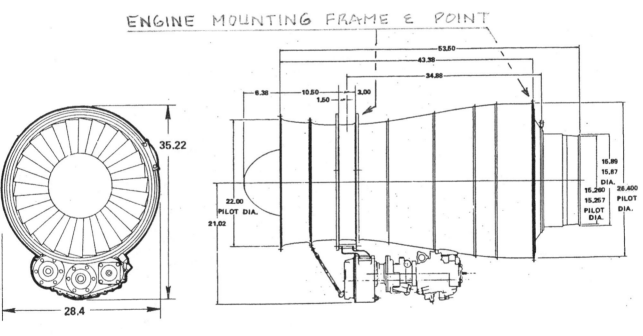

COURTESY: GARRETT

Figure 6.23 Uninstalled Performance and Geometry Data for Garrett TFE731-1042 Engine

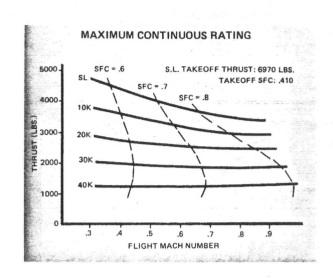
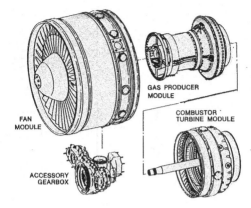

COURTESY: AVCO LYCOMING

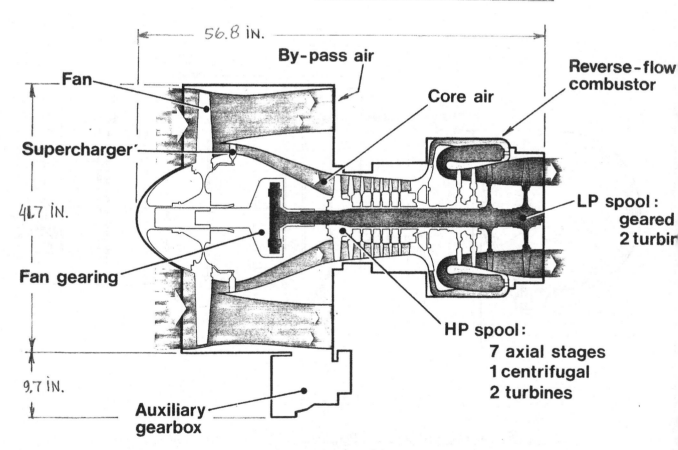

Figure 6.24 Uninstalled Performance and Geometry Data for AVCO Lycoming ALF-502R-5 Turbofan

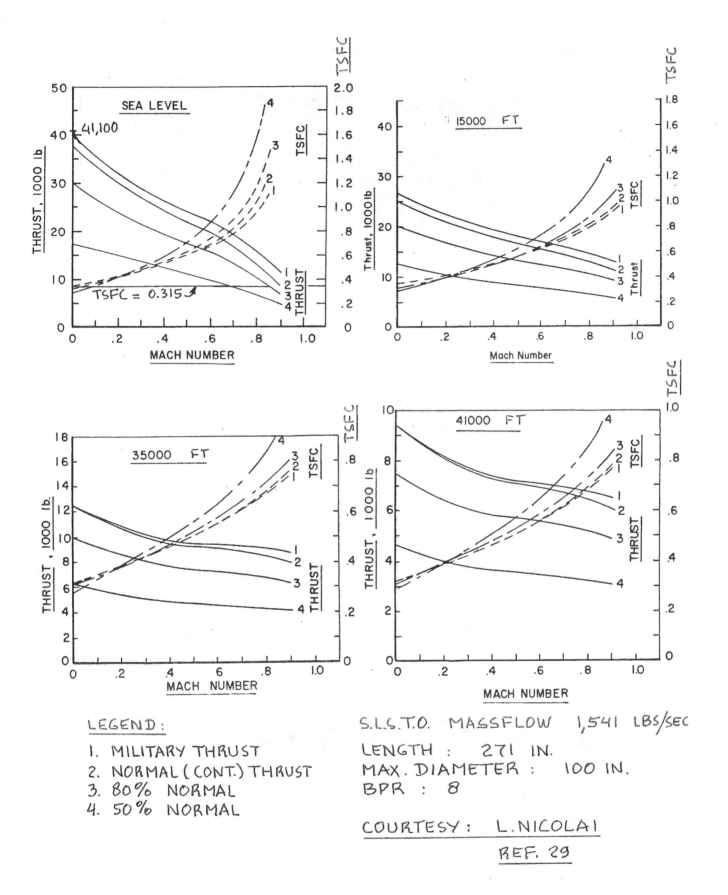

Figure 6.25 Estimated C5-A Installed Performance Data for General Electric TF-39-GE-1 Turbofan

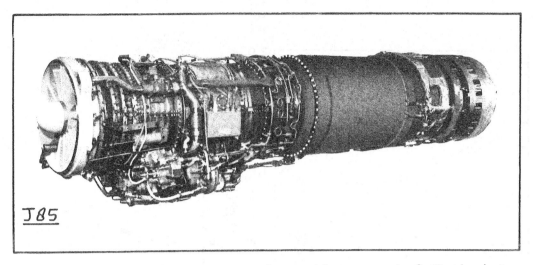

Figure 6.26 General Electric J-85 Augmented Turbojet

SPECIFICATIONS

	J85-13	J85-21A
Weight (lb)	597	684
Length (in.) (cold)	105.6	112.5
Max Dia (in.)(cold)	17.7	21
Max Radius (in.) (cold)	17.6	17.6
Comp/Turbine Stages	8/2	9/2
Thrust/Weight	6.8	7.3
Pressure Ratio	6.9	8.3
Air Flow (lb/sec)	44.0	53.0
RPM	16,500	16,600
T_4 (T O/°F)	1745	1800
EGT Limit (°F)	1325	1345
Max Thrust/HP (SLS)	4080	5000
SFC	2.22	2.13
MIL Thrust/HP (SLS)	2720	3500
SFC	1.03	1.00
Max Mach No/Alt	2.35/65K	2.35/65K
Cruise Mach No/Alt	.9/36K	.9/36K
Thrust	770	1200
SFC	1.29	1.23

SPECIFICATIONS

	J79-17	J79-17X*
Weight (lb)	3873	3847
Length (in) (cold)	208.69	208.69
Max. Dia. (in) (cold)	39.06	39.06
Max Radius	19.5	19.5
Comp/Turbine Stages	17/3	17/3
Thrust/Weight	4.60	4.9**
Pressure Ratio (MIL)	13.4	13.4
Air Flow (lb/sec)	170.0	170.0
RPM	7685	7839
T_4 (T.O./Cruise) °F	1810	1837
EGT Limit (°F)	1240	1300
Max Thrust (SLS)	17,820	18,730**
SFC	1.98	1.98
MIL Thrust (SLS)	11,810	11,810
SFC	0.85	0.85
Max Mach No./Alt.	2.4/45K	2.0/50K
Thrust (M2.0/35K)	18,600	20,840**
SFC (M2.0/35K)	2.07	2.05
Cruise Mach No./Alt	.9/35K	.9/35K
Thrust	2600	2600
SFC	0.98	0.98

*Various of these engine parameters are specifications and have not been verified by testing.
**Combat plus mode

Figure 6.27 General Electric J-79 Augmented Turbojet

COURTESY : GENERAL ELECTRIC

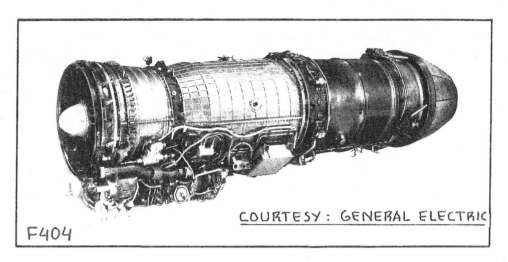

Figure 6.28 General Electric F404 Augmented Turbofan

SPECIFICATIONS

F404

Thrust	16,000-lb Class
Length (in.)	158.8
Maximum Diameter (in.)	34.8
Fan Stages	3
Compressor Stages	7
Compressor Pressure Ratio	Over 25:1 Class
Combustor	One Piece Annular
Turbine Stages	
High Pressure	1
Low Pressure	1
Control	Electrical-Hydromechanical
Variable Exhaust Nozzle	Hydraulically Actuated Converging-Diverging Type

F100-PW-100

MAXIMUM THRUST (FULL AUGMENTATION)	25,000-POUND (111.2 kN) CLASS
INTERMEDIATE THRUST (NON-AUGMENTED)	15,000-POUND (66.7 kN) CLASS
WEIGHT	3020 POUNDS (1371 kg)
LENGTH	191 INCHES (4.85 m)
INLET DIAMETER	36 INCHES (0.91 m)
MAXIMUM DIAMETER	46.5 INCHES (1.18 m)
BYPASS RATIO	0.6
OVERALL PRESSURE RATIO	24 to 1

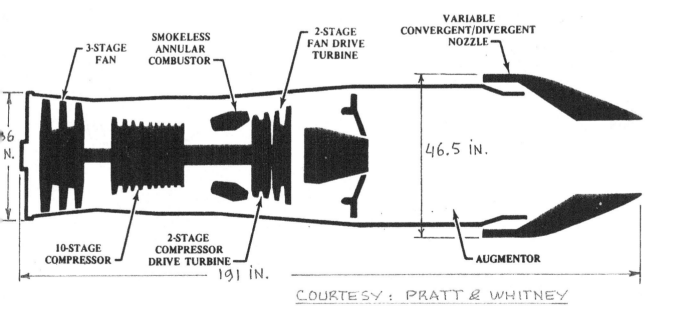

Figure 6.29a Pratt and Whitney F100-PW-100 Augmented Turbofan

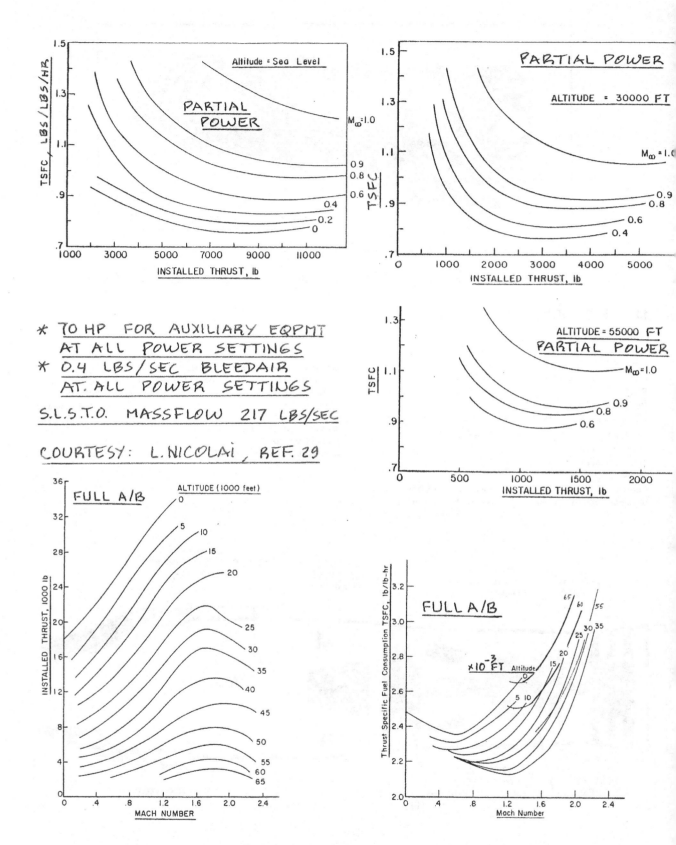

Figure 6.29b Estimated Installed Performance Data for Pratt and Whitney F100-PW-100 Augmented Turbofan

Table 6.8 Manufacturer Performance Data for Turbofan Engines

Type	Pratt and Whitney JT15D-1	JT15D-4C *	Garrett TFE731	ATF3-6 -2c	Teled. CAE 490-4**	Avco Lycoming ALF-502L-3
Max.T.O. Thrust (lbs)	2,200	2,500	3,500	5,440	2,965	7,500
T.O. Condition (static)	SLS	SLS	SLS	SLS	SLS	SLS
T.O. SFC (lbs/lbs/hr)	0.540	0.562	0.493	0.506	0.703	0.411
T.O. Massflow (lbs/sec)	69.4	77.7	112	162	61	256
T.O. BPR	3.3	2.6	2.82	2.55	1.13	5.0
Cruise Thrust (lbs) at 80 percent max.	2,065	2,125	755	1,047	1,400 max. contin.	2,100
Cruise Condition	SLS	SLS	0.8/40K	0.8/40K	0.8/20K	0.8/30K
Cruise SFC (lbs/lbs/hr)	0.537	0.556	0.815	0.816	1.00	0.750
Weight*(lbs)	514	575	725	1,125	640	1,270
Length (in), cold	56.6	63.3	49.7	33.6	51.2	56.8
Max. diam. (in)	27	27.3	39.1	102	h=28.3 w=23.2	41.7
Application	Cessna Citation	SIAI-M S211	Learjet M36	Falcon 200	Alphajet	BAe 146

* Military version ** Same as SNECMA/Turbomeca Larzac

Table 6.9 Manufacturer Performance Data for Turbojet and Turbofan Engines

Type	General Electric J79***	J85-21A ***	CF700	CJ610-5 (J85-4B)	F404***	F101***
Max.T.O. Thrust (lbs)	17,820	5,000	4,200	2,950	16,000	28,000
T.O. Condition (static)	SLS	SLS	SLS	SLS	SLS	SLS
T.O. SFC (lbs/lbs/hr)	1.98	2.13	0.66	0.980	NA	NA
T.O. Massflow (lbs/sec)	170	53	43/85**	44	NA	270
T.O. BPR	0	0	NA	0	NA	0.85
Cruise Thrust (lbs) at 80 percent max.	2,600 (dry)	1,200	1,060	870	NA	NA
Cruise Condition	0.9/35K	0.9/36K	0.8/36K	0.8/36K	NA	NA
Cruise SFC (lbs/lbs/hr)	0.98	1.23	0.98	1.15	NA	NA
Weight*(lbs)	3,873	684	725	402	2,000 (dry)	4,400 (dry)
Length (in), cold	209	113	53.6	51.1	159	181
Max. diam. (in)	39.1	21	33.1	17.7	34.8	55
Application	F4/F16	F-5E/F	Falcon	Learjet	F18	B1B

* No tailpipe, no thrust-reverser **generator/fan ***incl. afterburner

Table 6.10 Manufacturer Performance Data for Turbofan and Turbojet Engines
===

Type	Pratt and Whitney JT8D-217	2037= JT10D	JT9D-7	Rolls Royce RB211-22B	RB163 Spey	Ames TRS-18
Max.T.O. Thrust (lbs)	20,000	37,600	45,600	42,000	10,410	200
T.O. Condition (static)	SLS	SLS	SLS	SLS	SLS	SLS
T.O. SFC (lbs/lbs/hr)	0.562	NA	NA	NA	0.563	1.12
T.O. Massflow (lbs/sec)	483	1,340	1,509	1,380	203	N.A.
T.O. BPR	1.7	5.8	5.15	4.8	1.0	0
Cruise Thrust (lbs) max.continuous	5,350	NA	10,200	9,700	3,070	150
Cruise Condition	0.8/35K	0.8/35K	0.85/35K	0.8/35K	0.77/32K	0.3/5K
Cruise SFC (lbs/lbs/hr)	0.753	0.563	0.620	0.618	0.760	1.45
Weight*(lbs)	4,430	6,906	8,850	9,195	2,257	68
Length (in), cold	154	141.4	128.2	119.4	110	36.5
Max. diam. (in)	49.2	85.0	95.6	85.9	37	13
Application	MD-80	757	747-200	L-1011	BAC111	BD5J

Table 6.11 Manufacturer Performance Data for Turbofan Engines
===

Type	General Electric CF6-50C	CF6-50C1	CF6-6D	CF6-6K	CF6-32C1	CFM56-2	CF34
Max.T.O. Thrust (lbs)	51,000	52,500	40,000	41,500	36,500	24,000	8,650
T.O. Condition (static)	SLS	SLS	SLS	SLS	SLS	SLS	SL/59F
T.O. SFC (lbs/lbs/hr)	0.390	0.394	0.346	0.350	0.357		0.359
T.O. Massflow (lbs/sec)	1,450	1,470	1,303	1,328	1,104	830	307
T.O. BPR	4.26	4.24	5.72	5.67	4.9	6.0	6.3
Cruise Thrust (lbs) at 80 percent max.	8,720	9,080	7,160	7,270	6,630	N.A.	1,420
Cruise Condition	0.8/35K	0.8/35K	0.8/35K	0.8/35K	0.8/35K	0.8/30K	0.8/40K
Cruise SFC (lbs/lbs/hr)	0.628	0.626	0.616	0.616	0.609	0.650	0.690
Weight*(lbs)	8,731	8,731	7,896	7,896	7,140	4,610	1,580
Length (in), cold	173	173	177	177	140	95.7	100
Fan tip diam. (in)	86.4	86.4	86.4	86.4	76.3	68.3	49
Application	KC-10A	747 E4A	DC-10	DC-10	757	DC-8 (mod)	Challenger

* No tailpipe, no thrust-reverser

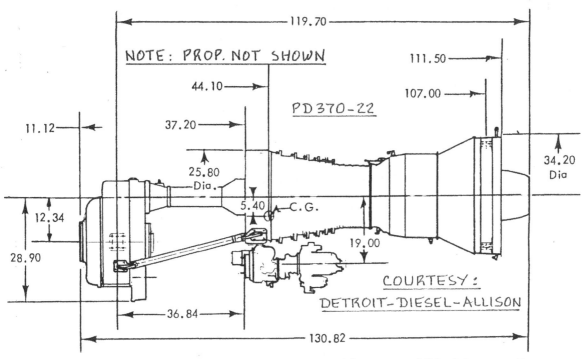

Figure 6.30 Detroit Diesel Allison PD370-22 Propfan

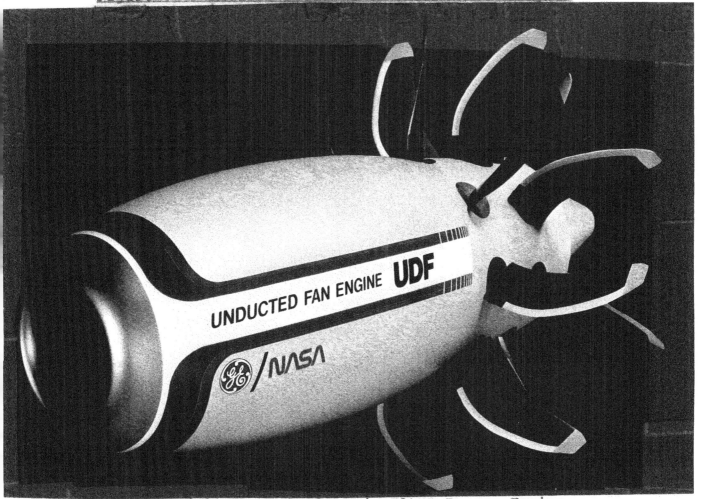

Figure 6.31 General Electric Ultra Bypass Engine

NOTE: THIS GASGENERATOR IS DESIGNED FOR DRIVING 8-BLADE OR 10-BLADE PROPELLERS (PROPFAN)

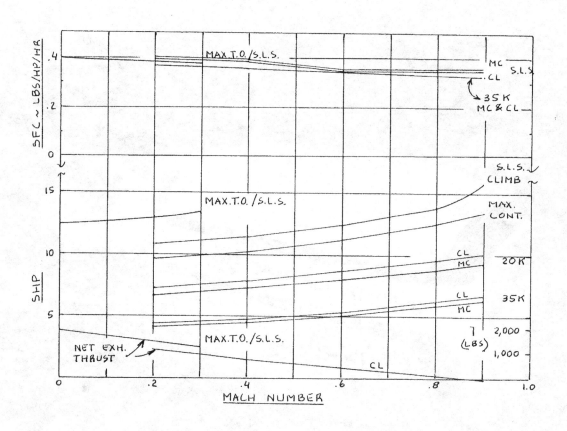

Figure 6.32 Uninstalled Performance Data for Detroit Diesel Allison PD370-22 Propfan

NO DATA AVAILABLE AT PRESS TIME. MAXIMUM S.L.S. T.O. THRUST ESTIMATED AT 30,000 LBS WITH T.S.F.C. OF 0.250

Figure 6.33 Guestimated Uninstalled Performance Data for General Electric Ultra Bypass Engine

6.2 RELATION BETWEEN FLIGHT ENVELOPE AND ENGINE TYPE

Matching engine type to the flight envelope of airplanes is of key importance to the ultimate succes (or lack thereof) of any airplane.

Figure 6.34 shows some fundamental relations between flight speed, mass flow rate, thrust and various efficiency parameters for piston engines, for turbojets and for turboprops.

Figure 6.35 shows the typical sfc values associated with different powerplants across the Mach range. For cruise range dominated airplanes it is usually the amount of fuel burned during the cruise phase of the mission which is decisive in selecting the type of powerplant. In such airplanes the achievable payload-range performance and the associated return-on-investment (ROI) play a dominant role in powerplant selection. Part VII shows how the payload range performance of an airplane can be predicted. Part VIII shows how the ROI characteristics of an airplane can be predicted.

There are however other considerations which may influence the powerplant decision. In military airplanes the acceleration potential, the radar cross section, the infrared signature and the physical size of the powerplant may be important considerations.

In commercial airplanes, the installed weight of the total powerplant package plus its fuel requirements, the cost of ownership and the reliability potential all may be deciding factors over and beyond pure efficiency considerations.

Figures 6.34 and 6.35 indicate that many areas of overlap exist. In these overlap areas it is not usually straightforward to select the 'proper' engine type. Customer preference, considerations of growth potential also play an important role in ultimately deciding the proper engine type.

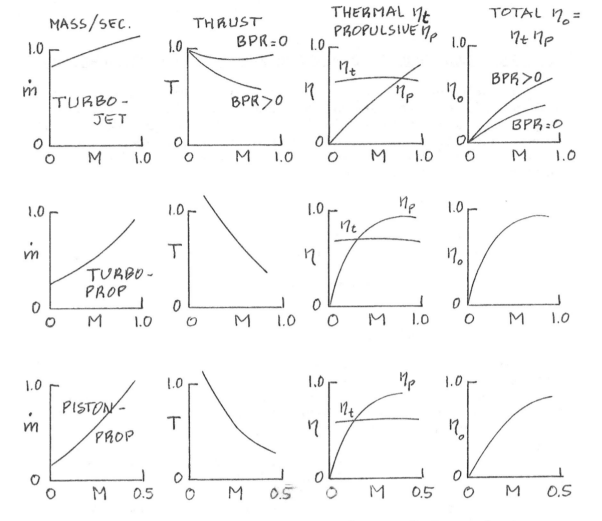

Figure 6.34 Variation of Fundamental Powerplant Parameters with Flight Speed

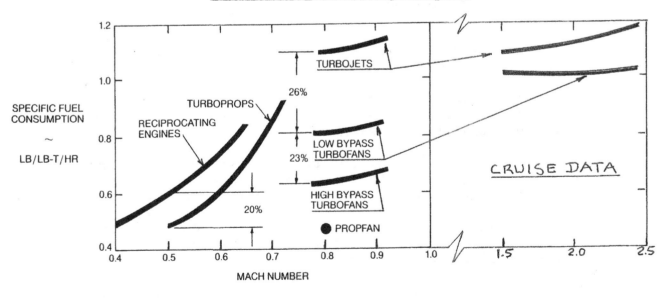

Figure 6.35 Effect of Mach Number on SFC for a Range of Powerplant Types

6.3 INSTALLED THRUST, POWER AND EFFICIENCY CONSIDERATIONS

To achieve the best combination of installed thrust and efficiency from any given powerplant installation requires a lot of attention to details. Requirements for power generation, airconditioning and other operationally required services all cause reductions in propulsive thrust and/or efficiency.

This section is organized as follows:

6.3.1 Power extraction
6.3.2 Propeller installations
6.3.3 Piston-engine installations
6.3.4 Subsonic and supersonic turbojet and turbofan installations

6.3.1 Power Extraction

To determine the amount of power extraction required from the engines it is necessary to make a list of all systems and services which take energy from the engines. Part IV contains a discussion of most system types found in civil and military airplanes. Once the system requirements for a given airplane have been defined (Step 17 in p.d. sequence II, p.18, Part II) such a list of power extraction requirements can be made.

The power extraction list should be subdivided into requirements for:

1. electrical power, in hp
2. bleedair mass flow, in lbs/sec

With this list it is possible to estimate the effect of power extraction on installed engine performance.

For typical cruise operations in passenger transports a total power loss of 2-5 percent should be used in preliminary design calculations.

In military operations the requirements for power extraction depend on the airplane mission: radar/search missions require large amounts of electrical power.

In fighters, typically 70 hp is needed to drive essential generators and auxiliary equipment while bleedair flow rates typically amount to 0.5 lbs/sec.

It is often desirable (notably in transports) to install auxiliary power units (APU) for power generation on the ground. This makes an airplane independent of ground based power. Whether or not the APU is designed to be used in flight depends on in-flight power requirements and on the question of system reliability for those systems which derive power from the propulsive installation.

6.3.2 Propeller Installations

For any given propeller design and for a given combination of rpm and delivered shaft horsepower, the thrust output of a propeller depends on the shape and the size of airplane components behind or ahead of it.

In tractor installations, the effect of the nacelle and/or wing directly behind the propeller is referred to as the 'blockage' effect. Figure 6.36 shows an example of a good and a poorly shaped installation.

In pusher installations, the effect of the nacelle and/or the wing directly in front of the propeller affects the propeller 'inflow field' and can affect propeller performance. Figure 6.37 shows an example of a good and a poorly shaped installation.

6.3.3 Piston-Engine Installations

The management of cooling air and of exhaust gasses plays a significant role in the ultimate propulsive efficiency of a given installation. Sub-section 6.9.1 contains further discussions of this subject.

6.3.4 Subsonic and Supersonic Turbojet and Turbofan Installations

Inlet design: The design of the inlet plays a very important role in determining overall installed engine performance. For a discussion of inlet types and their performance effects the reader should consult Refs 12, 14, 29, 60 and 65.

Part VI contains rapid methods for accounting for inlet effects on drag and thrust.

Nozzle design: The design of the nozzle (exhaust configuration) also plays a significant role in determining installed drag and thrust. References 12 and 65 should be consulted for details on exhaust (tailpipe) configurations. Reference 60 contains useful information on the design of exhaust nozzles. Examples of inlet and exhaust configurations are presented in Section 6.9.

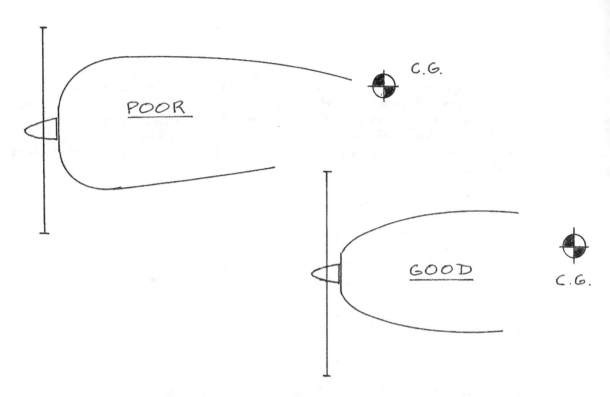

Figure 6.36 Example of 'Good' and 'Poor' Nacelle Shaping in a Tractor Installation

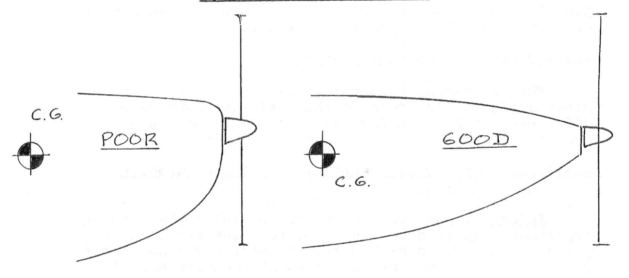

Figure 6.37 Example of 'Good' and 'Poor' Nacelle Shaping in a Pusher Installation

6.4 STABILITY AND CONTROL CONSIDERATIONS

The following stability and control effects should be considered:

6.4.1 Effect of one or more engines inoperative and effects of power transients

6.4.2 Tractor versus pusher

6.4.3 Effect of engine/propeller thrust line location and inclination

In preliminary design a good rule of thumb is: if the engine disposition differs significantly from that of existing, certified airplanes, considerable power effects on stability and control characteristics can be expected. In such cases the safest thing to do is to perform the necessary stability and control calculations before freezing the design. Part VII contains methods for doing this.

6.4.1 Effect of One or More Engines Inoperative and Effects of Power Transients

If the engines are disposed so that large yawing moment arms and/or large pitching moment arms prevail the effect of one or more engines becoming suddenly inoperative must be considered. Ref.37 and Part VII contain methods for computing engine-out effects on handling and on controllability.

6.4.2 Tractor Versus Pusher

Figures 6.36 and 6.37 define what is meant by a tractor and a pusher propeller installation. Part VI shows that tractor installations act to decrease longitudinal stability while pushers act to increase it.

The reader should not infer from this that pushers are good nor that tractors are bad. How much stabilizing effect or destabilizing effect is desirable depends entirely on the type of airplane and on its mission requirements.

Sub-section 6.4.3 presents a simple approximate equation from which the incremental stability of tractors and pushers may be estimated for preliminary layout purposes.

6.4.3 Effect of Engine/Propeller Thrust Line Location and Inclination

Figure 6.38 shows the geometry of propeller thrust line location with respect to the c.g. of the airplane. It is shown in Part VI that the static longitudinal stability of a propeller driven airplane varies roughly with z_T and with x_T in the following manner:

for vertical thrust line location:

$$(dC_m/dC_L)_T = 0.25(z_T/\bar{c})N_p \tag{6.2}$$

for horizontal thrust line location:

$$(dC_m/dC_L)_T = 0.02(x_T/\bar{c})N_p \tag{6.3}$$

These equations are valid for airplanes with power-to-weight ratios of about 0.1.

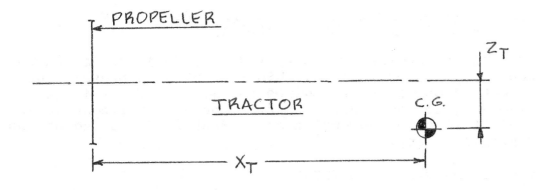

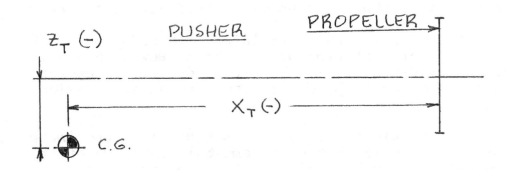

Figure 6.38 Geometry for Propeller Thrust Line Location

6.5 STRUCTURAL CONSIDERATIONS

The following structural considerations play an important role in the integration of the propulsion system into the airframe:

- 6.5.1 Transmission of thrust into the airframe
- 6.5.2 Lateral location of engines on the wing
- 6.5.3 Extension shafts and propeller blade excitation
- 6.5.4 Flutter

6.5.1 Transmission of Thrust into the Airframe

To transmit thrust forces into the airframe it is necessary to have a number of 'hard points' where the engine is physically attached to the airframe. The number of these hardpoints depends on the type of engine used.

Figure 6.39a shows an example of the principal method used to mount piston engines in airframes. Note that this is accomplished with a truss (usually made of welded steel tubes) or with a support cradle. Note that where the truss or cradle is attached to the airframe thrust and engine weight determine the attachment point loads.

The reader must realize that the attachment (mounting) points on the engine itself cannot be changed easily. Their location depends on the internal design of the engine and is normally determined by the engine manufacturer. Changing these attachment points is very expensive!

Since most piston engines transmit significant vibrations into the airframe it is essential to use some type of shock mount(s) to reduce these vibrations. Section 6.9 shows an example of such a shock mount in Figure 6.69.

Figure 6.39b shows the principal mounting arrangement used for jet engines. Note that there are usually two or three attachment points which are designed to transmit thrust as well as most of the engine weight. The rear attachment point is usually designed to allow for the considerable thermal expansion which a jet engine undergoes. This rear attachment point normally carries only part of the engine weight.

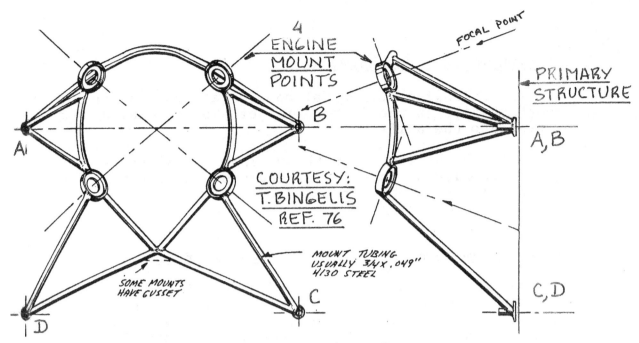

Figure 6.39a Method for Mounting Piston Engines in Airframes

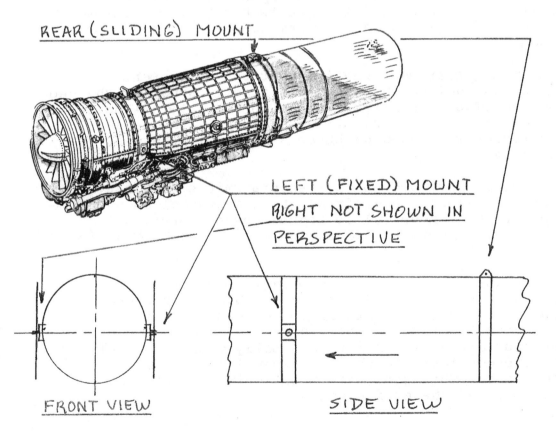

Figure 6.39b Method for Mounting Jet Engines in Airframes

Figure 6.39c shows a typical turboprop attachment. Note that the gasgenerator part of the engine is mounted in basically the same manner as a jet engine. However, in many turboprop engines the propeller and the gearbox must be separately supported. The transmission of thrust is normally done via the forward mounting points.

In jet engines as well as in turboprops, the engine attachment points cannot easily be changed. The engine manufacturer normally determines their location.

The reader should refer to Section 6.9 for more examples of engine installations.

6.5.2 Lateral Disposition of Engines over the Wing

Figure 6.40 shows head-on views of two types of four-engined jet transports. It is clear that a difference in design philosophy exists between these two designs.

Design 1 (compared to Design 2) will have a lower wing root bending moment due to the relieving bending moments induces by the engines. Design 1 (again compared to Design 2) will require a larger vertical tail and rudder to compensate for the larger asymmetric yawing moment which results from operation with one engine inoperative. Which design approach is best depends on how these trades affect the overall empty weight of the airplane.

6.5.3 Extension Shafts and Propeller Blade Excitation

Figure 6.41 shows examples of two airplanes with extension shafts between the engines and the propeller(s). This design approach has the advantage of giving excellent control over the c.g. location of an airplane: putting engines and propellers all the way back in an airplane can lead to stability problems which may be impossible to solve.

A serious problem associated with extension shaft installations is the vibration problems introduced by relatively flexible shafts. Bending vibrations can be eliminated by the use of more closely spaced bearings. Torsion vibrations can be reduced by a stiffer shaft. Both solutions cost extra weight: this must be considered before committing to such an installation. The airplanes of Figure 6.41 did suffer from serious vibration problems.

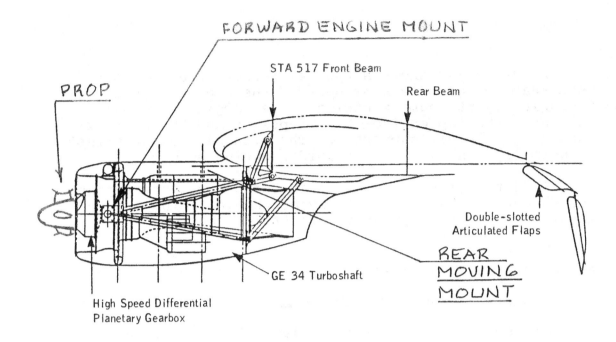

Figure 6.39c Method for Mounting Turboprop Engines in Airframes

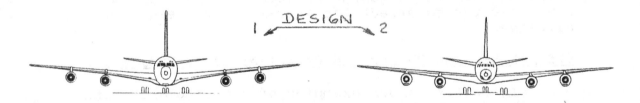

Figure 6.40 Examples of Lateral Disposition of Engines

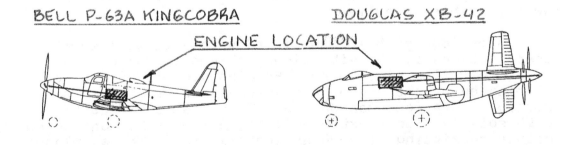

Figure 6.41 Examples of Application of Extension Shafts

Another structural problem may be caused by separated flow exciting the propeller blades whenever the propeller is mounted behind a wing, nacelle or a fuselage where flow separation can occur. Careful streamlining of the 'body' ahead of the propeller is essential. Fig.6.42 shows 'good' and 'poor' propeller locations behind a wing.

6.5.4 Flutter

The lateral and the longitudinal placement of engines relative to a slender wing is important from a wing flutter viewpoint. Detailed flutter analyses are required to ultimately decide on the best engine disposition.

In turbopropeller installations a so-called whirl-mode flutter may occur. Such a flutter mode usually consists of a combination of propeller-plus-gearbox coning motion excited by a wing and/or nacelle oscillation. It is of great importance to assure that the landing gear is installed such that structural damage due to hard landings cannot weaken key structural components in the whirl-mode scenario. Again, only detailed flutter analyses with and without assumed damage can identify the lightest ultimate installation.

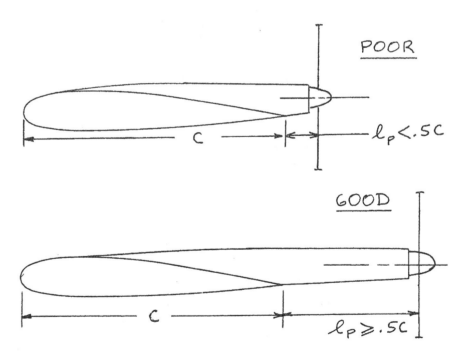

Figure 6.42 Example of 'Good' and 'Poor' Propeller Location From a Viewpoint of Blade Excitation

6.6 MAINTENANCE AND ACCESSIBILITY CONSIDERATIONS

Because engines need frequent servicing, inspection and component replacement, accessibility of the powerplant installation is a requirement in nearly all airplanes.

Figures 6.43 through 6.45 show how accessibility is provided in the case of a propeller driven single, a jet trainer and a small jet transport. Note the use of large clamshell doors. These doors must be designed so that they are supported in the 'open' position under prevailing wind conditions!

It must be possible to replace engines with relative ease. Figures 6.46 through 6.49 show how this requirement is addressed in a jet trainer, a jet fighter and in two jet transports respectively.

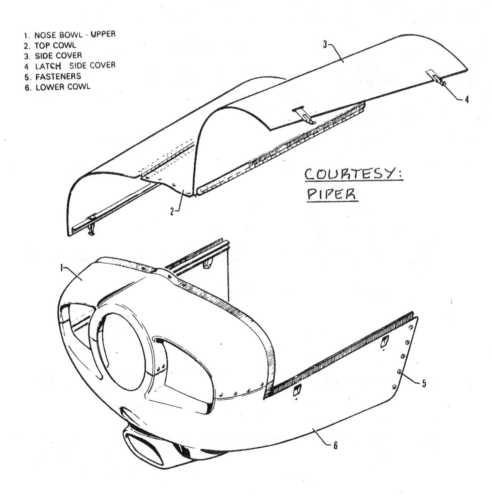

Figure 6.43 Accessibility of Engine Installation of the Piper PA-38-112 Tomahawk

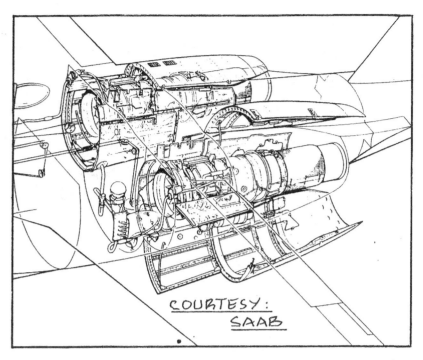

Figure 6.44 Accessibility of Engine Installation of the SAAB 105-XT

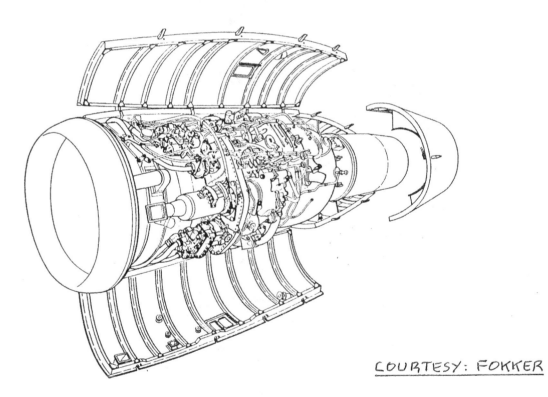

Figure 6.45 Accessibility of Engine Installation of the Fokker F28

Figure 6.46 Engine Removal for the SIAI Marchetti S211

Figure 6.47 Engine Removal for the Douglas A4 Skyhawk

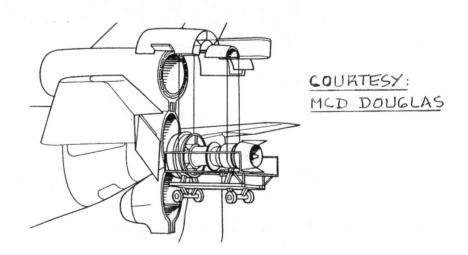

Figure 6.48 Engine Removal for the McDD DC10

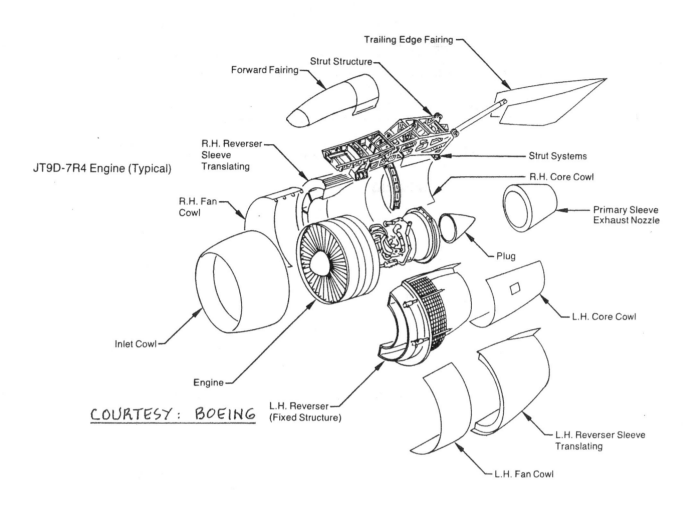
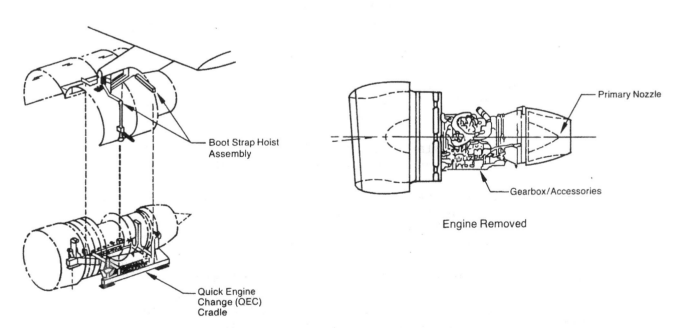

Figure 6.49 Engine Removal for the Boeing 767

6.7 SAFETY CONSIDERATIONS

The following safety aspects will be discussed:

6.7.1 Installation safety
6.7.2 Safety during ground operation
6.7.3 Foreign object damage (FOD)
6.7.4 Engine reliability and shutdown rates

6.7.1 Installation Safety

All engines and other heat generating equipment must be isolated from the rest of the airplane by means of firewalls and/or other suitable shrouds. This requirement is of great importance in isolating engines from fuel tanks. Rules to be followed in laying out fuel tanks and fuel systems are discussed in Part IV.

FAR 23.1191 and 25.1191 deal with this requirement for commercial airplanes. A similar requirement exists for military airplanes.

Figure 6.50 shows where insulating blankets can be applied in a jet engine nacelle. The material properties are also listed.

Figure 6.51 shows a typical firewall installation in a light airplane. Fire walls should be made out of stainless steel and/or titanium. DO NOT USE ASBESTOS!!!

The structural integrity of engines and propellers is historically less than that of airplane primary structure: the frequency of occurrence of disintegration type failures (although low) is one to two orders of magnitude greater than the likelihood of the primary structure failing. In deciding on any powerplant installation the following failures must be accounted for:

*Propeller blade failure
*Compressor and turbine wheel disintegration

Because of the very high kinetic energies associated with these items it is not acceptable to locate the crew, the passengers, flight crucial systems or primary structure in the path of such items. Figure 2.14 illustrates the so-called 5 degree cone criterion which is frequently used in layout design to satisfy this requirement with regard to humans. The designer should use the same criterion with regard to flight crucial systems and with regard to primary structure.

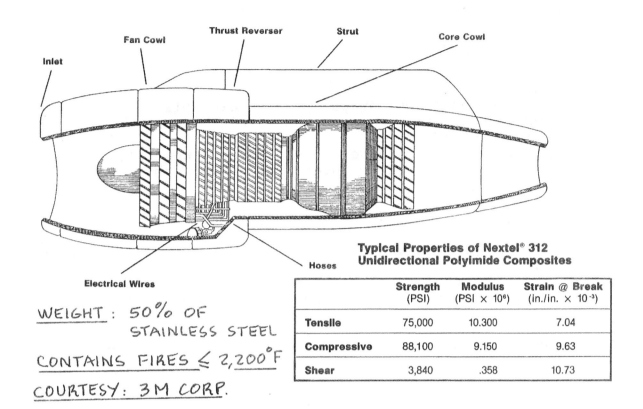

Figure 6.50 Application of Fire Insulating Blankets

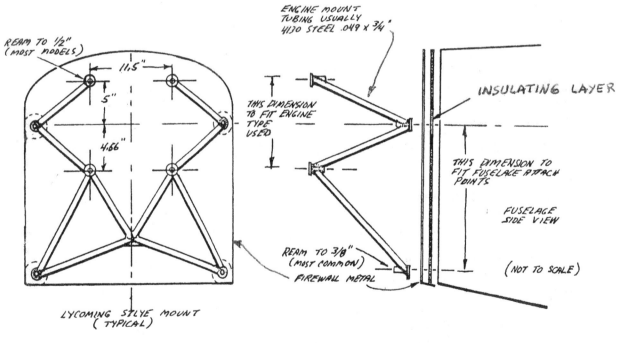

Figure 6.51 Fire Wall Installation in a Light Airplane

It should be obvious that powerplants must not be located in such a manner, that a disintegration type failure would result in serious damage to an adjacent powerplant. The reader should observe that all transport jets have the engines located to prevent this from happening!

6.7.2 Safety During Ground Operation

Propellers and jet engines create potential hazards for people who need to work in close proximity to airplanes while on the ground. It is not feasible to arrange for protection from running propellers and/or jet engines under most operating circumstances. Awareness of the potential danger and constant reminders must serve to deter serious accidents. Figure 6.52 illustrates the potential danger areas in the case of a large turbofan engine.

6.7.3 Foreign Object Damage (FOD)

Jet engines are sensitive to ingestion of many types of debris. Apart from the so-called bird strike requirement, it is necessary to assure that debris (gravel, slush, mud, snow add ice) thrown up by the landing gear cannot enter into inlets or cannot damage propellers or other critical components of the propulsive installation. To provide reasonable assurance in preliminary design that a proposed installation is not prone to landing gear FOD the reader should check the relative location of engine inlets and landing gears on existing airplanes and verify that critical angular relationships are not violated. Figure 6.53 illustrates a number of such relationships. By using special deflectors on the nose gear it is possible to prevent FOD despite a marginal angular relationship. To certify the 737 for operation from gravel runways, Boeing added such deflector shields to special versions of the 737.

Turbopropeller installations (with their relatively small air inlets) can be sensitive to ingestion of ice, dust particles, small birds and other particles. In such installations frequent use is made of so-called 'particle separators'. Figure 6.54 illustrates such a separator system. Note that even if the particle screen ices up a path must be provided for air to enter the engine even though some pressure loss may be associated with that.

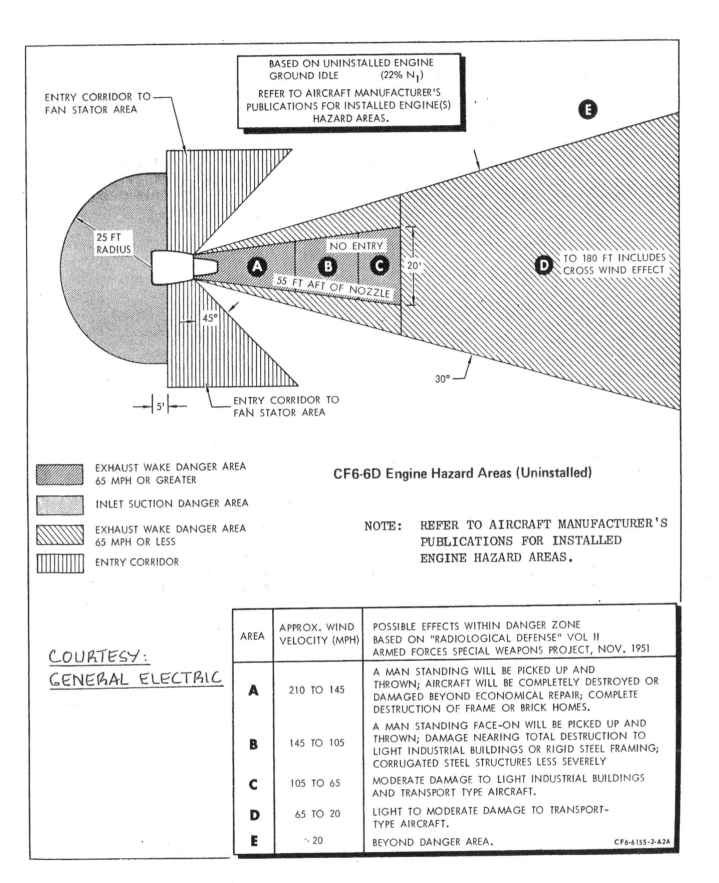

Figure 6.52 Danger Zone Behind a Turbofan

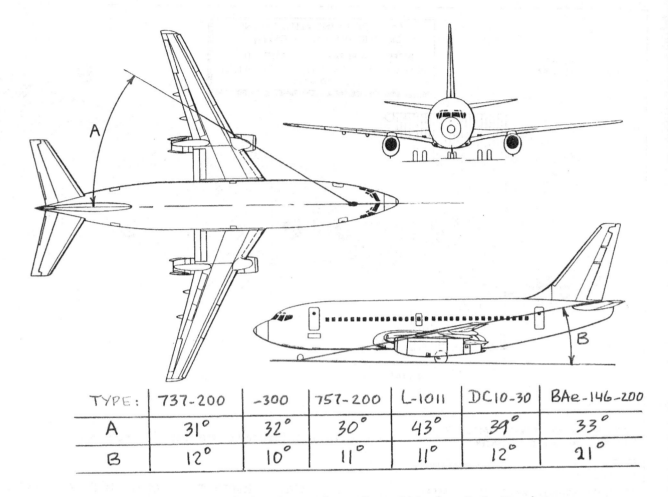

Figure 6.53 Critical Angles for FOD in Jet Engines

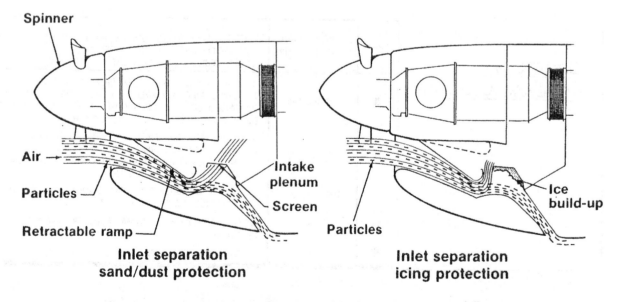

Figure 6.54 Particle Separator System for a Turboprop

6.7.4 Engine Reliability and Shutdown Rates

The following engine related scenarios need to be considered:

1. engine failure during take-off and/or go-around
2. engine failure during overwater flight

Case 1 is adequately addressed by the required one engine inoperative climb requirements which are imposed on all multi-engine airplanes. Part I, Chapter 3 has addressed these requirements in terms of basic airplane design parameters. Case 1 also can have stability and control consequences. These were addressed in Part II, Chapter 11. Part VII deals with this problem in detail.

Case 2 is of concern primarily in the case of twin engined passenger airplanes operating over long stretches of water. Figure 6.55 shows a recent study which indicates how long it takes to achieve certain mean diversion times at different engine cruise failure rates. The reader should monitor the flight safety articles published in 'Flight International' (British, weekly) at regular interval for recent information on actual engine shutdown rates. These shutdown rates depend not only on engine type but also on the type of airplane operations.

The concern over engine-out operation during extended overwater flights centers not merely on the engine itself but in particular on the effect on flight crucial systems. Part IV deals with this question in detail.

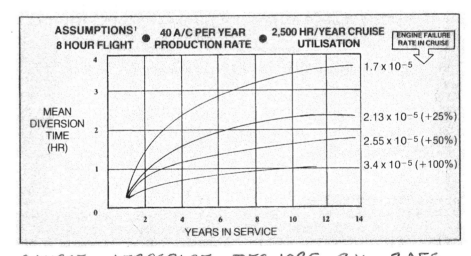

SOURCE: AEROSPACE - DEC. 1985, P.11, RAES.

Figure 6.55 Relation Between Mean Diversion Time and Calendar Time as a Function of Engine Shutdown Rate in Oceanic Twins

6.8 NOISE CONSIDERATIONS

Airplanes and their engines create a substantial amount of both interior and exterior noise. The interior noise levels should not be so high as to cause discomfort to the passengers or to make safe operation by the crew impossible.

The exterior noise levels should meet the requirements of FAR 36 (Ref.11). These requirements impose severe restrictions on the type of engine and/or propeller technology which can be utilized.

Figure 6.56 shows those noise sources which are of major importance to airplane designers.

The material in this section is organized as follows:

    6.8.1 Interior noise design considerations
    6.8.2 Exterior noise design considerations

6.8.1 Interior Noise Design Considerations

Reference 66 contains information about noise levels which are acceptable from a speech interference level and a hearing damage level viewpoint. Table 6.11 shows the noise levels permissible by OSHA standards. Table 6.12 relates noise levels to various activities.

Interior noise levels are of primary concern in propeller driven airplanes. Reference 67 reviews procedures for the prediction of cabin interior noise levels. Figure 6.57 shows typical interior noise levels measured inside the cabin of a light, propeller driven airplane.

Reductions of cabin noise levels can normally be achieved only by:

1. relocation of the propellers and/or powerplants
2. redesign of the propeller(s)
3. addition of sound damping materials

Examples of 1) are the Beech Starship I and the Piaggio 180 (Part II, Figures 3.42 and 3.47 respectively).

Item 2) normally amounts to lowering the tip Mach number of the propeller. This can be done by lowering propeller rpm and by lowering the propeller diameter.

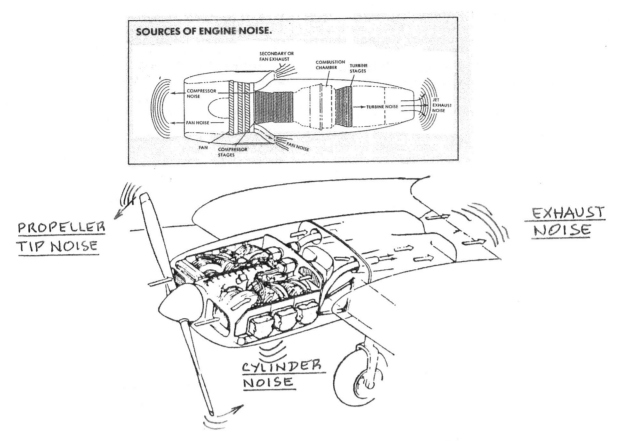

Figure 6.56 Sources of Engine Noise

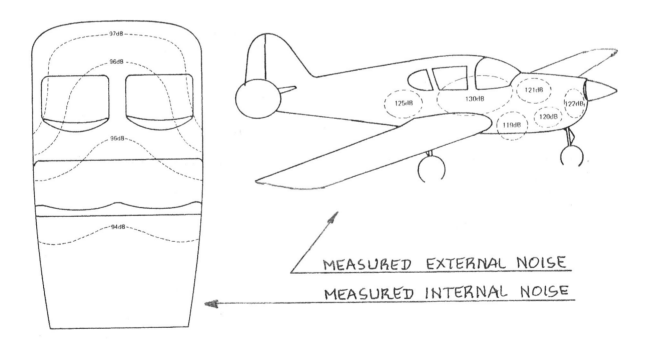

Figure 6.57 Typical Noise Contours in a Light Airplane

Table 6.11 Permissible Noise Exposure by OSHA Standard
===

Duration per day in hours:	Permissible Level in db:	Duration per day in hours:	Permissible Level in db:
8	90	1.5	102
6	92	1	105
4	95	0.5	110
3	97	0.25 or less	115
2	100		

Table 6.12 Noise Levels Associated with Activities
===

Overall Effect	Level in db	Activity
Extreme danger	155	Rifle blast, siren, jet engine (close to)
	140	Shotgun blast, drag strip (close to)
	120	Some rock music, rock drill (close to)
Probable permanent hearing damage	115-125	Drop hammers
	110-115	Routers, planers
	90-100	Subway, weaving mill
	90-95	Riveter, punch press
Possible hearing damage, depending on duration	80-95	Spinners, looms, lathes
	80	Heavy traffic
No hearing damage	65-75	Noisy typewriter
	70	Busy street
	60	Normal speech
	50	Average office
	20-30	Sleeping limit
	15	Leaf rustling

Changes in propeller airfoil to delay drag rise phenomena are also beneficial. Reference 68 contains a discussion of the penalties and the potential associated with propeller noise reduction.

Examples of item 3) are shown in Figure 6.58. Addition of sound damping materials adds to airplane empty weight and usually reduces airplane payload (or fuel) weight. In the F27 example discussed in Ref.69, the following weight penalties were identified:

For a 2 dbA reduction, 55 lbs for using 'tuned dampers' (Fig.6.58).

For a 5 dbA reduction, 66 lbs for using blankets (Fig.6.58)

6.8.2 Exterior Noise Design Considerations

The following specific exterior noise considerations are important to the design of airplanes:

1. Take-off noise
2. Approach noise
3. Sideline noise
4. Sonic boom generated noise

Regulations containing allowable noise levels and methods for demonstrating compliance are contained in Ref.11 (FAR 36). Figures 6.59a through 6.59d show the FAR 36 noise standards in relationship to airplane type and take-off weight. These figures also show the degree of compliance of several existing airplanes.

Refs 70 and 71 contain procedures for the prediction of fly-over noise from propeller driven airplanes.

Ref.72 presents a discussion on the noise from turbomachinery. Ref.73 presents two design conclusions:

1) If an airplane is designed to meet the take-off and approach performance requirements of Ref.11 by minimizing required thrust, it tends to also be optimum for minimizing noise within plus or minus 1 EPNdb.

2) The powerplant installation should be designed to allow for adequate sound attenuation treatment. About 10 to 12 EPNdb cen be expected from this source.

With regard to sonic boom overpressure, Reference 74 contains a simple procedure for its prediction.

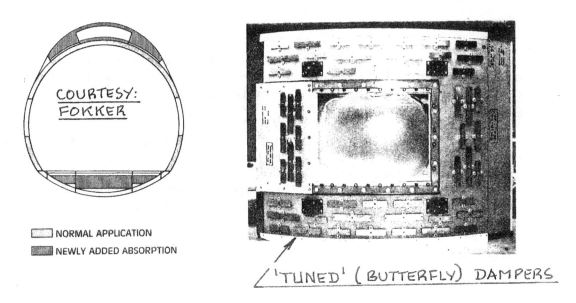

Figure 6.58 Examples of Sound Damping Treatment in the Fokker F27

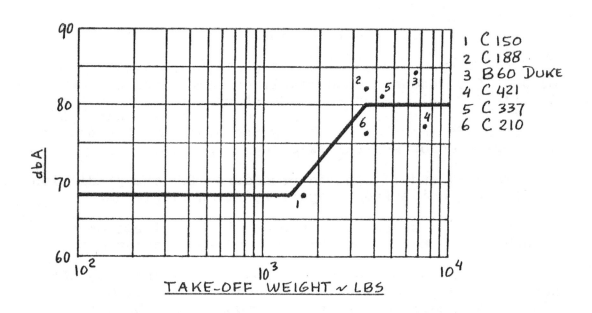

Figure 6.59a FAR 36 Noise Criteria For Light Airplanes

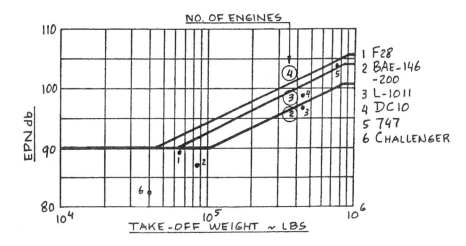

Figure 6.59b FAR 36 Take-off Noise Criteria for Jets

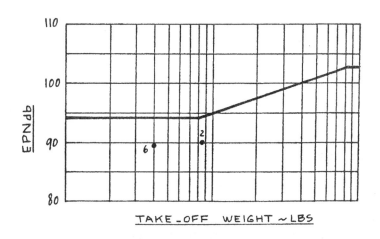

Figure 6.59c FAR 36 Sideline Noise Criteria for Jets

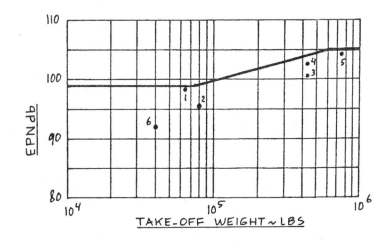

Figure 6.59d FAR 36 Approach Noise Criteria for Jets

6.9 EXAMPLE ENGINE INSTALLATIONS

Examples will be presented for the following types of engine installations:

6.9.1 Piston-propeller installations
6.9.2 Turbopropeller installations
6.9.3 Turbojet and turbofan installations
6.9.4 Propfan and ultra-bypass installations
6.9.5 Nozzles and thrust reversers

The reader should recognize that it is not clear at which point a propeller becomes a fan nor at which point a ducted fan becomes a ducted propeller.

6.9.1 Piston-Propeller Installations

Figure 6.60 shows a conventional piston-propeller installation, tractor fashion in a fuselage. Figs 3.1 through 3.6 in Part II contain many examples of this type of installation.

Significant design problems associated with piston/propeller installations are:

1. management of exhaust gasses
2. management of cooling air
3. reduction of airframe vibration
4. reduction of noise

1. Management of exhaust gasses: Note in Fig.6.60 that the exhaust stack is mounted perpendicular to the free stream. From a drag viewpoint this is very bad. Figure 6.61 shows how this can be improved somewhat. Reference 75 shows that poorly designed exhaust configurations can increase the zero lift drag coefficient of an airplane by 16 percent.

By directing the exhaust rearward, drag can be reduced and some thrust can be recovered as well: Ref.75 estimates that for highly powered general aviation twins the thrust recovery possible in a typical cruise flight condition amounts to some 15 percent of zero lift drag!

2. Management of cooling air: Figures 6.62 and 6.63 show the conventional ram-air cooling system as used in a tractor and a pusher arrangement respectively.

Warning: most piston engines are designed with the assumption that the cooling air will flow from top to bottom: downdraft. Figure 6.64 shows an updraft cooling

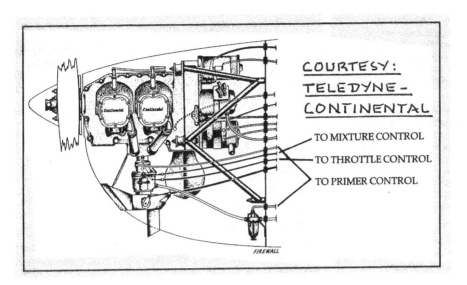

Figure 6.60 Light Airplane Piston Engine Installation

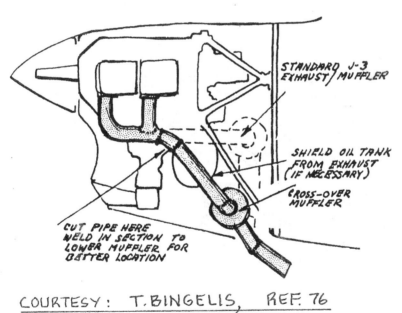

Figure 6.61 Improved Exhaust System Installation

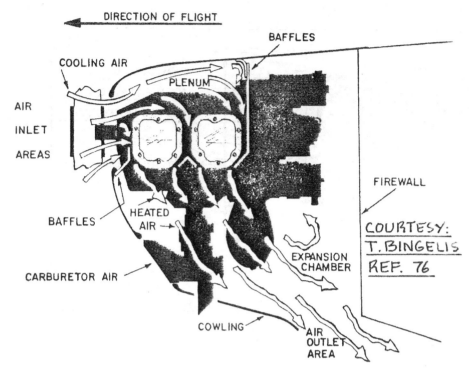

Figure 6.62 Conventional Ram Air Cooling for a Tractor Piston Engine Installation

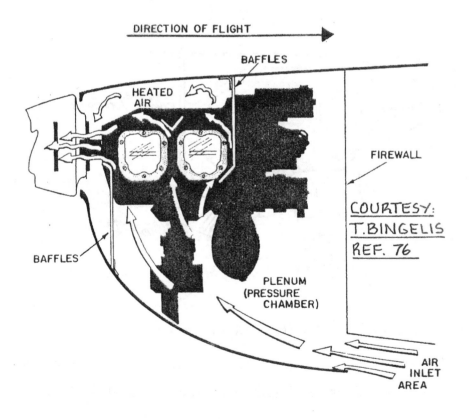

Figure 6.63 Conventional Ram Air Cooling for a Pusher Piston Engine Installation

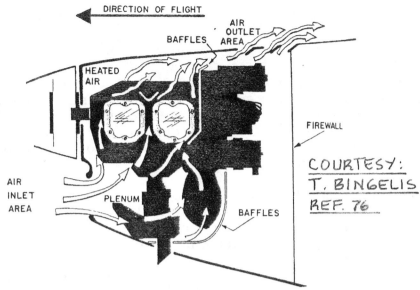

Figure 6.64 Updraft Cooling for a Tractor Piston Engine Installation

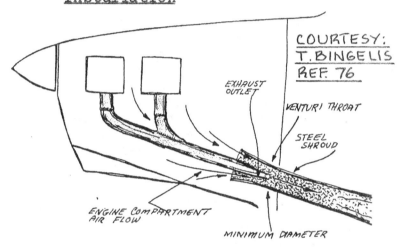

Figure 6.65 Example Ejector Installation in a Single

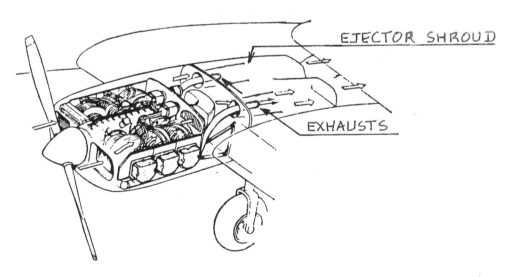

Figure 6.66 Example Ejector Installation in a Twin

arrangement. This may look good, but can result in the cooling air being heated by the exhaust stack thereby reducing cooling effectiveness! In pusher installations such as that of Fig.6.63 this can be a problem. Special engine driven cooling fan(s) may have to be installed to solve this problem.

Mismanagement of cooling air can cause drag increases up to 9 percent of zero lift drag according to Ref.75. Significant reductions in cooling air drag and exhaust configuration drag can be obtained by using so-called ejector systems. Figures 6.65 and 6.66 show examples of ejector installations. However: ejectors will increase weight and initial cost: this must be evaluated against the potential benefits.

Most of todays piston engines are of the horizontally opposed type. For installed power requirements above roughly 750 hp it has been found that the in-line or radial piston engines offer lighter and less 'draggy' installations. Examples of in-line engine installations are found in WWII fighters. Fig.6.67 shows an example of a modern radial engine installation. Note the 'tight' cowling system used in this arrangement. Figure 6.67 also shows the extra cowl flaps used to improve engine cooling during take-off.

Piston engines tend to loose power rapidly at altitude. Most high performance piston/propeller driven airplanes use supercharging to increase power available at altitude. Figures 6.12 and 6.68 show typical arrangements used in supercharger installations.

   3. Reduction of airframe vibration: Most piston engines transmit significant vibrations into the airframe to which they are attached. To reduce such vibrations, piston engines are normally mounted on shock absorbing engine mounts. Figure 6.69 shows an example. References 76 and 77 contain many more examples of shock mount installations.

Section 6.8 presents a discussion of airplane noise. Piston-propeller combinations tend to be very noisy unless special steps are taken to reduce noise. It is possible to achieve major reductions in powerplant noise by using ducted fans. Figure 6.70 shows a potential application of a ducted fan driven by a piston engine. Whether the additional weight and wetted area is worth the reduction in noise remains to be seen.

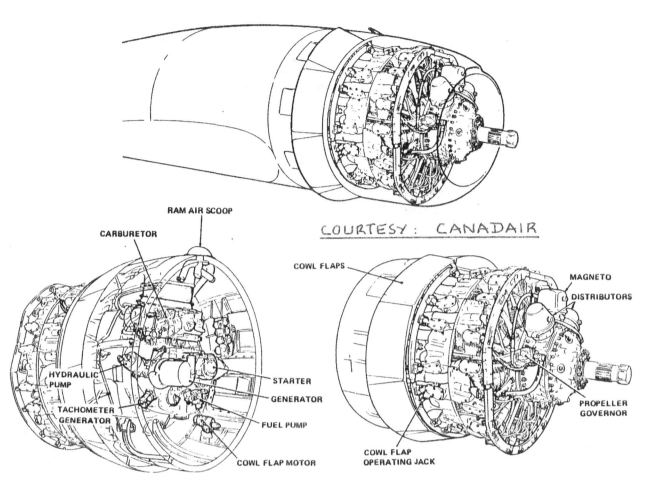

Figure 6.67 Radial Piston Engine Installation

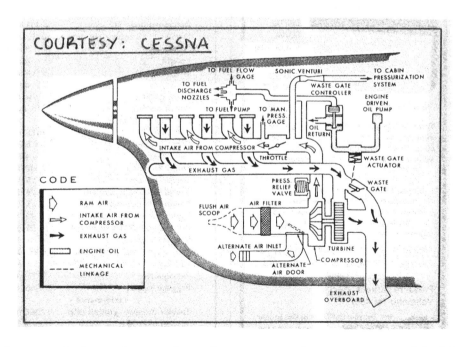

Figure 6.68 Schematic of a Supercharger Installation

Part III Chapter 6 Page 361

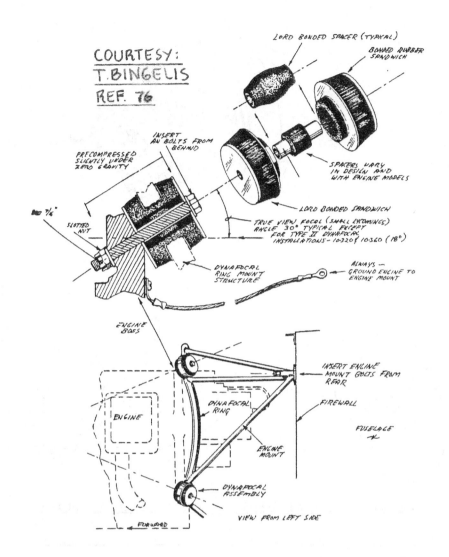

Figure 6.69 Shock Mount Installation

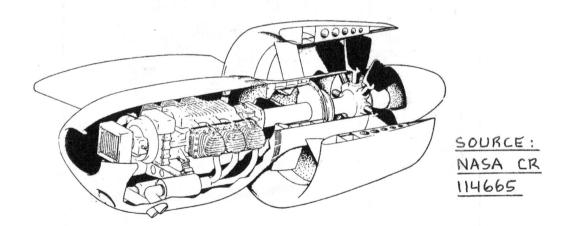

Figure 6.70 Piston Engine Driving a Ducted Fan

4. Reduction of noise: The problem of exterior and interior noise was discussed in Section 6.8. Major noise reductions due to the powerplant installation can be obtained by using a ducted propeller. An example of such an arrangement is shown in Figure 6.70. The reader should carefully evaluate the weight and drag penalties associated with this type of installation before deciding on its use. Figure 3.22d in Part II shows an airplane which uses a similar ducted propeller installation.

6.9.2 Turbo-Propeller Installations

Figure 6.71 shows a typical single engine turboprop tractor installation. Note the engine truss mount connecting to the firewall bulkhead.

Examples of 'over-the-wing' and 'under-the-wing' turboprop installations are shown in Fig.6.72. A counter-rotating turboprop installation is shown in Fig.6.73.

A cutaway drawing of a complete turboprop/nacelle/landing-gear installation is provided in Figure 6.74.

Figure 6.75 shows a pusher propeller installation driven by two gas generators. This installation was flown on the Learfan, also shown in Figure 6.75. An advantage of this installation is the absence of any yawing moment following an engine failure.

Many turboprop engines are sensitive to bird and/or (sand) particle ingestion. For that reason particle separators and/or screens are included in many turboprop installations. Figure 6.54 shows an example of such a particle separator.

6.9.3 Turbojet and Turbofan Installations

In transports and bombers the so-called 'buried' engine installation has been used in airplanes such as the DeHavilland Comet, the Vulcan and the Valiant: see Figure 6.76.

A major advantage of buried installations is the low installed drag. Disadvantages include: interruption of spars, relatively poor accessibility and long tailpipes.

In most of todays transports the engines are installed in external pods: wing or fuselage mounted. Fig.6.77 is a typical example of a wing mounted installation. Observe the forward and aft engine mounting points: recall the material in Section 6.5 on engine mounting.

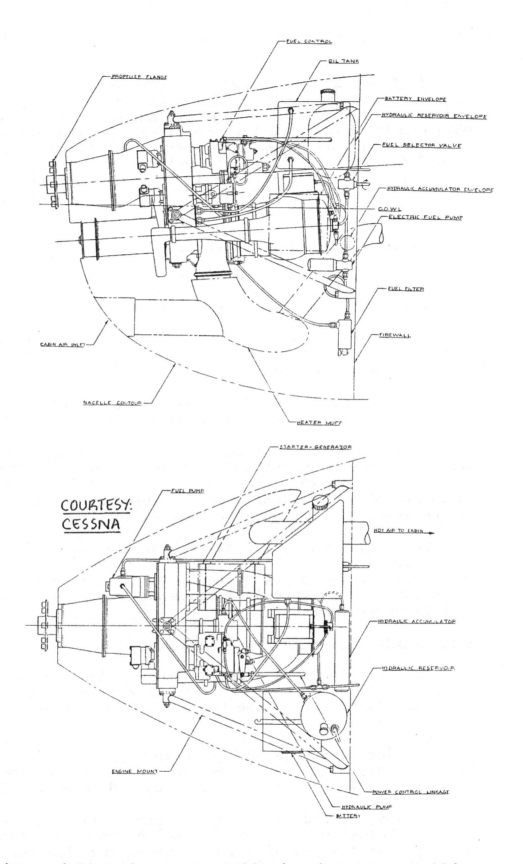

Figure 6.71 Turboprop Installation in a Cessna 406

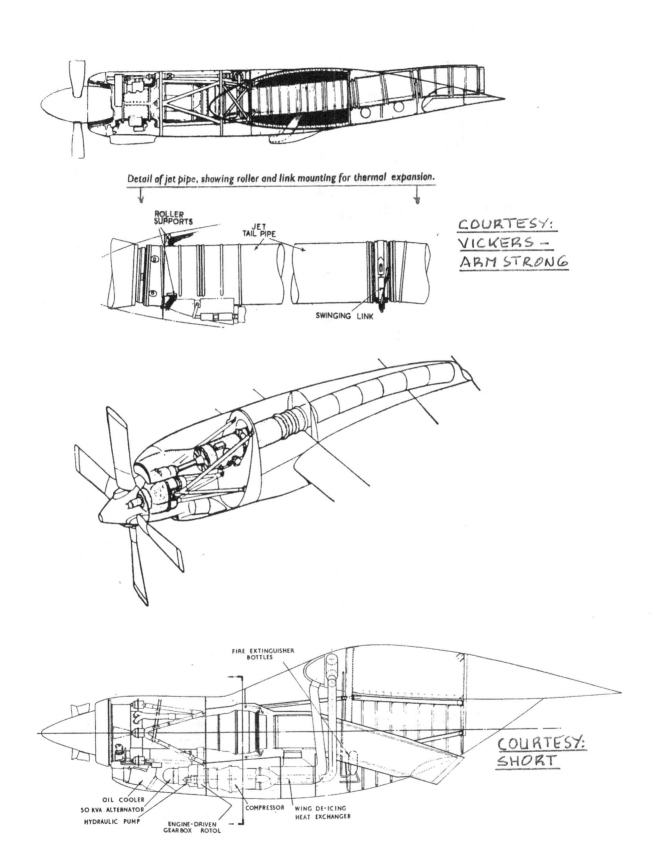

Figure 6.72 Examples of Over-Wing and Under-Wing Turbo-Prop Installations

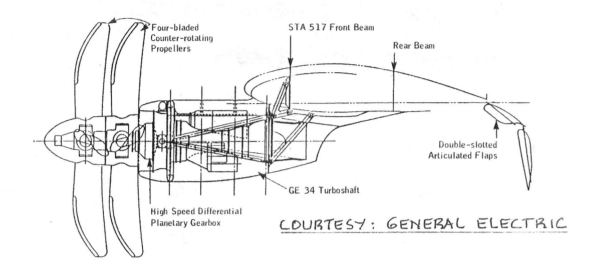

Figure 6.73 Counter Rotating Turboprop Installation

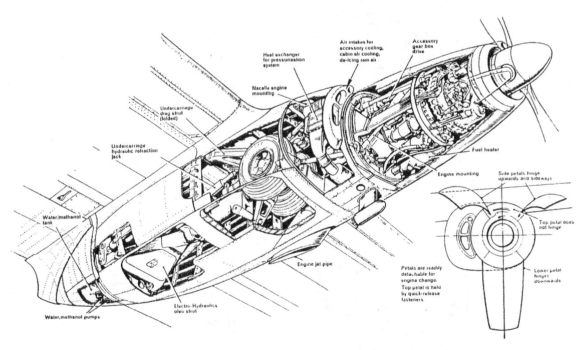

Figure 6.74 Cutaway of Turboprop Installation in the Handley Page Dart Herald

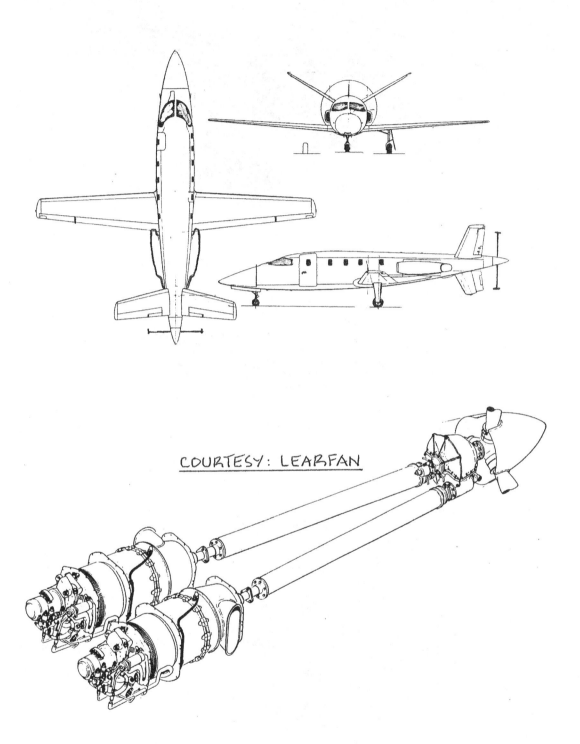

Figure 6.75 Turboprop and Drive Shaft Installation in the Learfan

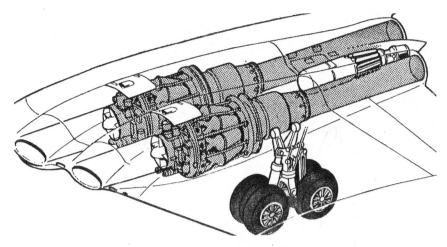

Figure 6.76 Example of Buried Turbojet Installation

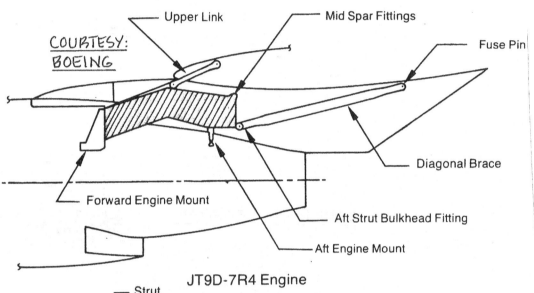

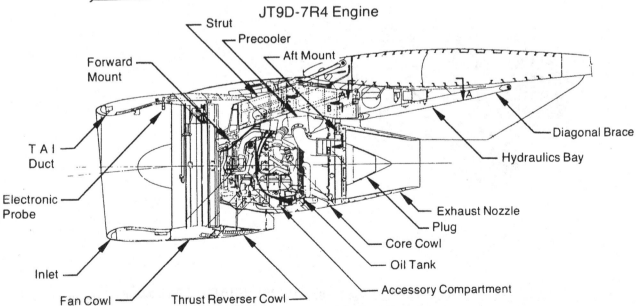

Figure 6.77 Nacelle/Engine Installation in the Boeing 767

Examples of fuselage mounted nacelles are found in the B727 and the McDD DC9. Figure 6.78 shows the 727 installation. Note the buried center engine: this requires a long duct which causes some loss in pressure recovery as well as a weight increase. Even high bypass ratio fans can be installed in this manner. The L1011 installation of Figure 6.79 is an example. An interesting contrast to this arrangement is that of the center engine of the McDD DC10 shown in Figure 6.80.

The DC10 center engine installation eliminates the need for a long inlet duct but complicates the design of the vertical tail.

In fighter airplanes the engines are normally buried in the fuselage. The A10 of Figure 3.27c in Part II is an exception.

Figures 6.81 and 6.82 are examples of subsonic fighter engine installations.

In supersonic fighters the installation is complicated because of the requirement of inlet shock management. Figures 6.83 through 6.85 show examples of supersonic installations. For best efficiencies at Mach numbers of 2.0 and above variable geometry inlets are required. The F15 inlet of Figure 6.84 has a variable ramp inlet.

Supersonic transports above Mach 2.0 also require vaiable geometry inlets. The Concorde inlet of Figure 6.86 illustrates such a system.

References 65 and 78 contain discussions of the characteristics of various types of inlets. Part VI also addresses this question.

6.9.4 Propfan and Ultra-bypass Installations

These installations are currently in the preliminary design and development stage. It appears likely that 'range-payload' driven airplanes will be strong candidates for these engines.

Figures 6.87 show several proposed propfan installations arranged as tractors. Counter rotating arrangements are also possible: Figure 6.88 shows a tractor and a pusher installation.

Figure 6.89 puts the large propfan diameter in

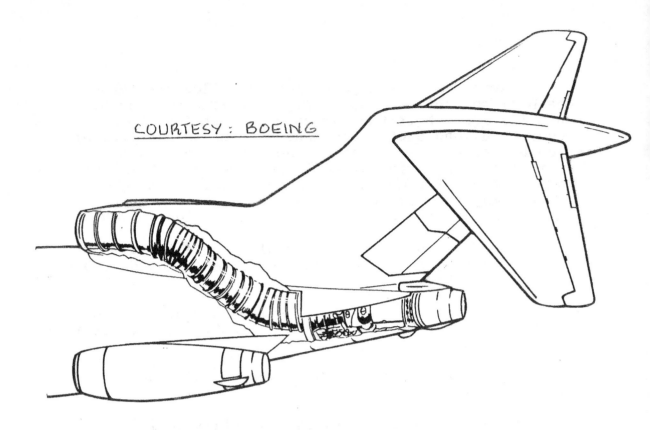

Figure 6.78 Center Engine Installation in the Boeing 727

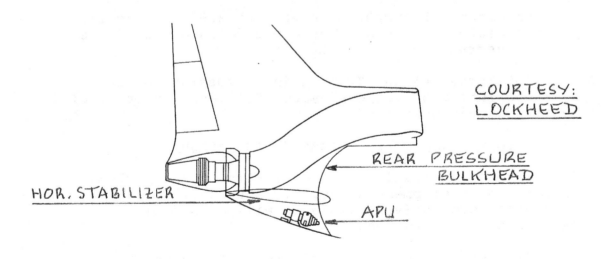

Figure 6.79 Center Engine Installation in the Lockheed L-1011

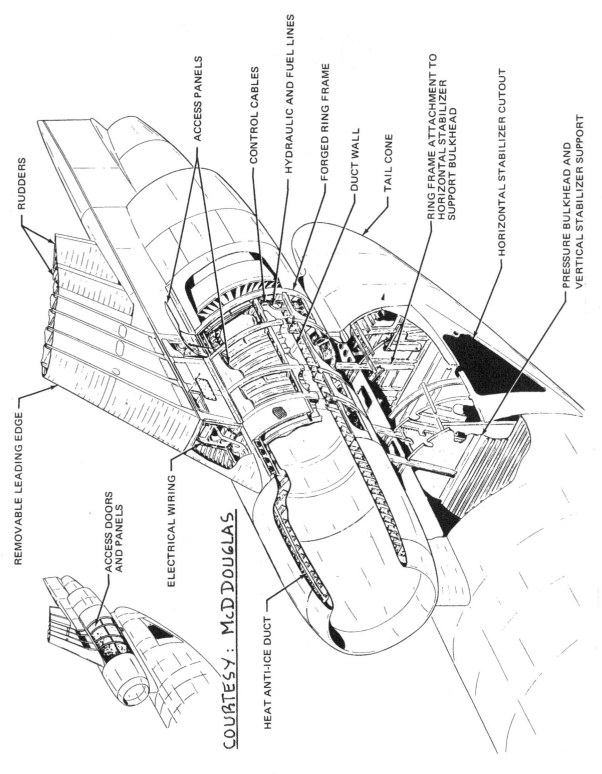

Figure 6.80 Center Engine Installation McDD DC10

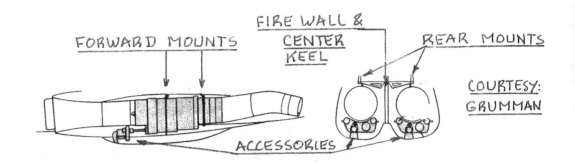

Figure 6.81 Subsonic Engine Installation: Grumman A6

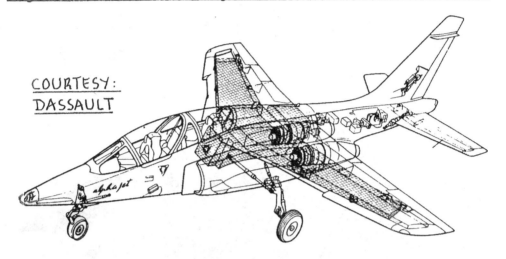

Figure 6.82 Subsonic Engine Installation: DBD Alphajet

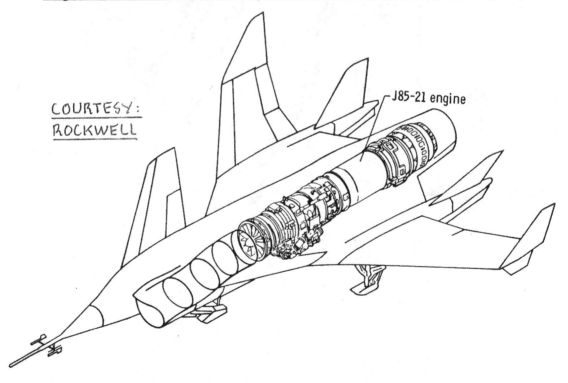

Figure 6.83 Subsonic Engine Installation: Rockwell HiMat

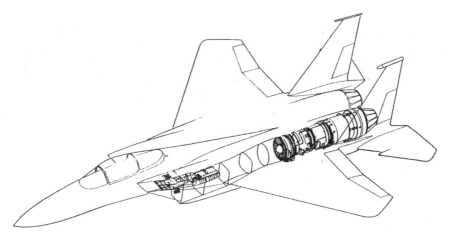

COURTESY:
McDONNELL DOUGLAS

Figure 6.84 Supersonic Engine Installation: McDD F15

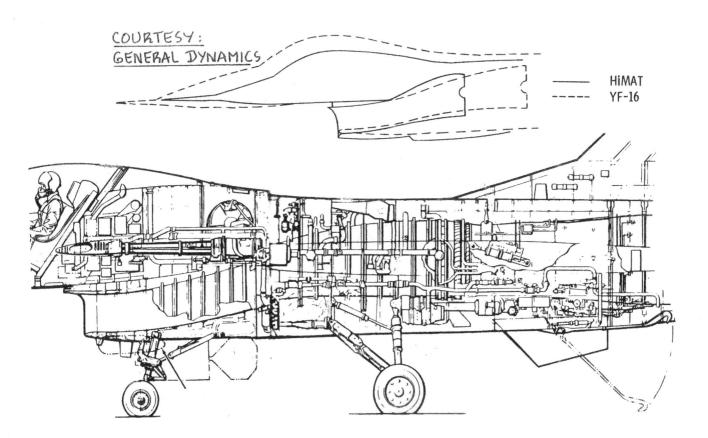

Figure 6.85 Supersonic Engine Installation: GD F16

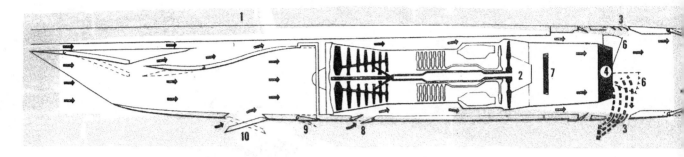

LEGEND:

1. WING
2. TURBINES
3. THRUST REVERSERS
4. PRIMARY NOZZLE
5. SECONDARY NOZZLE
6. THRUST REV. BUCKET
7. AFTERBURNER
8. COOLING AIR FLAP
9. BYPASS FOR EXCESS AIR
10. INLET FOR EXTRA AIR

Figure 6.86 Supersonic Engine Installation: Concorde

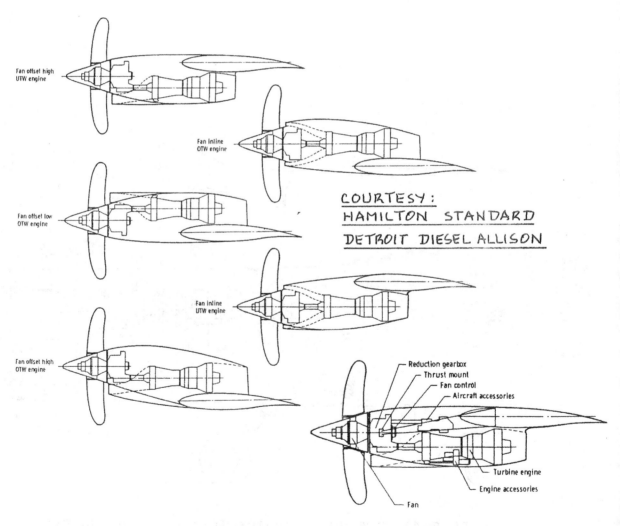

COURTESY:
HAMILTON STANDARD
DETROIT DIESEL ALLISON

Figure 6.87 Single Rotating Propfan Installations

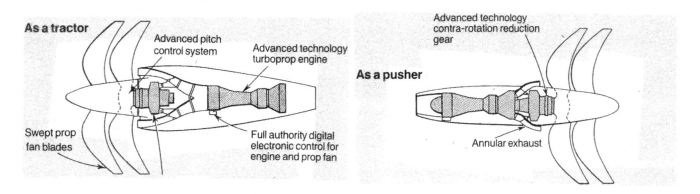

COURTESY: PRATT & WHITNEY

Figure 6.88 Counter Rotating Propfan Installations

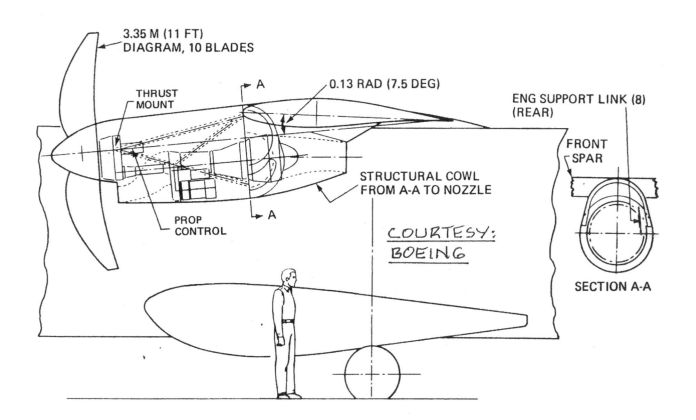

Figure 6.89 Propfan Installation in a High Wing Commuter

perspective in a small commuter transport proposal.

A variant of the propfan is the General Electric ultra-bypass-engine (UBE), shown in the proposed Boeing 7J7 in Figure 6.90. Figure 6.91 shows a tractor and a pusher arrangement for the UBE powerplant.

6.9.5 Nozzles and Thrust Reversers

All jet engines must have exhaust nozzles to generate thrust. Figures 6.77 through 6.86 all show such nozzles. A close-up view of a typical nozzle installation is given in Figure 6.92.

A desirable feature of most exhaust systems is that they be reversible for landing operations. Although this adds complexity to the nozzle installation such reversers are incorporated in most transports to provide improved slowdown capability. The latter is essential when operating on slick surfaces such as icy runways. Figures 6.93 through 6.95 show example installations on transports.

Even some fighters use thrust reversers: the SAAB 37 Viggen is an example. Figure 6.96 shows its thrust reverser installation.

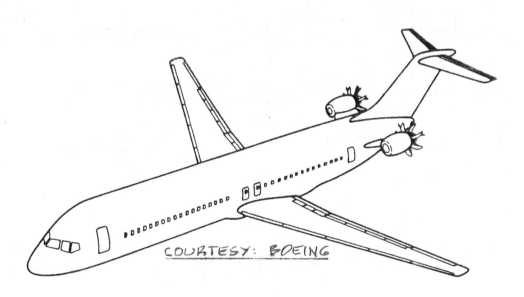

Figure 6.90 Ultra Bypass Engine Installation: Boeing 7J7

COURTESY: GENERAL ELECTRIC

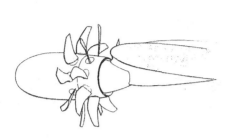
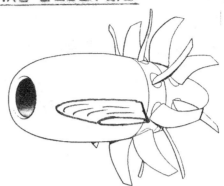

TRACTOR
STING MOUNT

PUSHER
PYLON MOUNT
(SEE FIG. 6.90)

Figure 6.91 Ultra Bypass Engine Arranged as Pusher or as Tractor

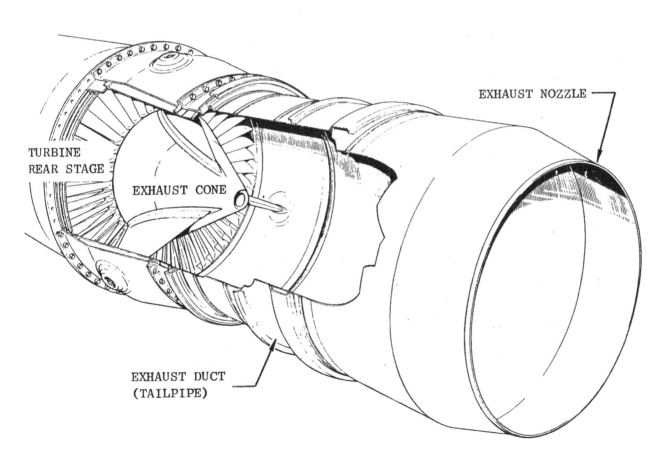

Figure 6.92 Typical Turbojet Exhaust Nozzle Arrangement

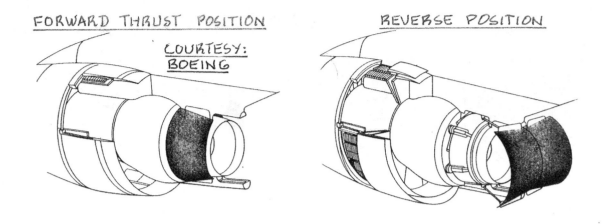

Figure 6.93 Thrust Reverser Installation: Boeing 747

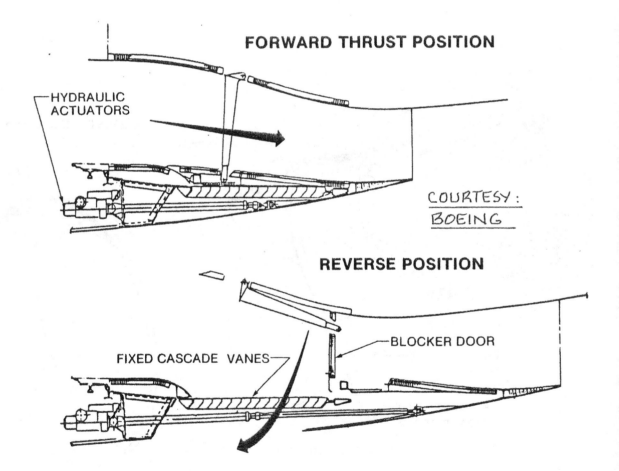

Figure 6.94 Thrust Reverser Schematic: Boeing 757

Figure 6.95 Thrust Reversers: Lockheed Jetstar

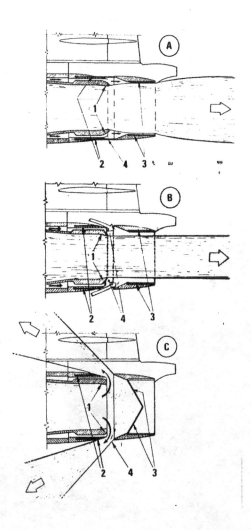

A   MAXIMUM THRUST

B   CRUISE THRUST

C   REVERSE THRUST

1. VARIABLE PRIMARY NOZZLE
2. THRUST REVERSER SLIDE
3. THRUST REVERSER CLAMSHELL DOORS
4. THRUST REVERSER NOZZLES

Figure 6.96 Thrust Reverser: SAAB 37 Viggen

7. PRELIMINARY STRUCTURAL ARRANGEMENT, MATERIAL SELECTION AND MANUFACTURING BREAKDOWN

The purpose of this chapter is threefold:

1. to provide a logical method for arriving at a reasonable structural arrangement: Step 19 in Chapter 2 of Part II.

2. to provide a preliminary basis for the selection of structural materials.

3. to provide data for deciding on a manufacturing breakdown.

Sections 7.1 through 7.3 present discussions of these subjects.

7.1 PREPARING A PRELIMINARY STRUCTURAL ARRANGEMENT*

Before a structural arrangement can be decided upon, it is required that the following is available:

1. A dimensioned threeview, including an indication where all moving surfaces are.

From Step 15 in p.d. sequence I (p.17, Part II) a dimensioned threeview of the proposed design is available. For examples, see Chapter 13 of Part II.

2. A weight and balance calculation and the resulting c.g. excursion diagram.

From Steps 10-13 in p.d. sequence I (p.15-16, Part II) a weight and balance analysis is available.

With the threeview and the weight and balance analysis at hand, it is suggested that the following steps be taken:

Step 1: On all lifting surfaces, identify the moving control surfaces and/or the flaps. Locate the rear and the front spar of each lifting surface so that a sufficient amount

---

* This section is based on suggestions made to the author by Dr. H.W. Smith of the Department of Aerospace Engineering of The University of Kansas.

of room is available for hinges and/or for tracks. Draw in these spar locations.

Note: the information needed to do this, is available after carrying out Steps 6,7 and 8 of p.d. sequence I (p.12-14, Part II). Preliminary design procedures for the wing, for the layout of high lift devices, for the empennage and for control surface layouts are presented in Chapters 6,7 and 8 in Part II.

Step 2: Carry the spars, which form the so-called torque boxes of the structure of the lifting surfaces through to where they intersect with the fuselage. At the point of intersection of these spars with the fuselage, draw in the required fuselage frames or bulkheads.

Note: in most airplanes the torque box actually passes through the fuselage (at the top or at the bottom). Several wing design examples which illustrate this are given in Chapter 4.

Step 3: Determine the location of major structural cutouts for such items as doors and emergency exits. Draw in the necessary frames. If these cutouts are in the wing or in another lifting surface then ribs need to be drawn in at those locations.

Note: in fuselages there are specific requirements for doors and for exits. These requirements MUST be adhered to! However, they strongly impact the location and spacing of fuselage frames and longerons.

Examples of door and exit requirements and specific layouts which satisfy these requirements are given in Chapter 3.

Step 4: Determine the location of the landing gears.

Note: this information is available after carrying out Step 9 of p.d. sequence I (p.14-15, Part II). A preliminary design procedure for landing gear design is presented in Chapter 9 of Part II.

a) If the gears are attached to the fuselage, then it will be necessary to support the attachment points with frames or bulkheads. Draw these in.

b) If the gears are attached to the wings, then it

will be necessary to draw in local ribs to support the attachment points. Draw these in.

Part IV contains more details on the subject of landing gear design as well as specific examples of landing gear arrangements and layouts.

> Step 5: Determine the location of the powerplants. From engine manufacturer's data, locate where on the engine the attachment pads are. It is at these points that the engines will be attached to surrounding structure.

Note: this information is available after carrying out Step 5 of p.d. sequence I (p.12 of Part II). A procedure for the preliminary integration of the propulsion system into the airframe is contained in Chapter 5 of Part II.

a) If the powerplants are attached to the wing then locally, ribs for attachment are required. Draw these in.

b) If the powerplants are attached to the fuselage, for example with the help of pylons, then locally frames or bulkheads are required. Draw these in.

Chapter 6 contains examples of many types of engine installations.

> Step 6: Determine where major masses (other than powerplants) are located.

Examples of major masses are:

a) Guns, pods and bombracks: these normally require extra ribs and/or frames. Typical wing 'hard point' requirements are discussed in Chapter 4.

b) Radar antennas: depending on their size, these may require extra frames or bulkheads for attachment.

c) special fuel tanks and/or hoppers: these may require extra ribs, spars, frames and/or bulkheads.

d) passenger seats: these require tracks to move on and to attach to. These tracks in turn are mounted on stiffened floors. See Ch.3 for floor design examples.

f) concentrated payloads such as large vehicles and/or machine equipment: these require special floor

Part III Chapter 7 Page 383

strengthening and special tie-down provisions to prevent accidental 'sliding' of the payload. All this usually results in the need for special structural provisions in the form of longerons, frames, ribs and/or spars. See Chapter 3 for cargo floor design examples.

g) auxiliary power units (APU): these require special installation provisions and firewalls. Part IV deals with APU installations.

Draw in the required ribs, frames, bulkheads and floor beams to account for ALL major masses.

For data on typical spacings, see the following chapters:

> Chapter 3 for fuselage framedepths, frame spacings and longeron spacings.
>
> Chapter 4 for typical wing rib spacings.
>
> Chapter 5 for typical empennage rib spacings.

<u>Step 7:</u> Locate the pressure bulkheads and draw those in.

Note: pressure bulkheads are required in most transports and fighters. See Chapter 3 for examples of pressure bulkhead arrangements.

<u>Step 8:</u> Draw in the major fuselage cross sections.

Note: major cross sections, from a structural arrangement viewpoint are those where intersections with other major structural components occur. Major cross sections are also those, where major cutouts, such as doors and windows, occur.

In each of these cross sections, locate where longerons need to be placed and draw them in.

Examples of fuselage cross section design are found in Chapter 3.

<u>Step 9:</u> Take a good look at the resulting 'skeleton structure' and see if any items can be relocated such that structural synergism can be achieved.

The following examples illustrate what is meant by 'structural synergism'.

Part III　　　　　　　　Chapter 7　　　　　　　Page 384

<u>Synergism Example 1</u>: Where the vertical tail and horizontal tail spars intersect the fuselage, four frames could be required. It would save a lot of weight if they could both attach to the same two frames. It is not always possible to achieve this, but it certainly is a desirable target to shoot for. If the front spars of both tails can be made to intersect the fuselage at the rear pressure bulkhead, additional weight savings will accrue.

<u>Synergism Example 2</u>: Placement of the landing gear was dictated by a number of nonstructural considerations, as seen in Chapter 7. It would save weight, if the landing gear locations could be made to coincide with already existing major structural components.

<u>Step 10:</u> In the spaces between the frames, bulkheads, spars, ribs and longerons, it is normally necessary to install minor similar items. The spacings for these items differ from one airplane type to another.

The following general rules can be used in a preliminary layout:

For fuselage frames typical frame spacings are: 18 inches for transports and 24 inches for general aviation airplanes.

For wing ribs typical rib spacings are: 24 inches for transports and 36 inches for general aviation airplanes.

For fighter and military trainer airplanes the spacings vary considerably, depending on the mission and on the design flight envelope.

Draw in these additional structural items.

This completes the initial structural arrangement.

For additional guidance in preparing a structural arrangement, the reader is encouraged to consult the inboard profile drawings in Chapter 3 (p.154-162), the structural arrangement drawings in Chapter 4 (p.244-248) and the cut-away drawings in Chapter 8.

Next, attention needs to be focussed on material selection and on manufacturing breakdown. These topics are covered in Sections 7.2 and 7.3 respectively.

7.2 PRELIMINARY SELECTION OF STRUCTURAL MATERIALS

The following considerations determine which materials are selected in the structural design of an airplane:

1. Design strength to weight ratio
2. Fatigue characteristics
3. Crack propagation behavior
4. Dominant failure mode(s)
5. Damage and corrosion tolerance and/or resistance
6. Familiarity and experience with a given material
7. Existing manufacturing facilities
8. Cost of material
9. Manufacturing cost
10. Customer requirements

These considerations are not listed in an intended order of importance.

Table 7.1 presents a list of typical airplane structural materials which are in use. The reader should consult the following materials for information on the use of structural materials:

For light airplanes: pages 132, 133, 219, 238, 276 and 286.

For fighter airplanes: Figures 7.1 and 7.2a-b.

For transport airplanes: Figures 7.3a-d.

References 12,20,31,79,80 and 81 contain data which provide further insight into the problem of airplane material selection.

Table 7.1 Examples of Use of Airplane Materials (Data Source: Ref.82)

Material	Aluminum	Steel	Titanium	Composites
Shapes and Forms	Sheet Extrusions Plate Forgings Tubes	Sheet Extrusions Plate Forgings Tubes	Sheet Extrusions Plate Forgings Tubes	Sheet Extrusions Plate Forgings Tubes
Typical Airplane usage	Skins Spar webs and Spar caps Stiffeners Control surfaces and flaps Fittings	Engine and/ or fuselage trusses Landing gear, system and engine parts	Skins Stringers and spars Stiffeners Fittings	Skins Spar webs and spar caps Stiffeners Control surfaces and flaps Fairings
Major design consideration	Light weight Corrosion resistant	High strength Suited for high temp.	High strength Suited for high temp. Expensive and difficult to manufacture	Light weight Can be layed up in any shape Low temp. Needs anti-lightning
Commonly used Alloys	2024-T3, T8, T81 T861, 2124-T3, T8 6061-T6 7075-T73, T76 7175-T73, 76 7050-T736, T76 T73, T736	Stainless Steels 5Cr-Mo-V AlSi4340 300M	Ti-140A, Ti-155A Ti-4Al-3Mo-V Ti-8Al-Mo-V Ti-6Al-25N- 4Zr-2Mo	Graphite epoxy Kevlar Fiberglass Boron Aluminum

Part III Chapter 7 Page 387

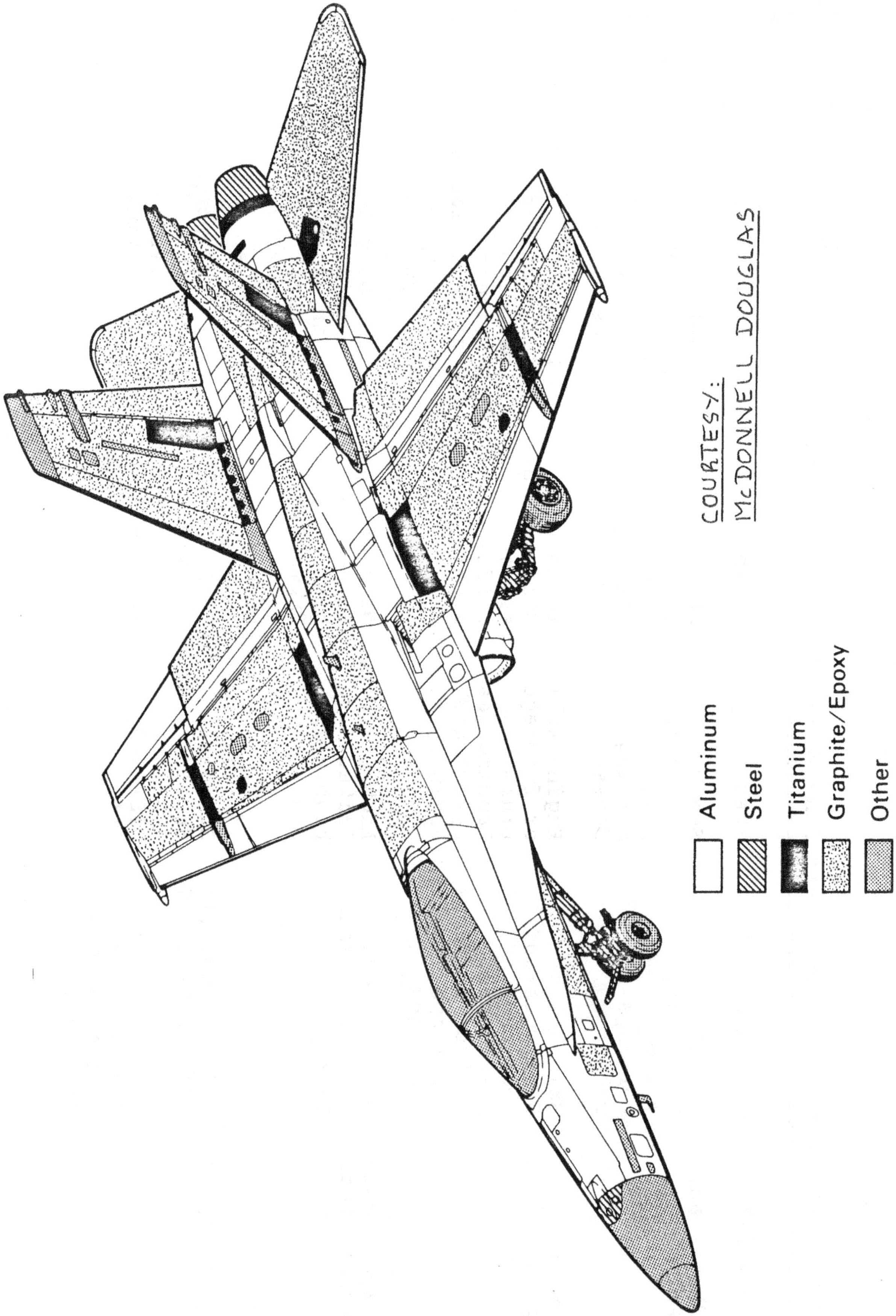

Figure 7.1 Material Distribution: McDonnell Douglas F-18

COURTESY:
McDONNELL DOUGLAS

Aluminum
Steel
Titanium
Graphite/Epoxy
Other

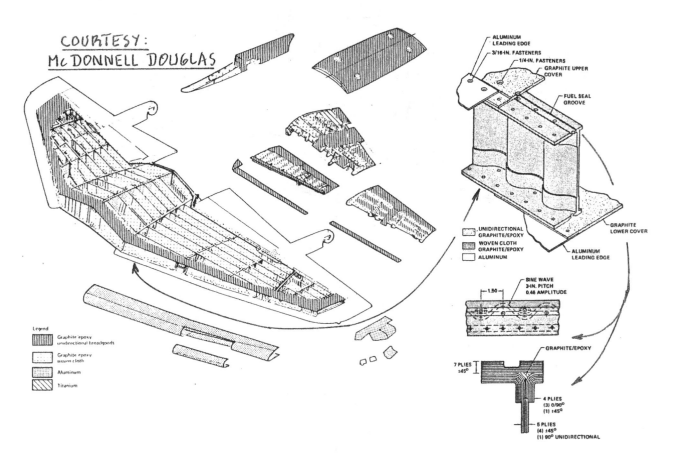

Figure 7.2a Material Distribution, Wing: McDD AV-8B

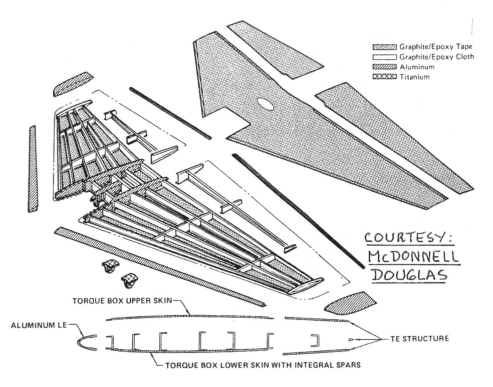

Figure 7.2b Material Distribution, Hor. Tail: McDD AV-8B

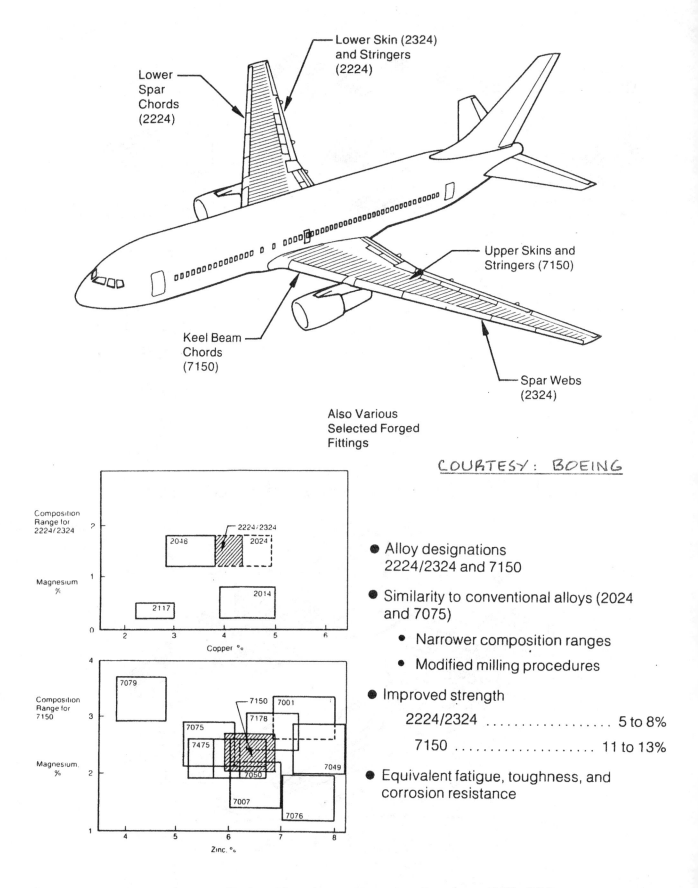

Figure 7.3a Aluminum Usage: Boeing 767-200

Part III   Chapter 7   Page 390

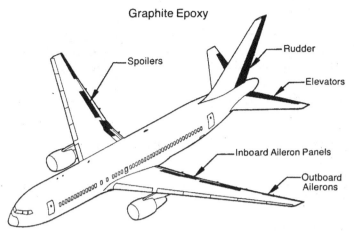
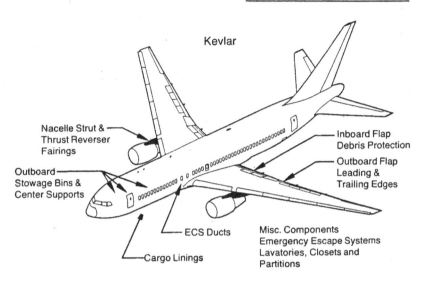
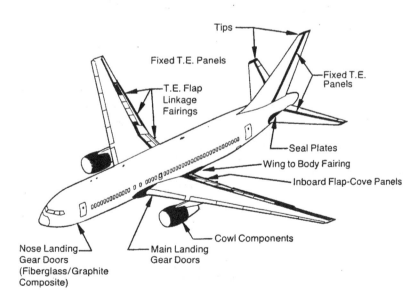

Figure 7.3b Advanced Composites Usage: Boeing 767-200

Material	Application
Graphite/Epoxy	Inboard Aileron Outboard Aileron Inboard and Outboard Spoilers Rudder Elevator
Hybrid (Kevlar/Graphite)	Fixed Panels-Wing T.E. Cowls-Thrust Reverser, Inlet and Fan Fairing-T.E. Flap Linkage Cove Panels-Inbd T.E. Flap Wing/Body Fairing Landing Gear Doors-Body Fixed T.E. Panels, Tip-Empennage Seal Plates-Stabilizer
Hybrid (Fiber- Glass/Graphite)	Nose Landing Gear Doors (Graphite Weight Only)
Kevlar	ECS Ducts Cargo Liner Outboard Stowage Bins & Center Stowage Supports Emergency Escape System Lavs, Closets, & Partitions Fairings-Engine Pylon Outboard Flap-L.E. and T.E. Inboard Flap-Debris Protection Fairing-Thrust Reverser

Composite Applications Example
Main Landing Gear Door
(Kevlar/Graphite Composite)

COURTESY: BOEING

Figure 7.3c Composites Applications: Boeing 767-200

COURTESY: BOEING

Fiberglass Usage

- Durable
- Lightweight
- Sonic and buffet resistant
- Easily repairable
- Corrosion resistant

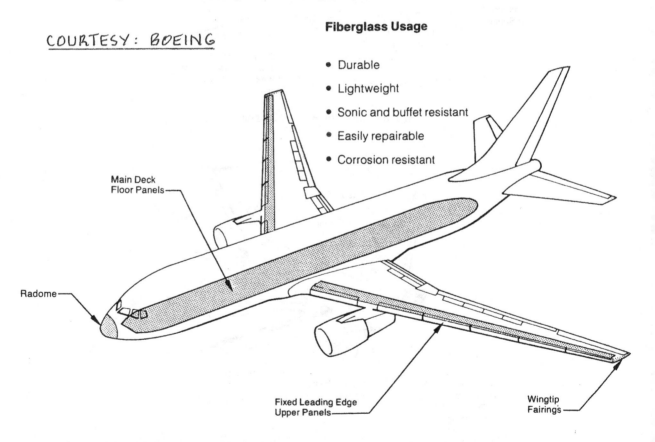

Figure 7.3d Fiberglass Usage: Boeing 767-200

7.3 PRELIMINARY SELECTION OF MANUFACTURING BREAKDOWN

The following considerations play a role in deciding on the manufacturing breakdown for an airplane:

1. Number of units to be produced

2. Available sub-assembly and final assembly facilities

3. Agreements for sub-assemblies to be performed in other areas or in other countries (international programs)

4. Future growth potential

5. Questions of production optimization: how many workers can perform their jobs without interfering with one another

The material presented in this chapter does not address these considerations. Part VIII contains methods for predicting manufacturing cost which depends to a large extent on the manufacturing materials and the manufacturing breakdown selected. Figures 7.4 through 7.9 present examples of manufacturing breakdowns used in a range of airplanes:

For a light airplane: Figure 7.4 Beech Musketeer

For a military trainer: Figure 7.5 SIAI Marchetti S-211

For a jet transport: Figure 7.6 Boeing 767-200

For fighters: Figure 7.7 Fairchild Republic A-10

        Figure 7.8 McDonnell Douglas F-15

        Figure 7.9 McDonnell Douglas F-18

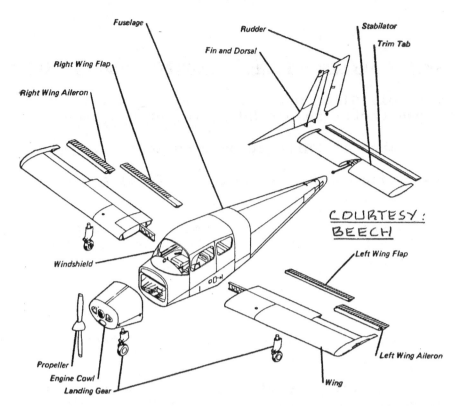

Figure 7.4 Manufacturing Breakdown: Beech Sport

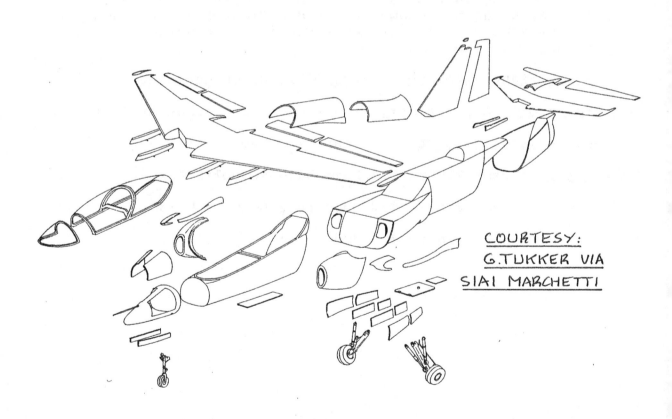

Figure 7.5 Manufacturing Breakdown: SIAI Marchetti S-211

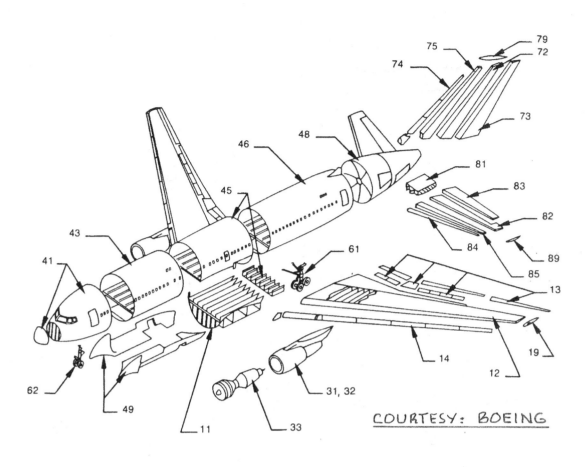

Section Number	Description	Section Number	Description
11	Wing Center Section	49	Multi-Section (Wing/Body Fairings)
12	Wing Outboard Structural Box	61	Main Landing Gear
13	Wing Trailing Edge (Includes Ailerons, Flaps, and Spoilers)	62	Nose Landing Gear
		72	Vertical Tail – Aft Torque Box
14	Wing Leading Edge (Fixed and Movable)	73	Vertical Tail – Trailing Edge (Fixed and Movable)
19	Wingtip		
31	Propulsion – Wing Installation (Strut and Cowling)	74	Vertical Tail – Leading Edge
		75	Vertical Tail – Forward Torque Box
32	Propulsion – Center Installation (Mount, Inlet, Ducting & Thrust Reverser	79	Vertical Tail – Tip
		81	Horizontal Tail – Center Section
33	Propulsion – Built-Up Engine	82	Horizontal Tail – Aft Torque Box
41	Fuselage Nose Section (Includes Radome)	83	Horizontal Tail – Trailing Edge (Fixed and Movable)
43	Forward Fuselage Section		
45	Fuselage Section, Wing Join	84	Horizontal Tail – Leading Edge
46	Aft Fuselage Section	85	Horizontal Tail – Forward Torque Box
48	Fuselage Tail Section	89	Horizontal Tail – Tip

Figure 7.6 Manufacturing Breakdown: Boeing 767-200

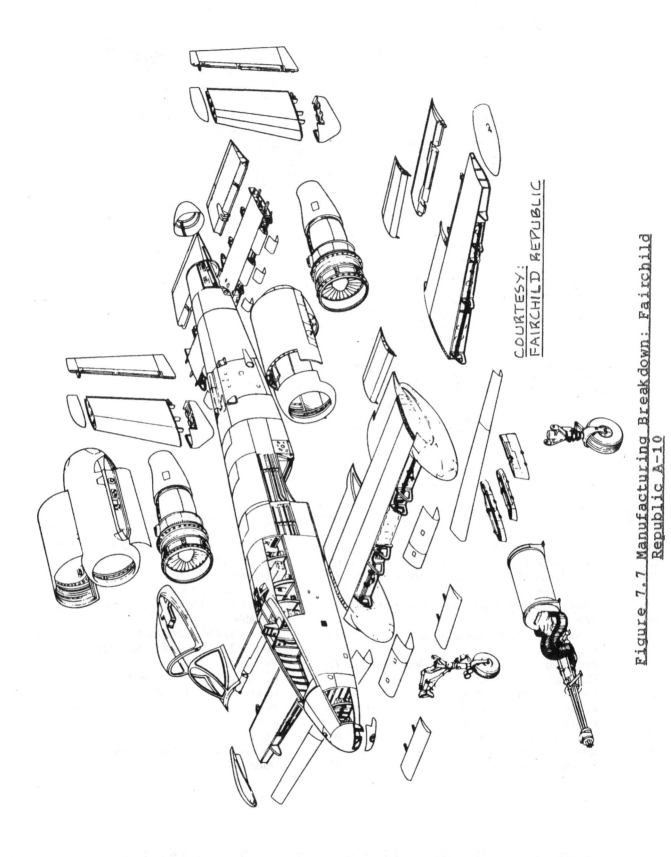

Figure 7.7 Manufacturing Breakdown: Fairchild Republic A-10

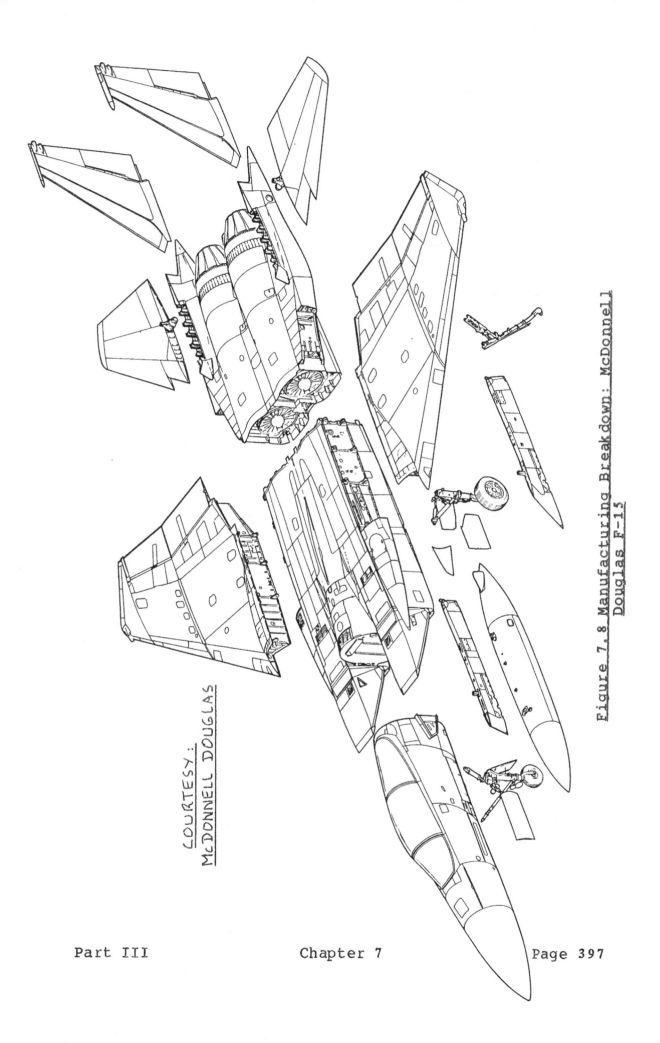

Figure 7.8 Manufacturing Breakdown: McDonnell Douglas F-15

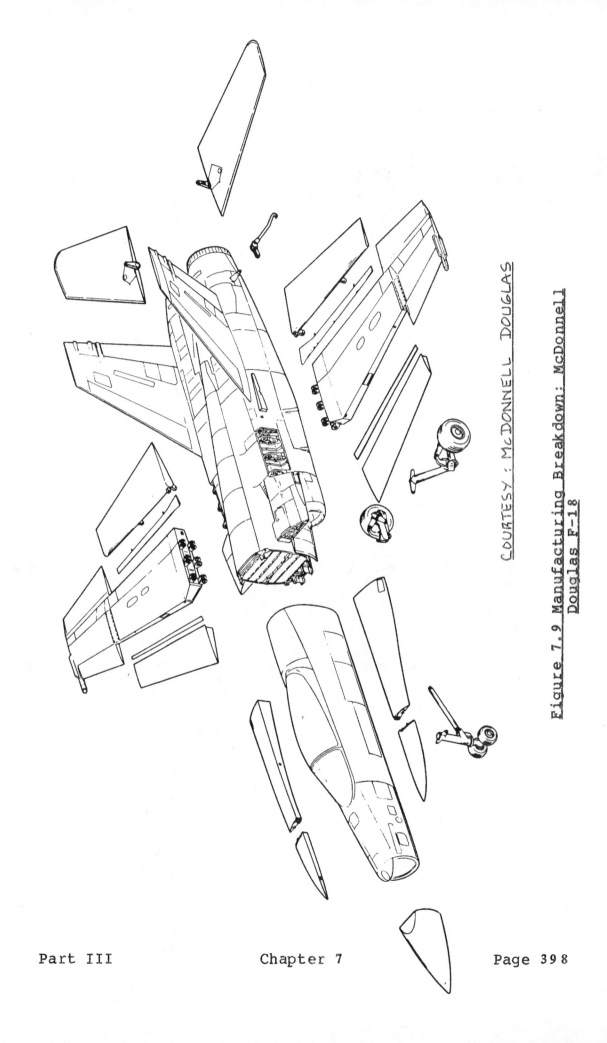

Figure 7.9 Manufacturing Breakdown: McDonnell Douglas F-18

Courtesy: McDonnell Douglas

8. COLLECTION OF CUTAWAY DRAWINGS

The purpose of this chapter is to provide a number of airplane cutaway drawings. Such cutaway drawings show the overall internal structural, systems and propulsion layout of an airplane. They also serve to provide the aeronautical engineering student with the opportunity to 'look' at airplanes from a 3-dimensional viewpoint.

Cutaways are presented for modern airplanes and for some older airplanes. The reason for the latter is to provide some historical perspective. The indicated years are approximate. The following cutaways are included:

1. Homebuilt Airplanes:

    Piston-Propeller Driven:

    Figure 8.1 Scaled Down Focke-Wulf 190 (1943)

2. Single Engine Propeller Driven Airplanes:

    Piston-Propeller driven:

    Figure 8.2 Percival Prentice (1938)
    Figure 8.3 Fokker Promotor (1948)
    Figure 8.4 Cessna 150 (1962)
    Figure 8.5 Cessna 172 (1964)
    Figure 8.6 Beech 77 Skipper (1978)

    Turbo-Propeller driven:

    Figure 8.7 Pilatus Turbo-porter (1980)

3. Twin Engine Propeller Driven Airplanes:

    Piston-Propeller driven:

    Figure 8.8 Italair F.20 Pegaso (1970)
    Figure 8.9 Beech 76 Duchess (1978)

    Turbo-Propeller Driven:

    Figure 8.10 Beech Super King Air 200 (1980)

4. Agricultural Airplanes:

    Piston-Propeller Driven:

    Figure 8.11 Hollandair HA-001 Libel (1960)

5. Business Jets:

   Figure 8.12 Gulfstream Peregrine (1980 proposal)
   Figure 8.13 Gates Learjet 25 (1972)
   Figure 8.14 Aerospatiale Corvette (1976)
   Figure 8.15 Gates Learjet 28 (1980)
   Figure 8.16 Gulfstream GIII (1984)

6. Regional Propeller Driven Airplanes:

   Piston-Propeller Driven:

   Figure 8.17 Scottish Av. Twin Pioneer (1950)
   Figure 8.18 Britten Norman Trislander (1970)

   Turbo-Propeller Driven:

   Figure 8.19 Armstrong Whitworth Apollo (1952)
   Figure 8.20 Handley Page Dart Herald (1956)
   Figure 8.21 Short Skyvan (1958)
   Figure 8.22 Fokker F-27 Friendship (1958)
   Figure 8.23 Piper T-1040 (1982)

7. Transports and Transport Jets

   Piston-Propeller Driven:

   Figure 8.24 Fokker F22 (1938)

   Jet Driven:

   Figure 8.25 McDonnell Douglas DC9 (1962)
   Figure 8.26 Fokker F28 (1964)
   Figure 8.27 McDonnell Douglas DC10 (1972)

8. Military Trainers

   Piston-Propeller Driven:

   Figure 8.28 Fokker S12 (1948)

   Turbo-Propeller Driven:

   Figure 8.29 Beech T-34C-1 (1978)

   Jet Driven:

   Figure 8.30 Fokker S14 (1952)
   Figure 8.31 SAAB XT-105 (1958)
   Figure 8.32 British Aerospace Hawk (1980)

Figure 8.33 Microturbo Microjet 200 (1984)

9. Fighters

   Piston-Propeller Driven:

   Figure 8.34 Fokker G.1 (1938)

   Turbo-Propeller Driven:

   Figure 8.35 Piper PA-48 Enforcer (1980)

   Jet Driven:

   Figure 8.36 Gloster Meteor (1945)
   Figure 8.37 Douglas A4 Skyhawk (1960)
   Figure 8.38 General Dynamics F-16 (1980)
   Figure 8.39 McDonnell Douglas F-4E (1970)
   Figure 8.40 McDonnell Douglas F-15 (1978)
   Figure 8.41 McDonnell Douglas F-18 (1982)
   Figure 8.42 McDonnell Douglas AV-8B (1984)

10. Military Patrol, Bomb and Transport Airplanes:

    Piston-Propeller Driven:

    Figure 8.43 Fokker T.IX (1939)

    Turbo-Propeller Driven:

    Figure 8.44 CASA C212A (1978)

11. Flying Boats, Amphibious and Float Airplanes

    Piston-Propeller Driven:

    Figure 8.45 Fokker T.VIII-W (1940)
    Figure 8.46 Riviera Amphibian (1965)

12. Supersonic Cruise Airplanes

    No cutaways available

13. NACA/NASA Experimental Airplanes:

    Figure 8.47 Bell X2 (1952)
    Figure 8.48 Ryan X13 (1956)
    Figure 8.49 North American X15 (1959)

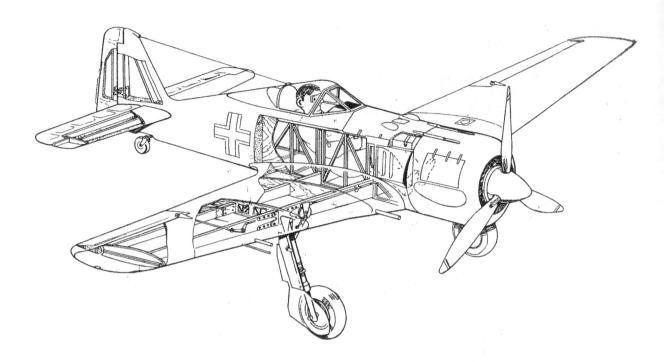

Figure 8.1 Scaled Down Focke-Wulf 190

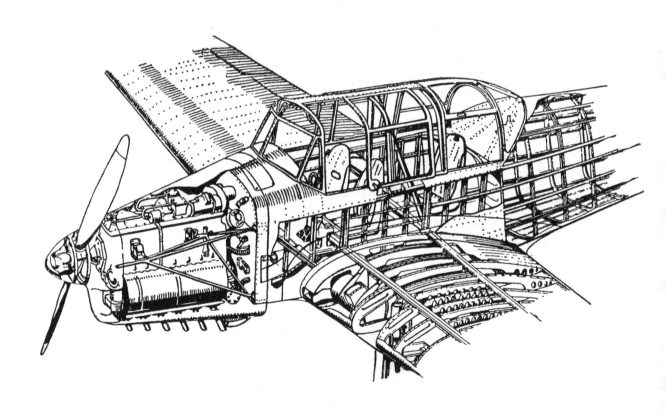

Figure 8.2 Percival Prentice

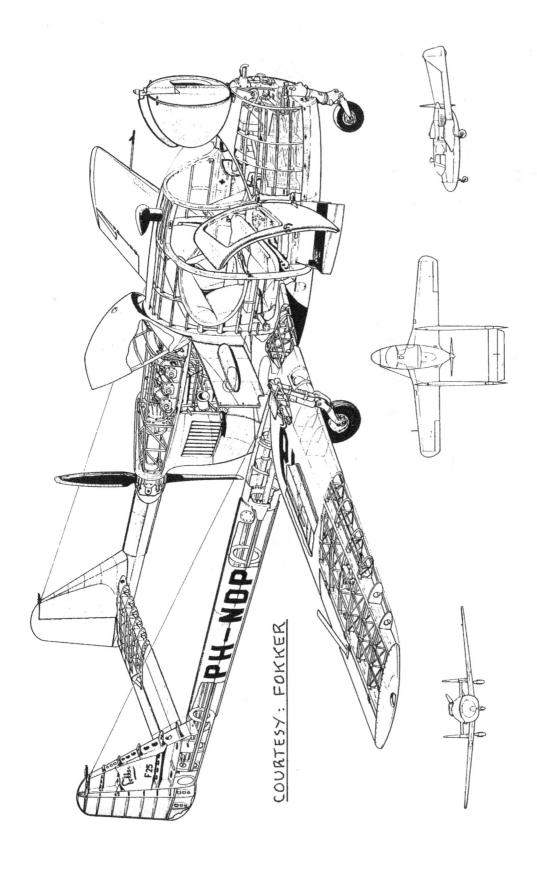

Figure 8.3 Fokker Promotor

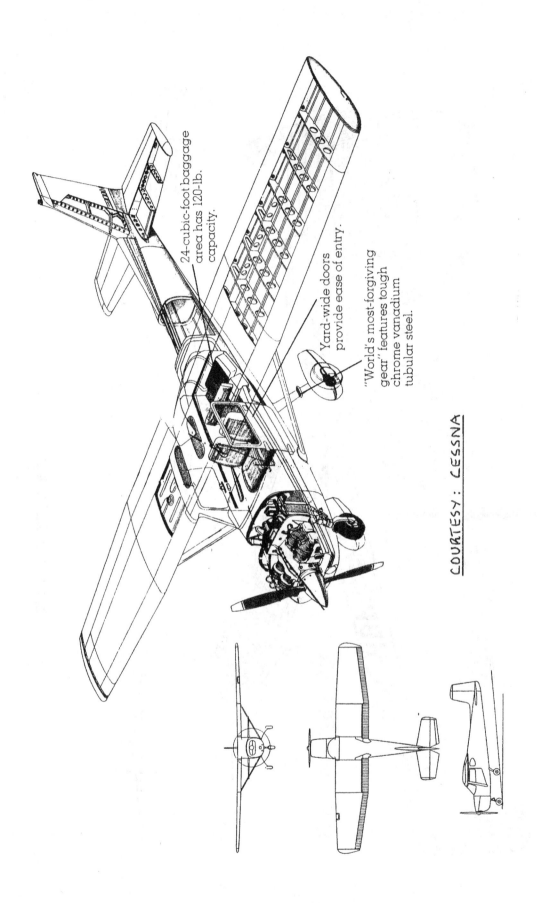

Figure 8.4 Cessna 150

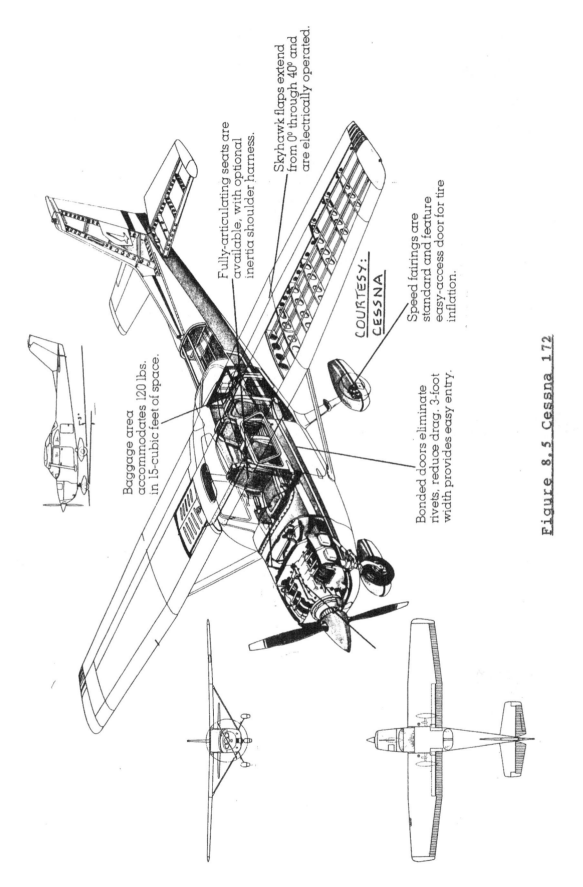

Figure 8.5 Cessna 172

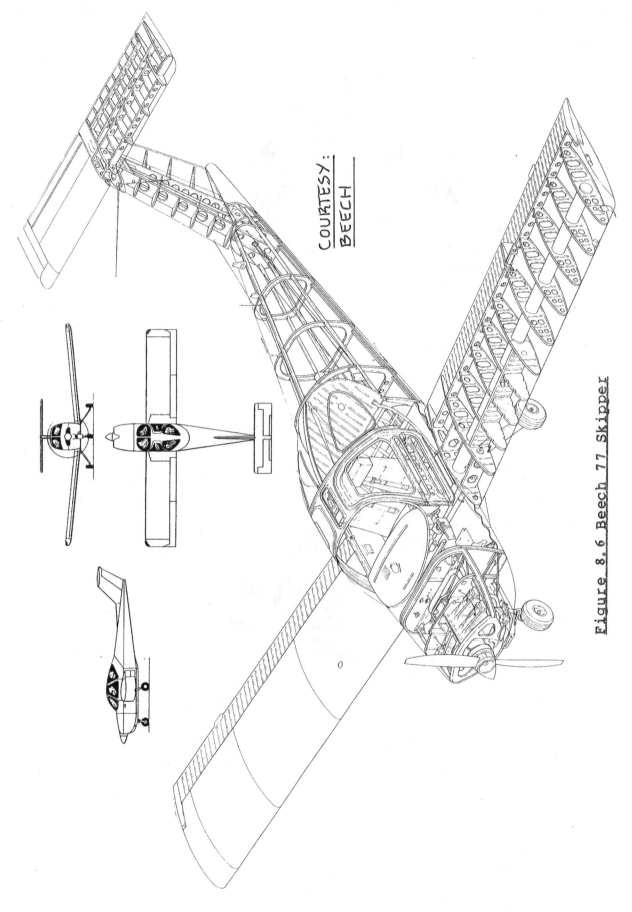

Figure 8.6 Beech 77 Skipper

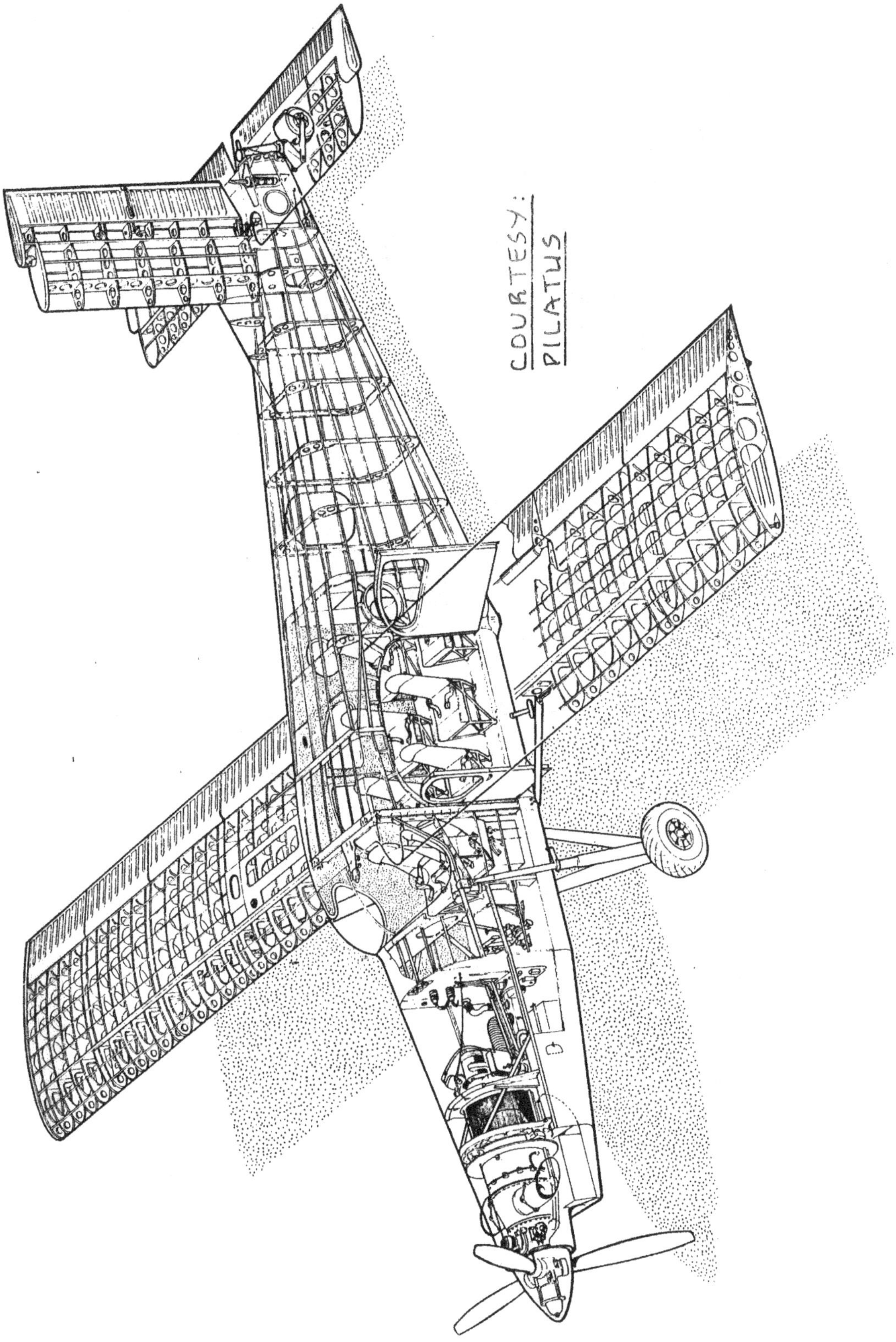

Figure 8.7 Pilatus Turbo-porter

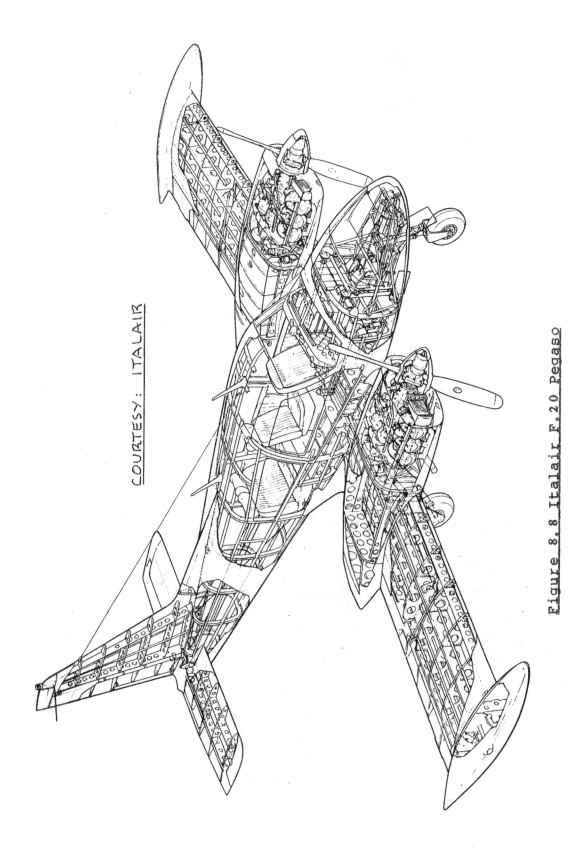

Figure 8.8 Italair F.20 Pegaso

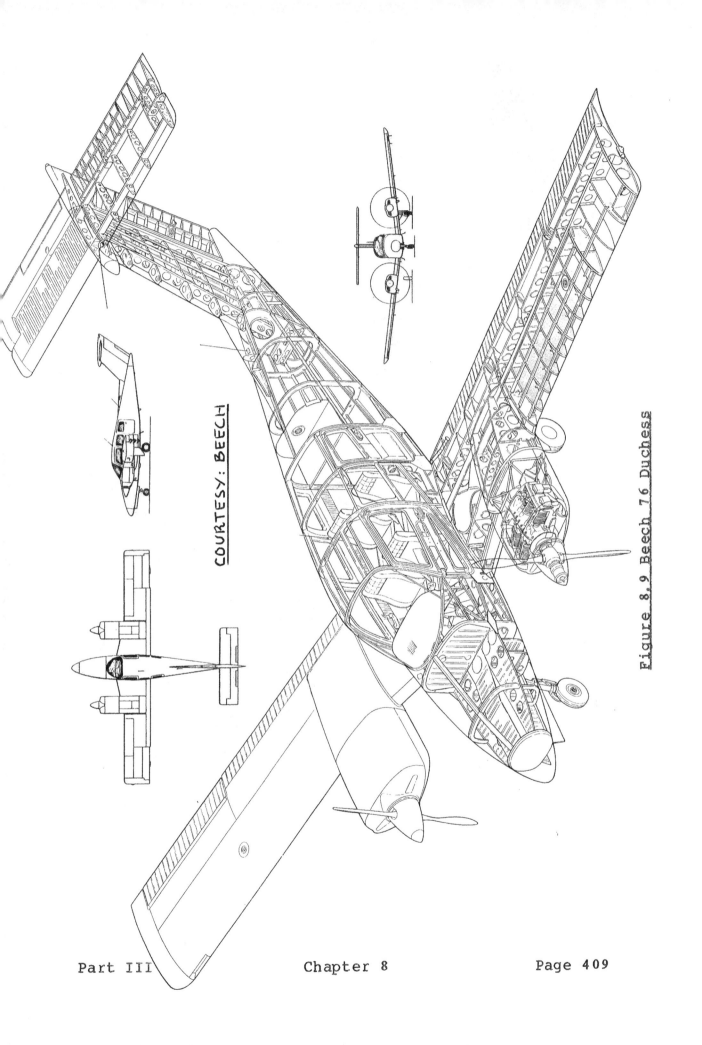

Figure 8.9 Beech 76 Duchess

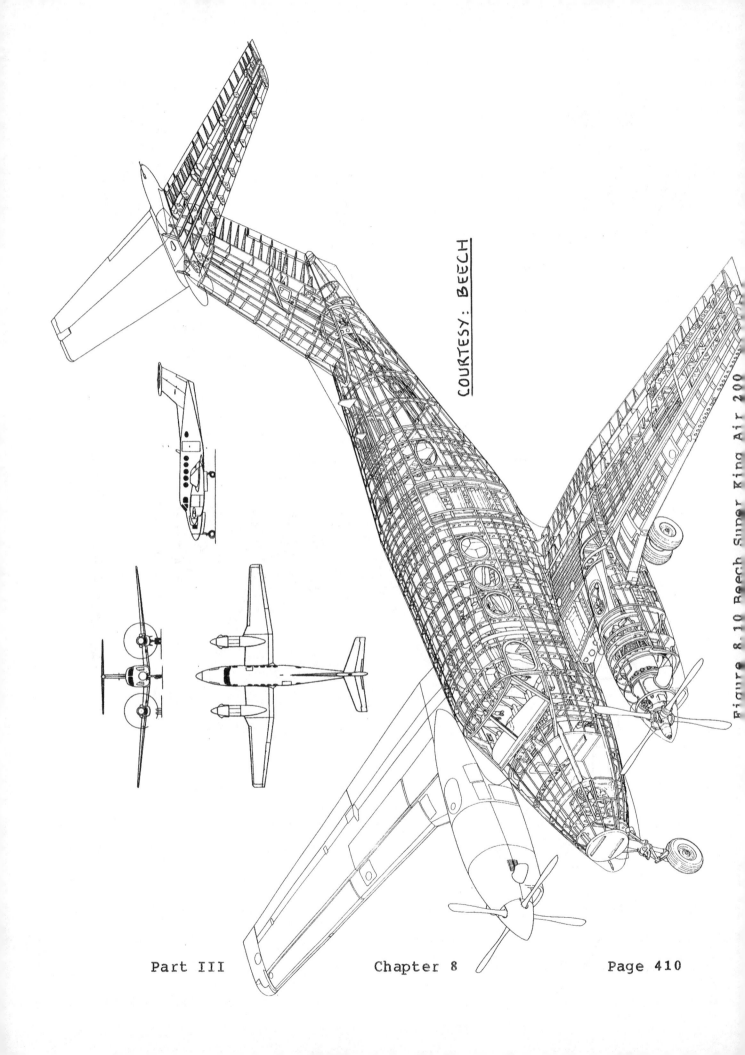

Figure 8.10 Beech Super King Air 200

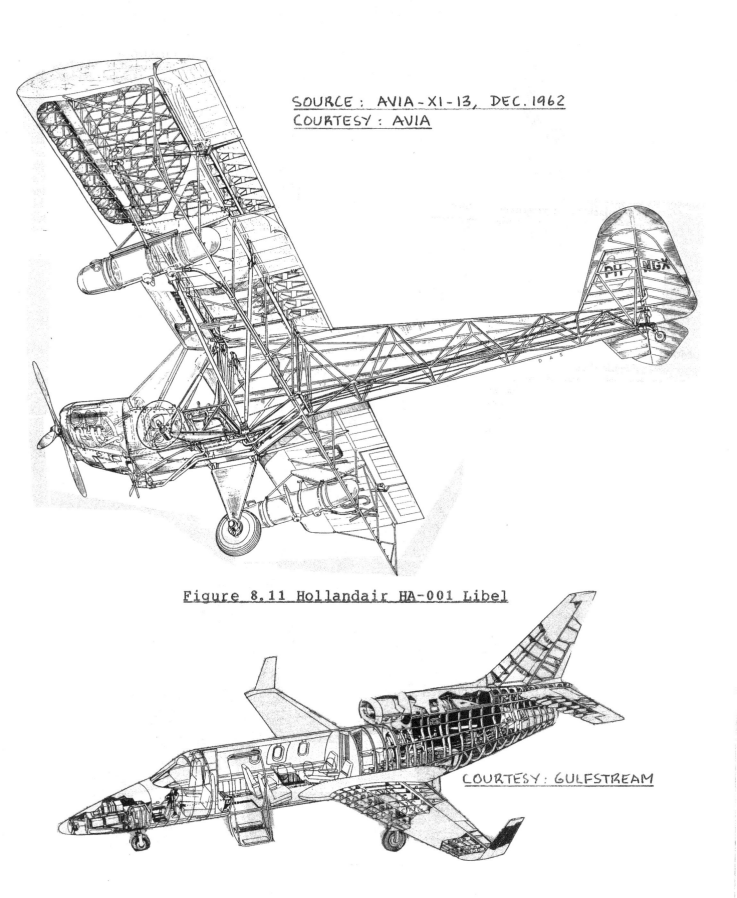

Figure 8.11 Hollandair HA-001 Libel

Figure 8.12 Gulfstream Peregrine

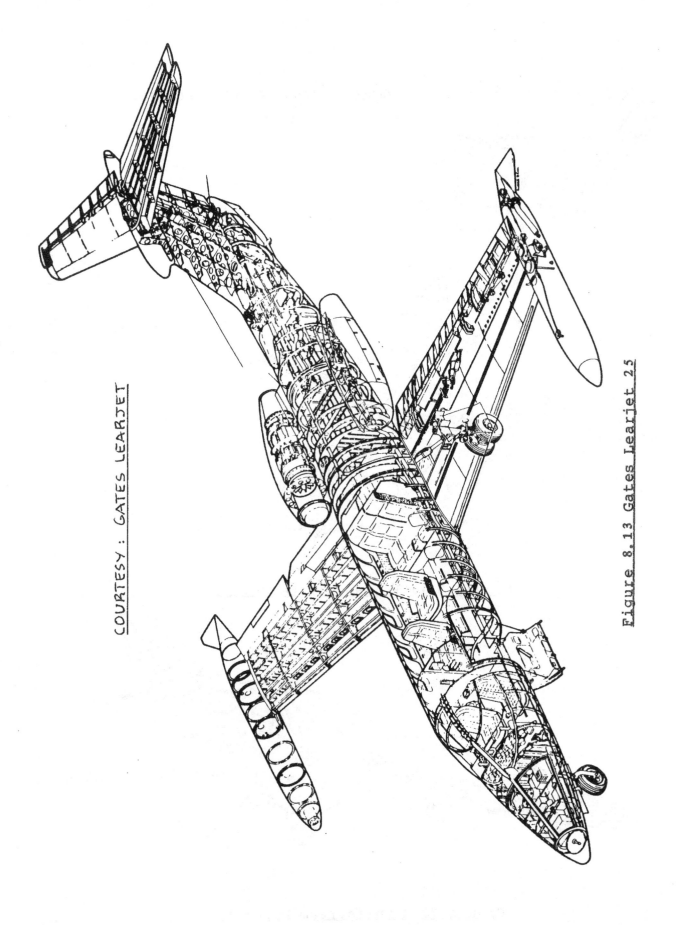

Figure 8.13 Gates Learjet 25

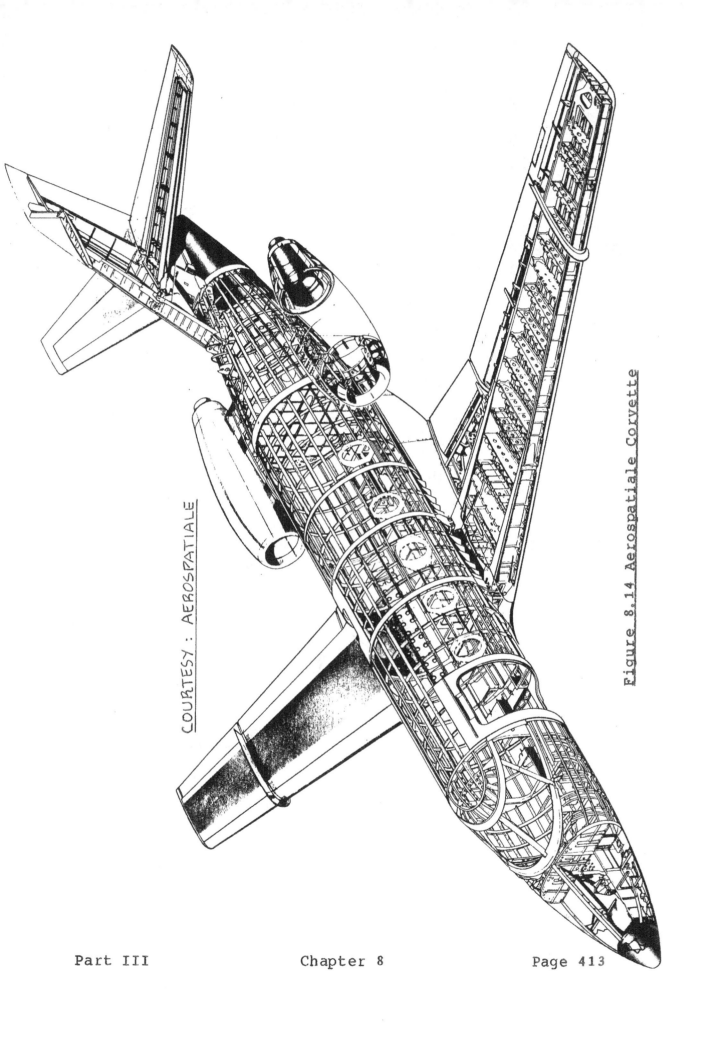

Figure 8.14 Aerospatiale Corvette

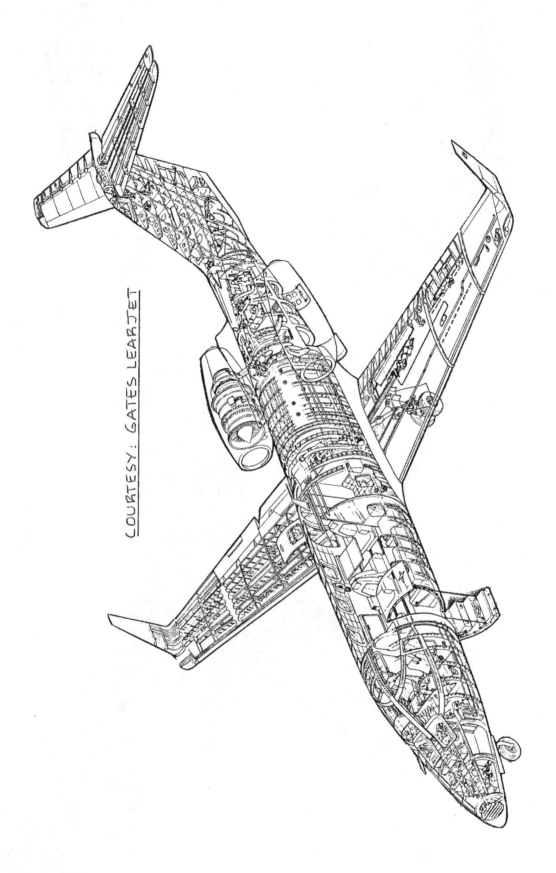

Figure 8.15 Gates Learjet 28

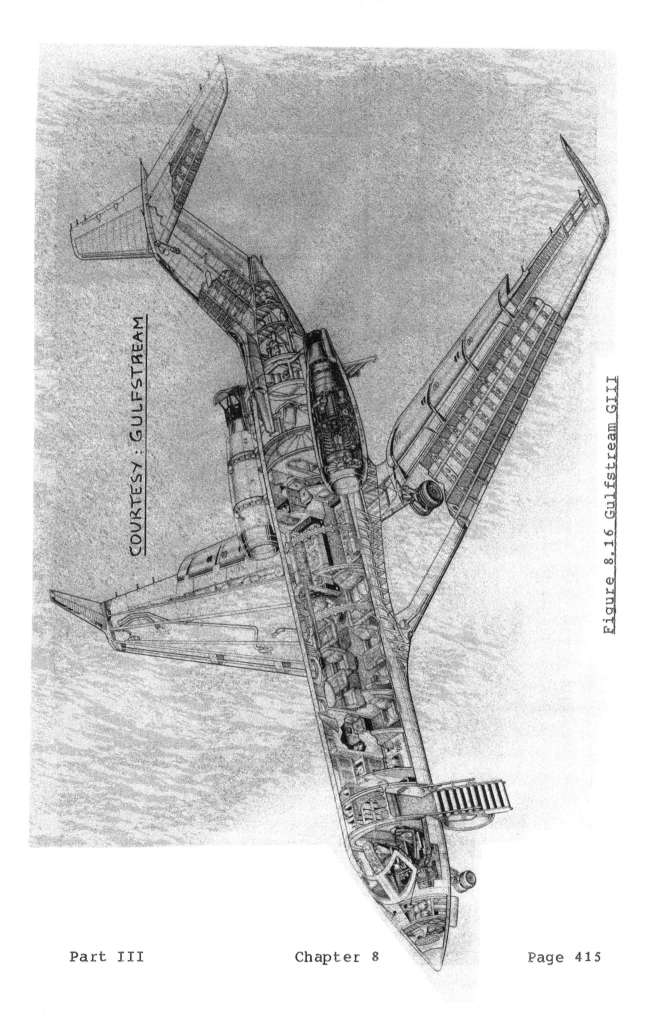

Figure 8.16 Gulfstream GIII

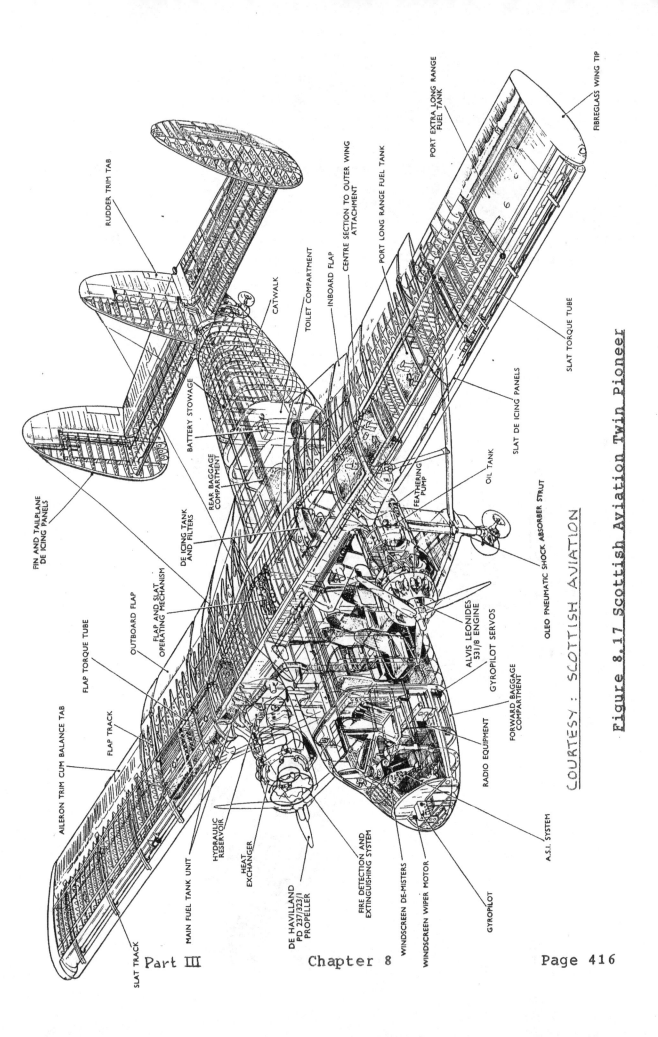

Figure 8.17 Scottish Aviation Twin Pioneer

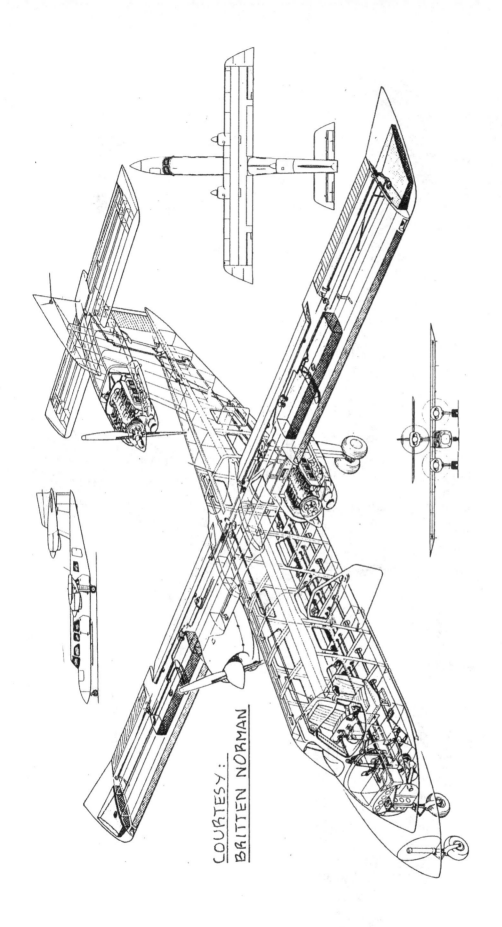

Figure 8.18 Britten Norman Trislander

Figure 8.19 Armstrong Whitworth Apollo

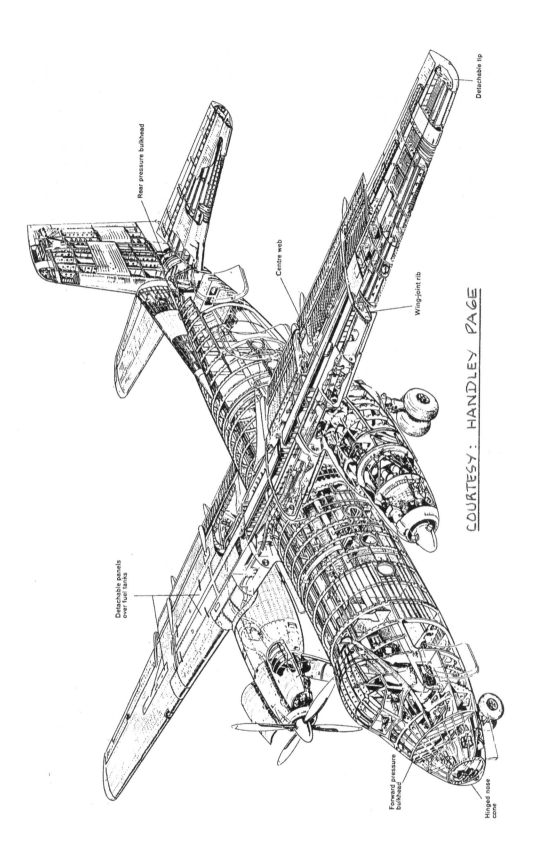

Figure 8.20 Handley Page Dart Herald

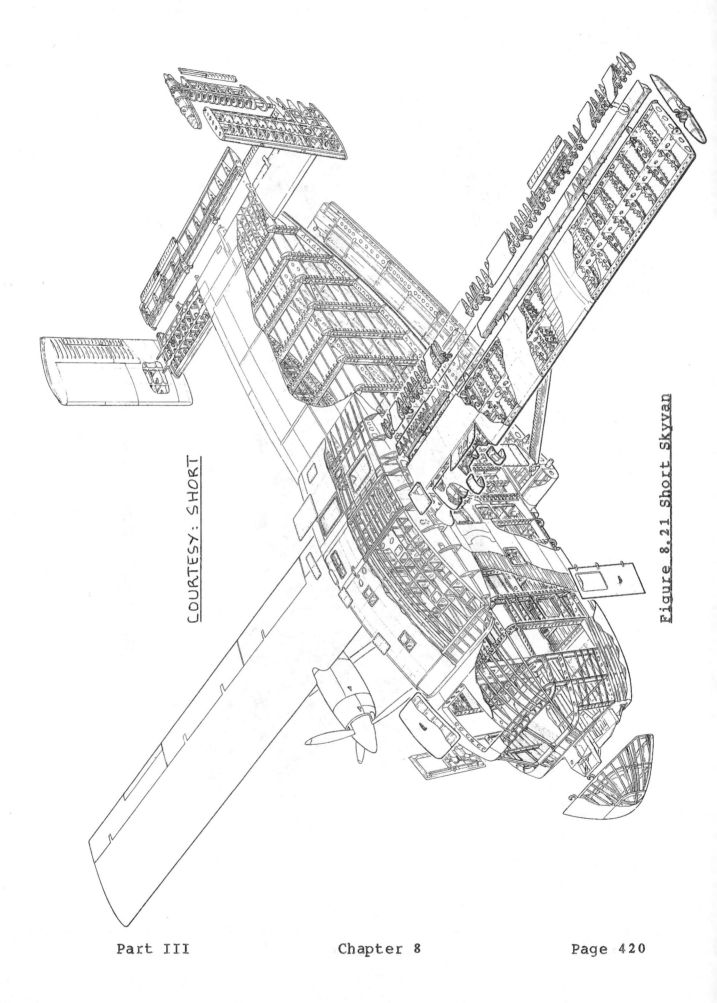

Figure 8.21 Short Skyvan

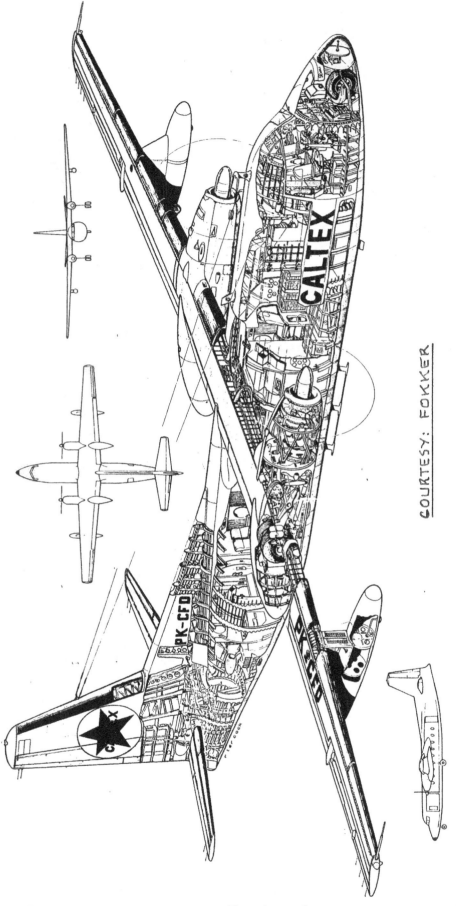

Figure 8.22 Fokker F-27 Friendship

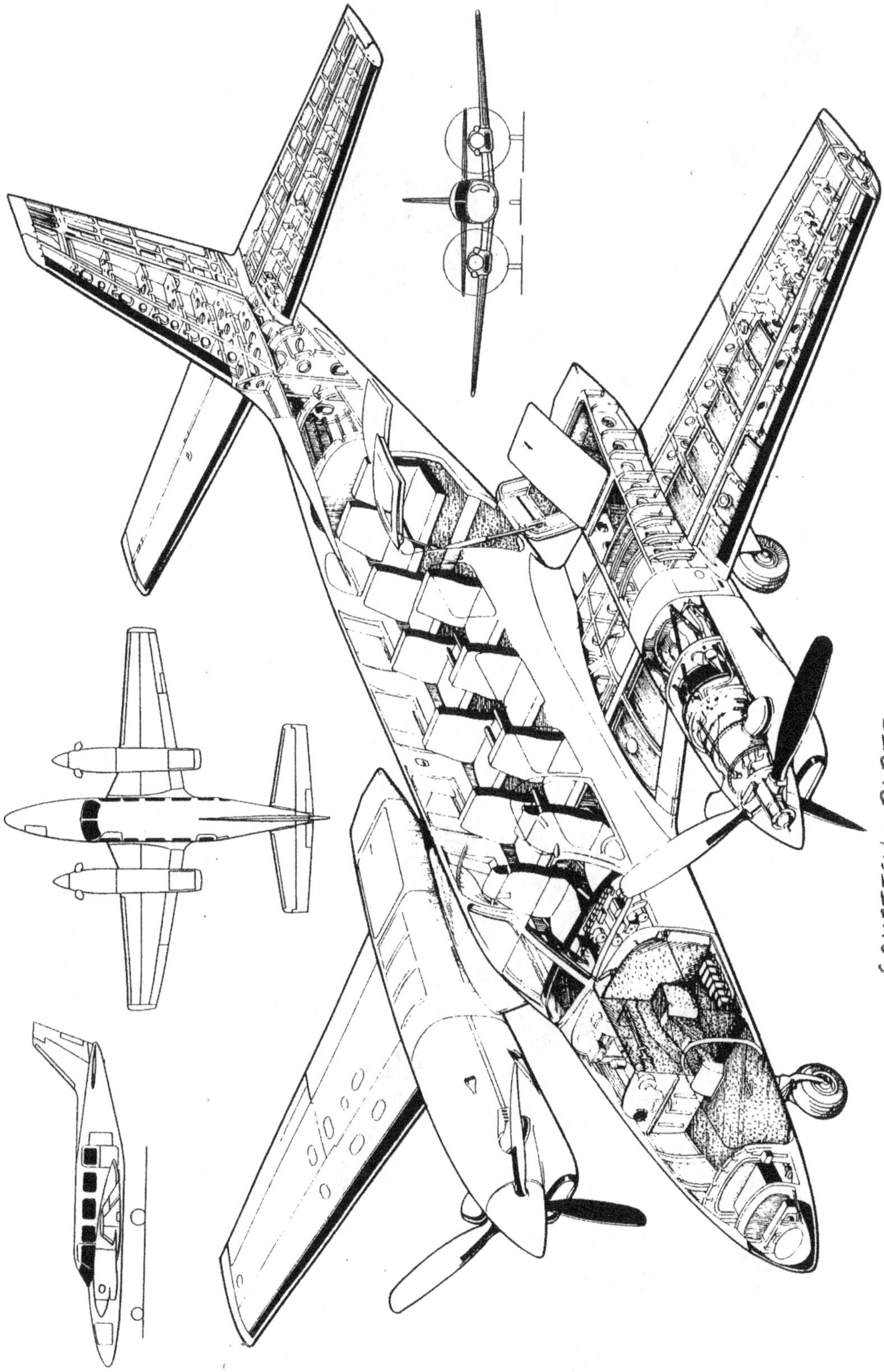

Courtesy: PIPER

Figure 8.23 Piper T-1040

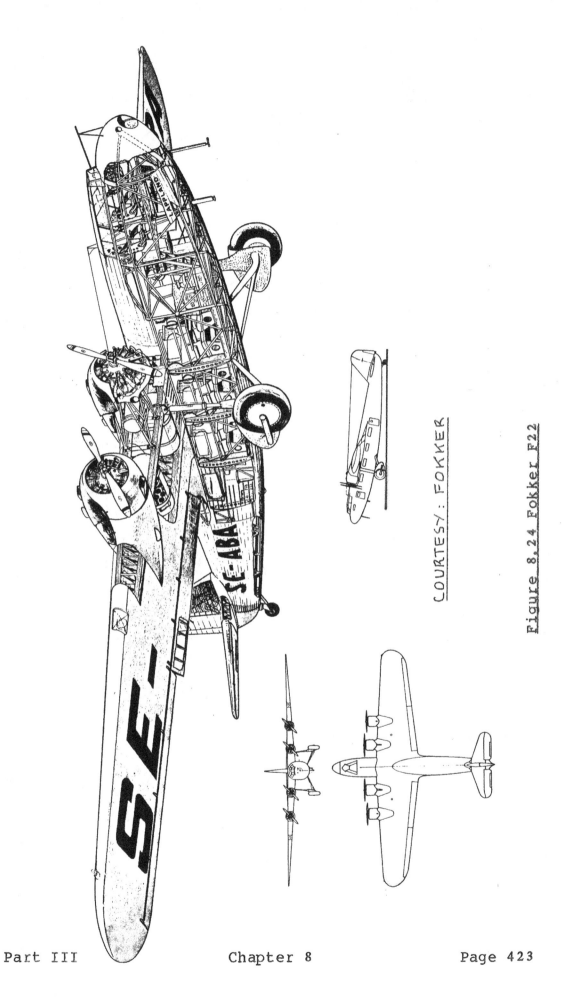

COURTESY: FOKKER

Figure 8.24 Fokker F22

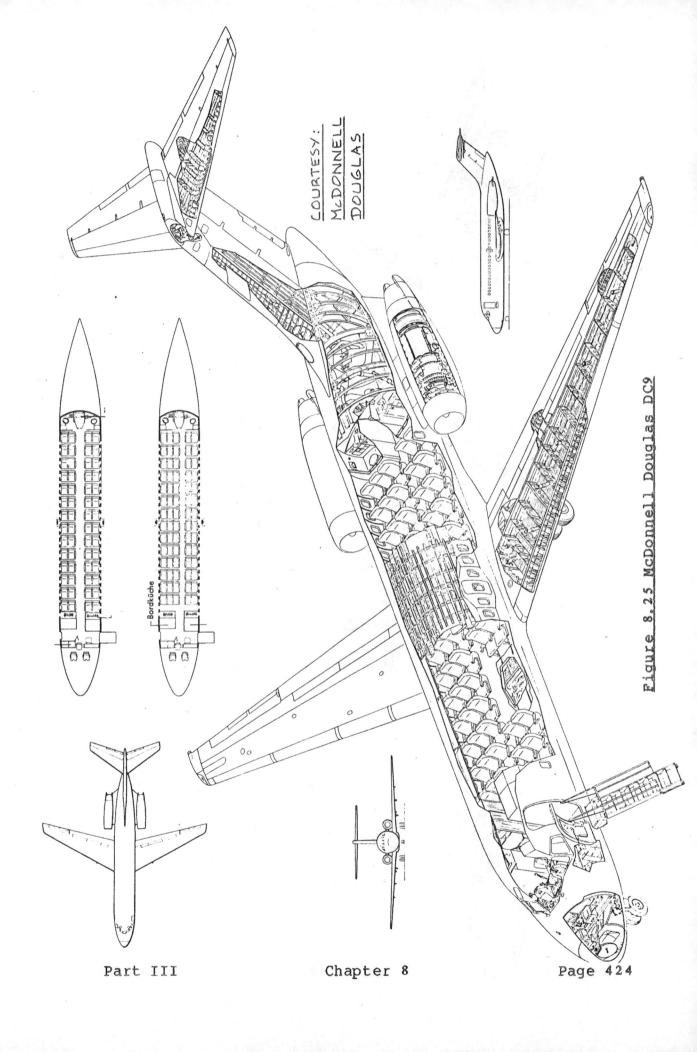

Figure 8.25 McDonnell Douglas DC9

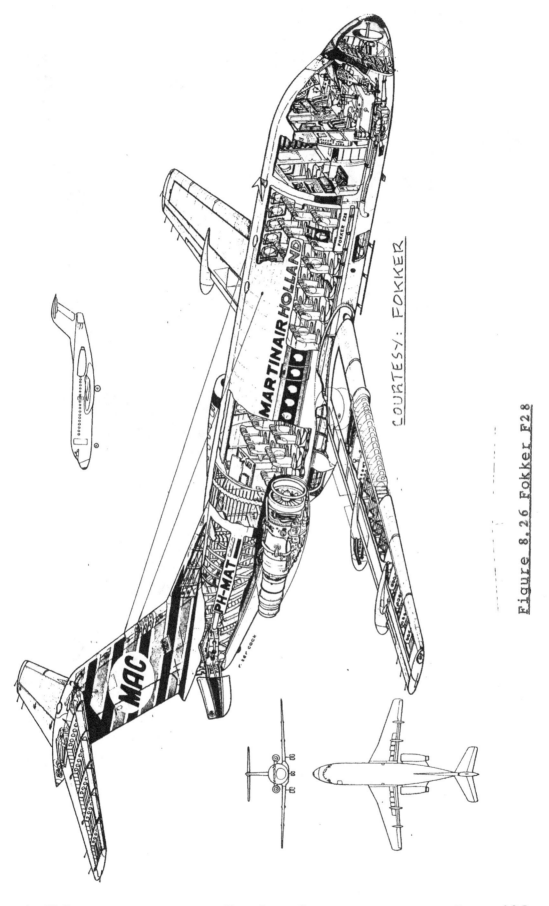

Figure 8.26 Fokker F28

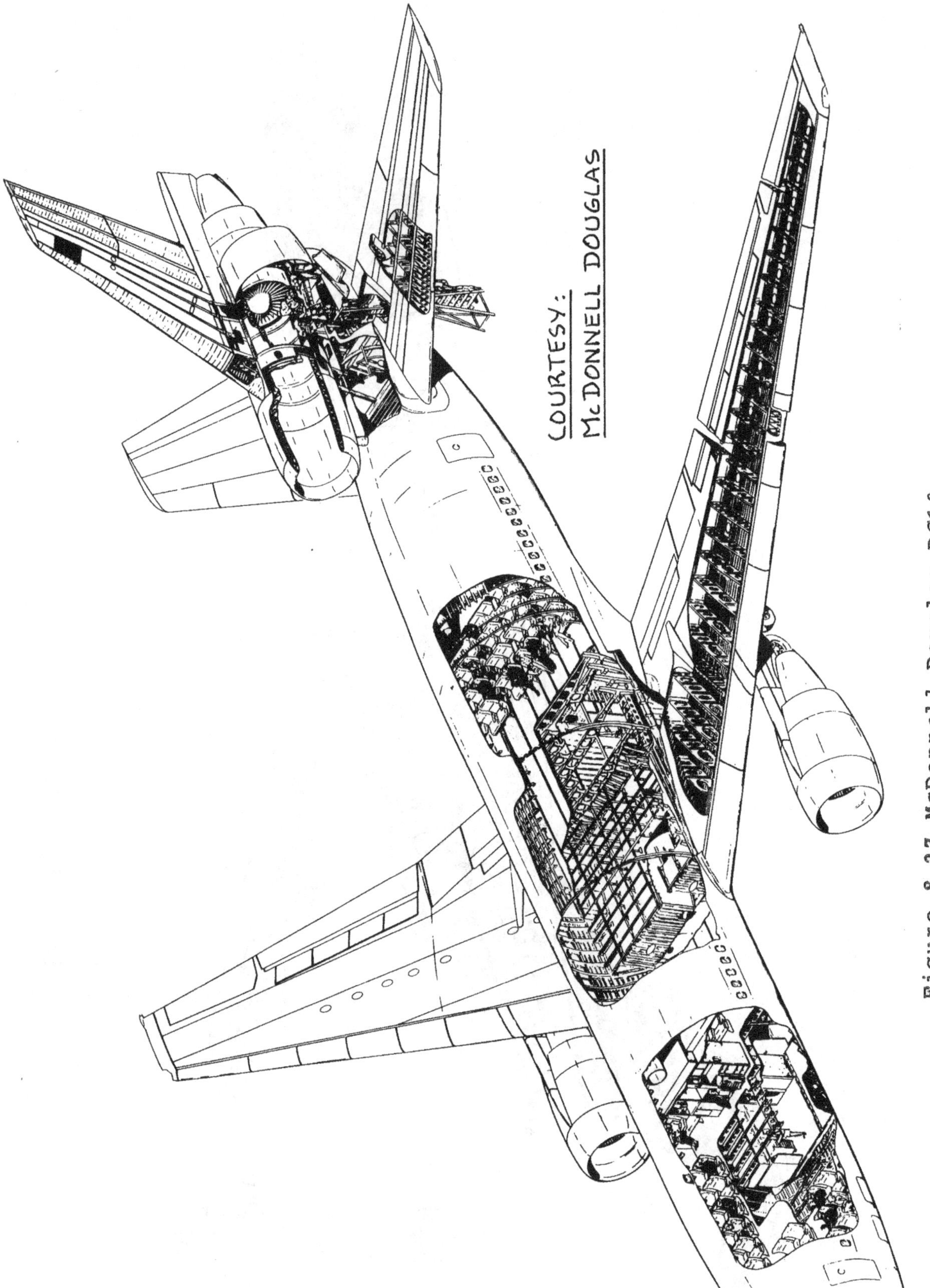

Figure 8.27 McDonnell Douglas DC10

Courtesy: McDonnell Douglas

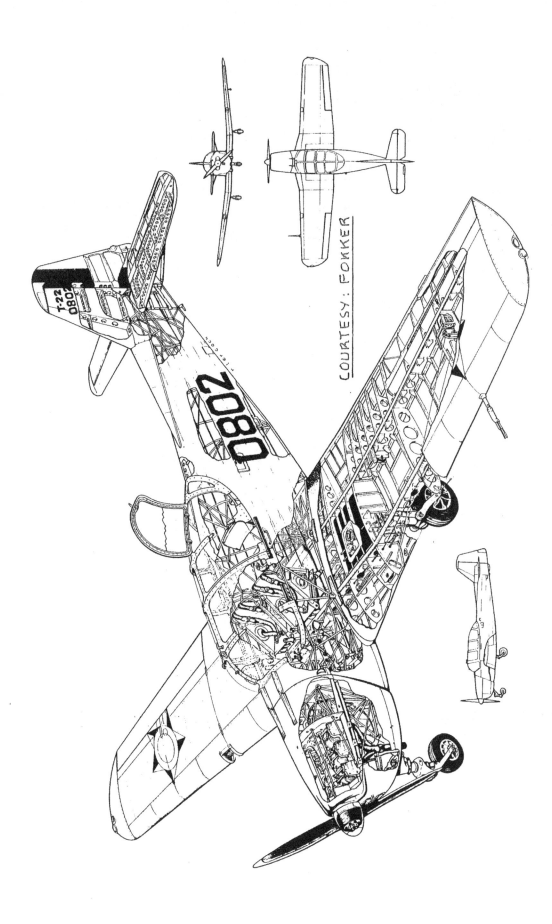

Figure 8.28 Fokker S12

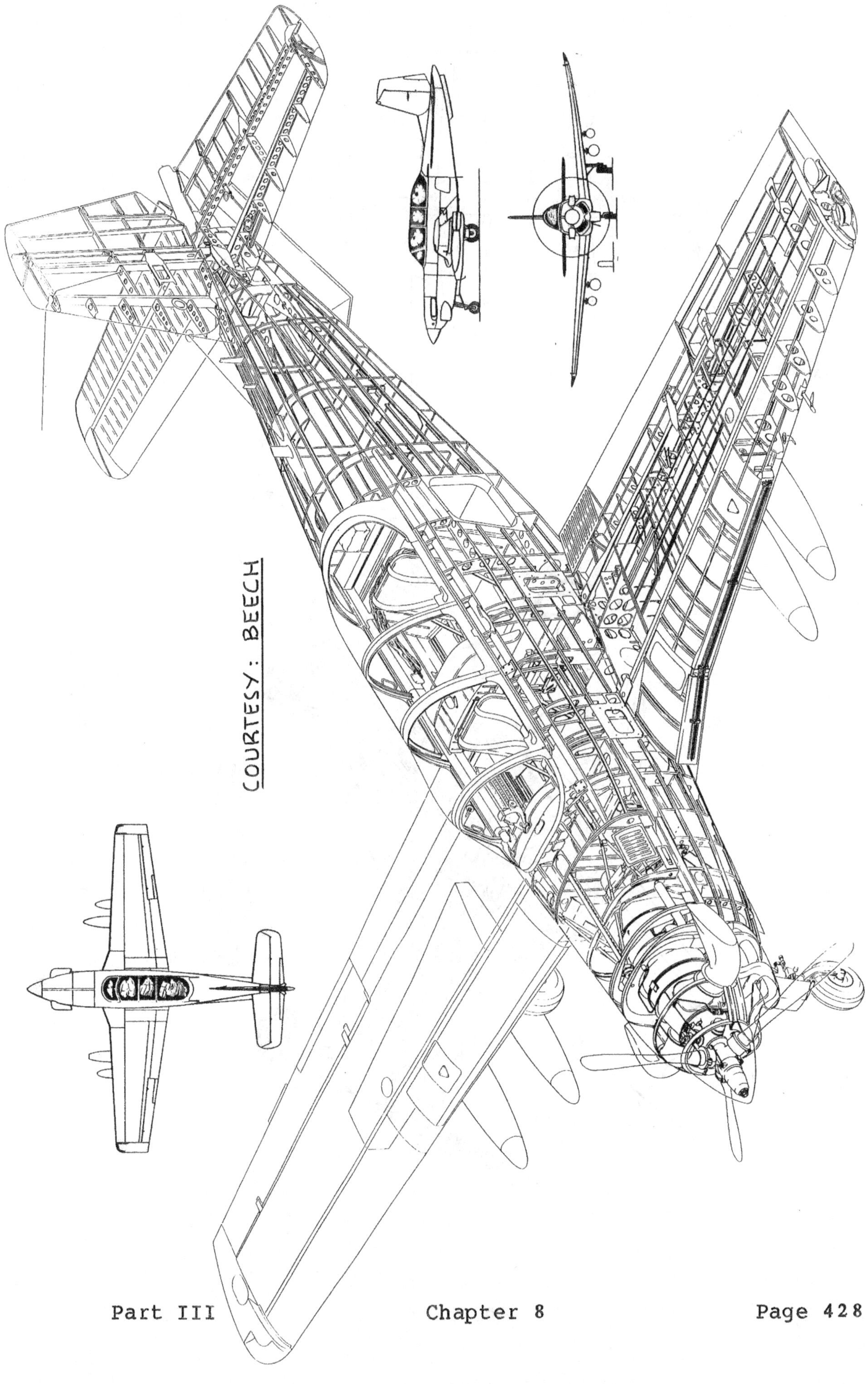

Figure 8.29 Beech T-34C-1

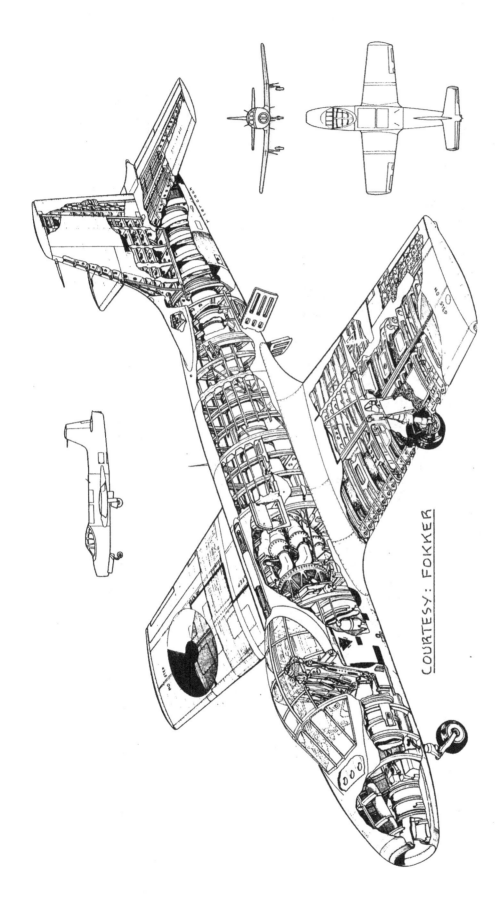

Figure 8.30 Fokker S14

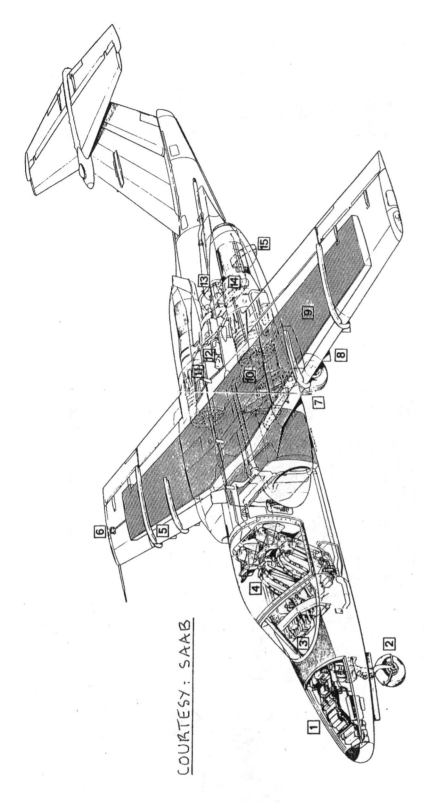

1. Forward equipment compartment housing nav/com equipment.
2. Nose landing gear, steerable, with shimmy damper retracts forward.
3. Instrument panel, with duplicated instruments and controls, under bird-proof windshields and acrylic hood.
4. Pressurized cabin seating two in ejection seats or four in standard seats. Light cargo and other cabin arrangements are also possible.
5. Wing, thin, moderately swept and with negative dihedral. Two-spar, one piece design.
6. Pressure refuelling point. Gravity refuelling also possible.
7. Main landing gear, retracting into fuselage, hydraulically operated, in emergency gravity lowered. Hydraulic anti-skid disk brakes.
8. Six weapon pylons for up to 2000 kg (4400 lb) stores.
9. 2050 litres (540 US gallons) fuel tanks. 920 litres (243 US gallons) in fuselage and 1130 litres (297 US gallons) in wing.
10. Lower equipment compartment.
11. Rear equipment and electronics compartment. 28 V DC and 115/200 V 400 c/s AC electric systems. Engine start on internal or external power.
12. Hydraulic compartment. Hydraulic system driven by both engines, operational on one idling engine.
13. Cabin pressure and air-conditioning system.
14. Two General Electric J85-17B jet engines giving 1293 kg (2850 lb) thrust each.
15. Air brakes, hydraulically operated, continuously variable with position indicated in cockpit. Air brakes permit controlled vertical dive.

Figure 8.31 SAAB XT-105

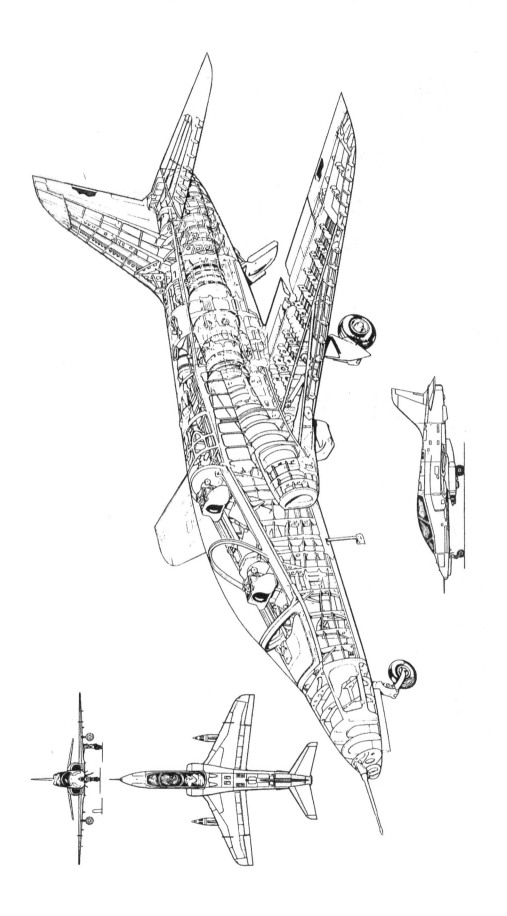

Figure 8.32 British Aerospace Hawk

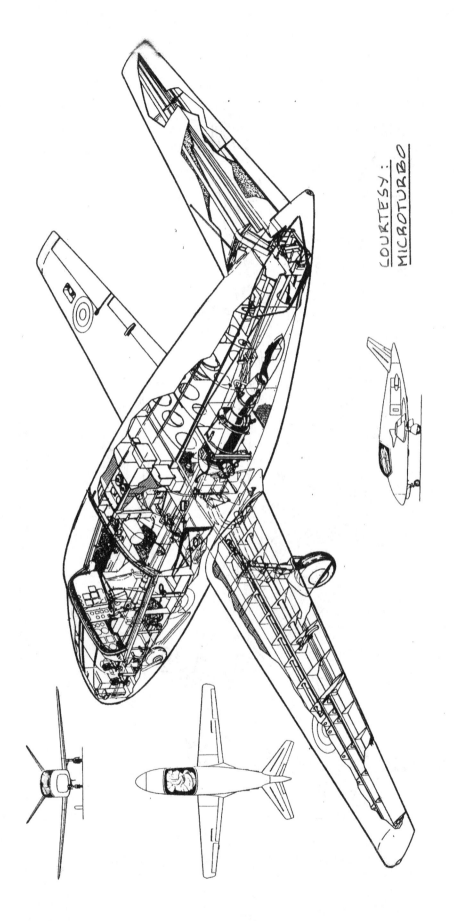

Figure 8.33 Microturbo Microjet 200

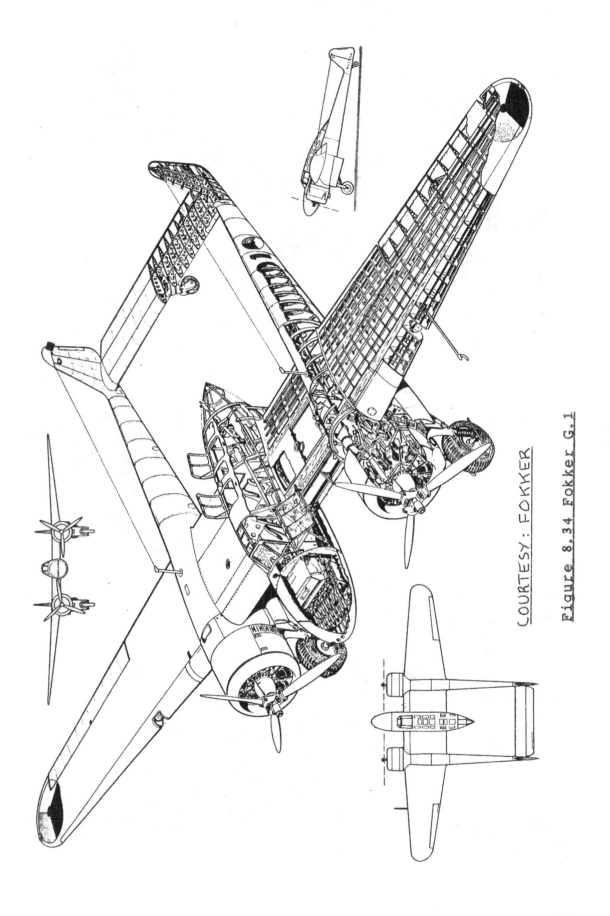

COURTESY: FOKKER

Figure 8.34 Fokker G.1

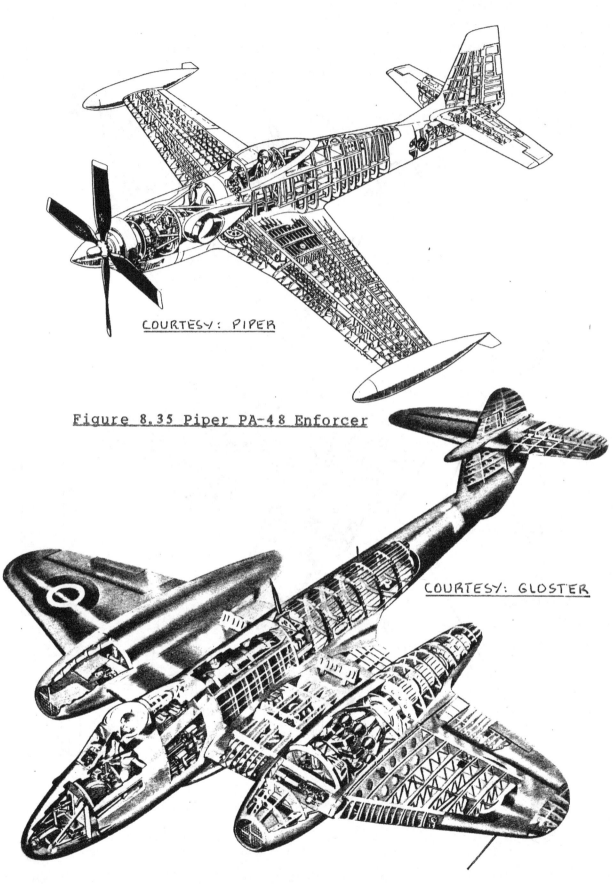

COURTESY: PIPER

Figure 8.35 Piper PA-48 Enforcer

COURTESY: GLOSTER

Figure 8.36 Gloster Meteor

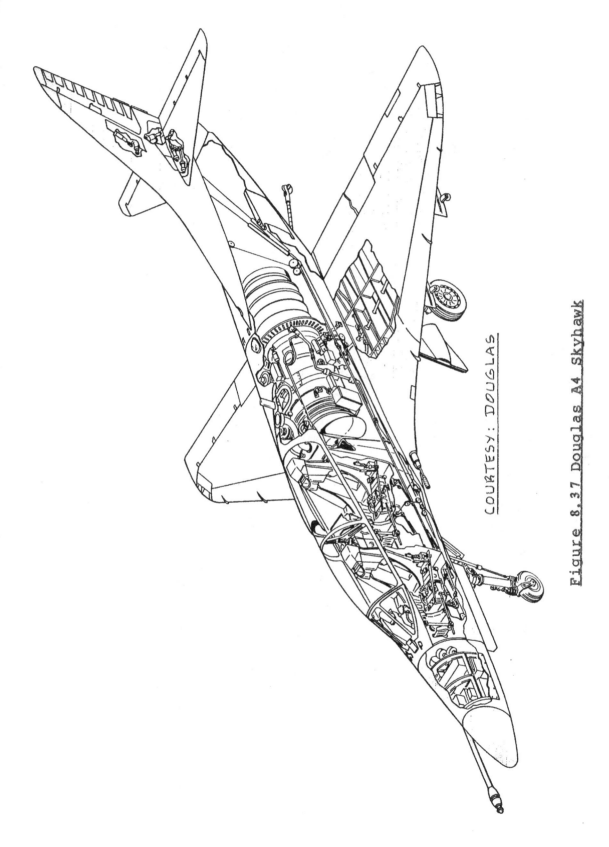

Figure 8.37 Douglas A4 Skyhawk
Courtesy: Douglas

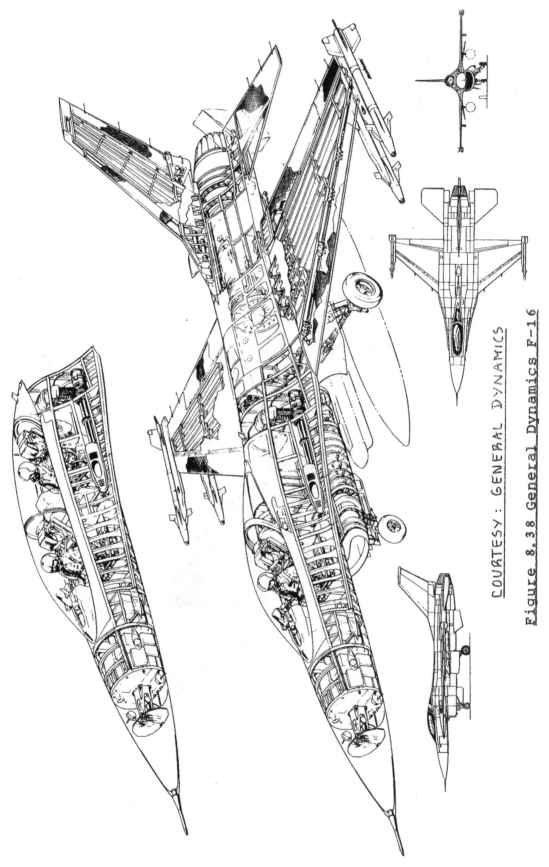

Courtesy: General Dynamics
Figure 8.38 General Dynamics F-16

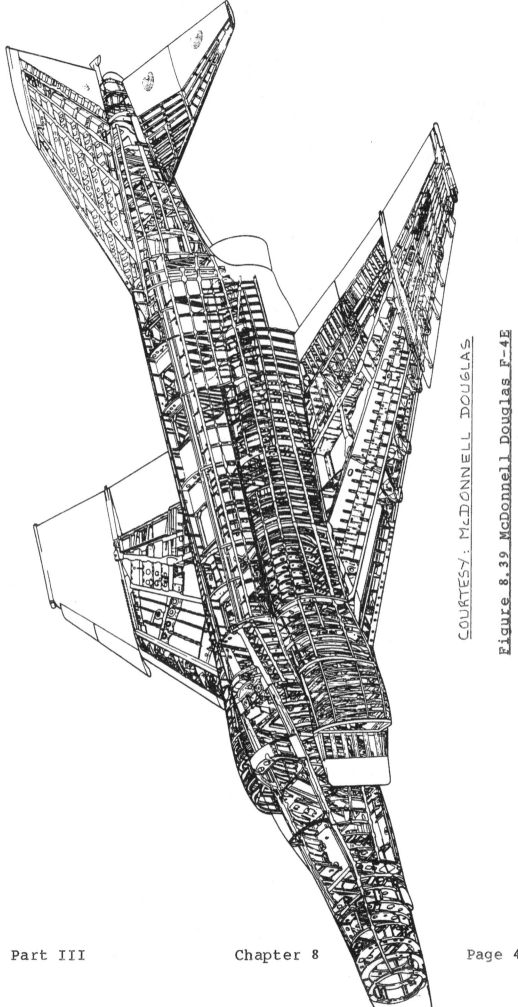

COURTESY: McDONNELL DOUGLAS

Figure 8.39 McDonnell Douglas F-4E

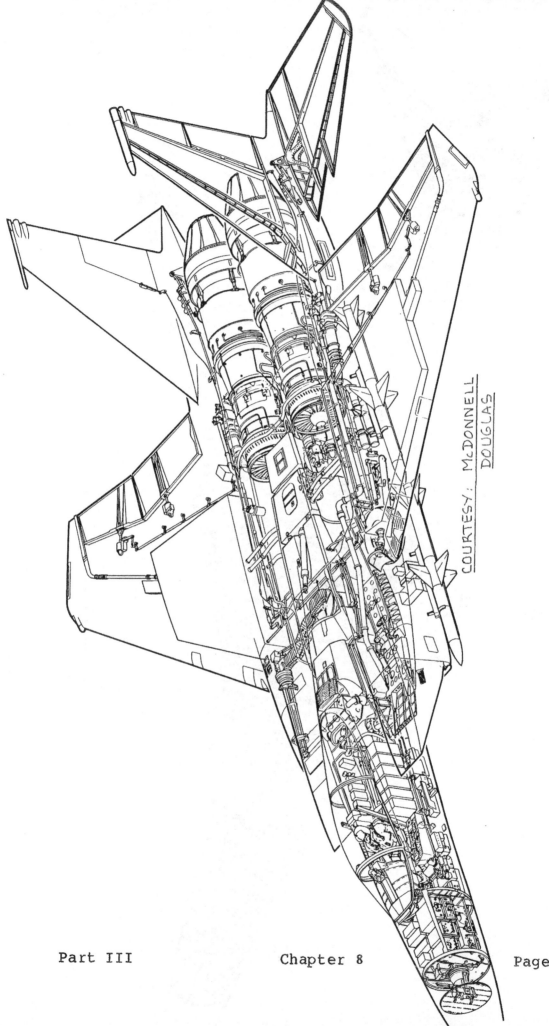

Figure 8.40 McDonnell Douglas F-15

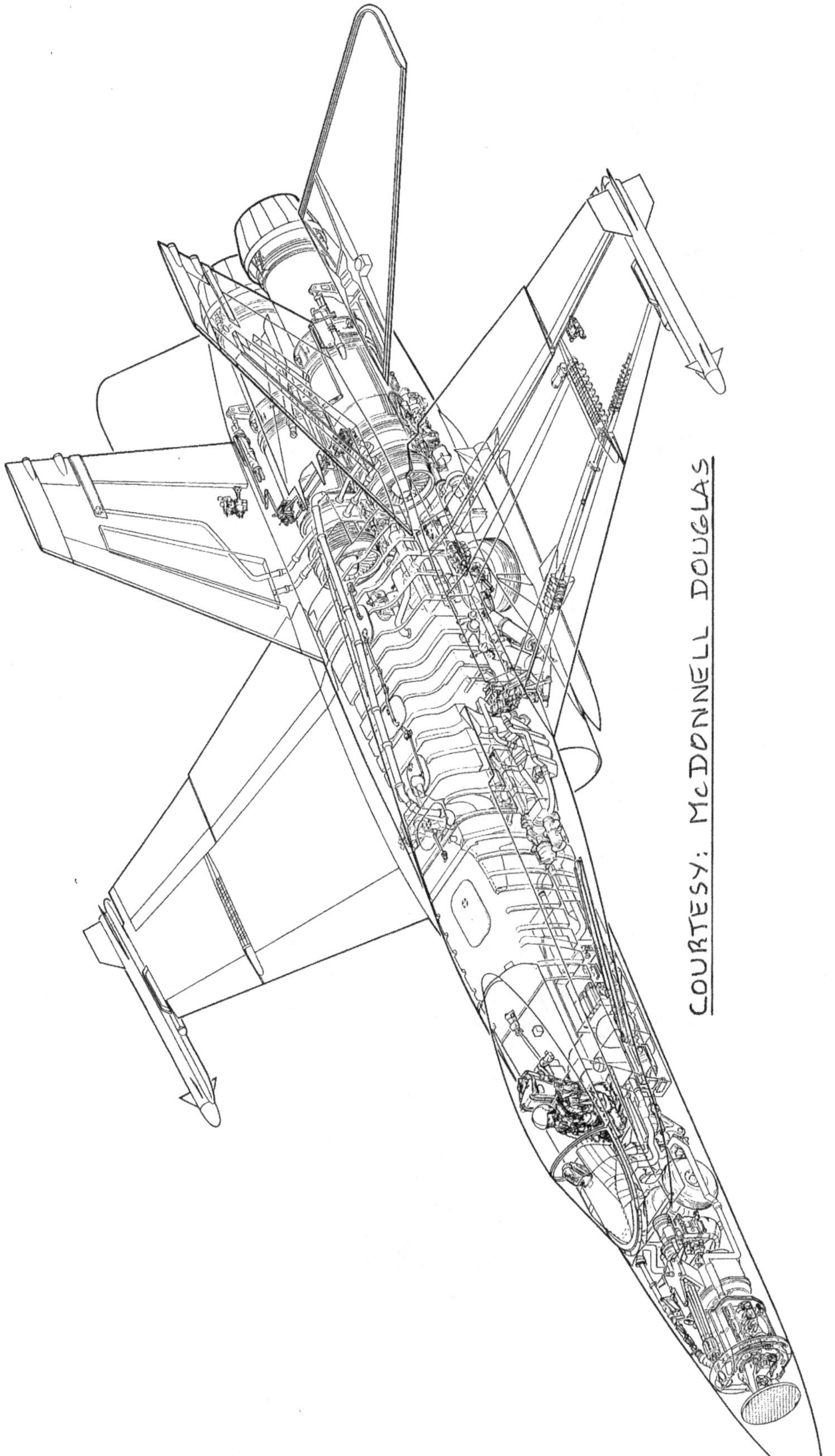

Courtesy: McDonnell Douglas

Figure 8.41 McDonnell Douglas F-18

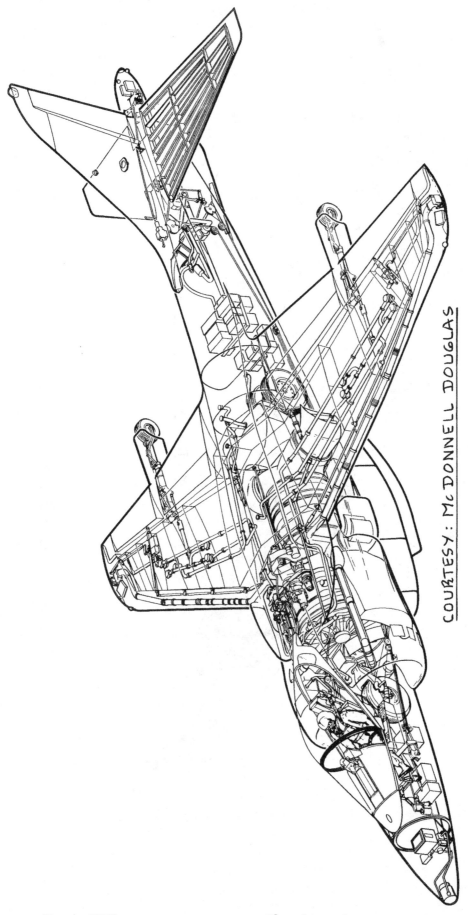

Figure 8.42 McDonnell Douglas AV-8B

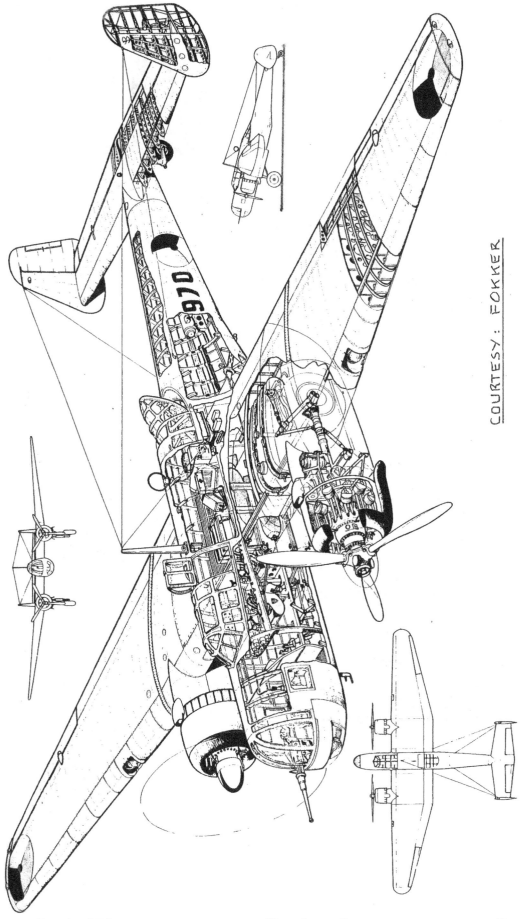

Figure 8.43 Fokker T.IX

COURTESY: FOKKER

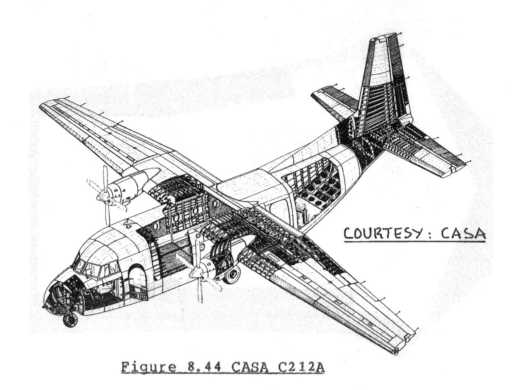

Figure 8.44 CASA C212A

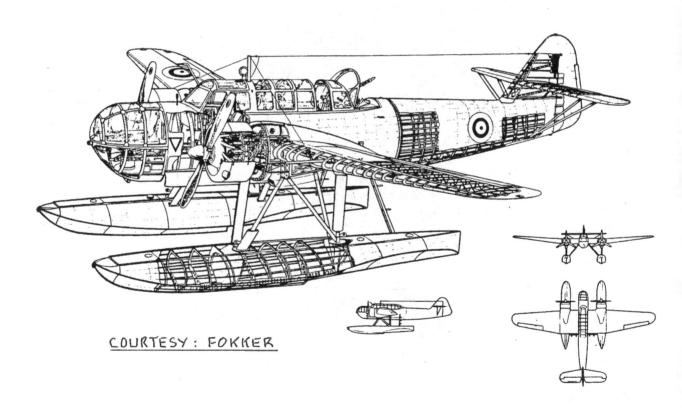

Figure 8.45 Fokker T.VIII-W

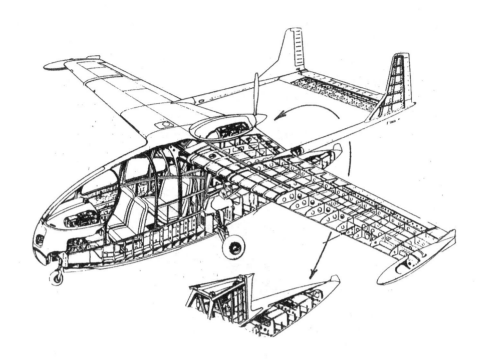

Figure 8.46 Riviera Amphibian

COURTESY: BELL AEROSPACE

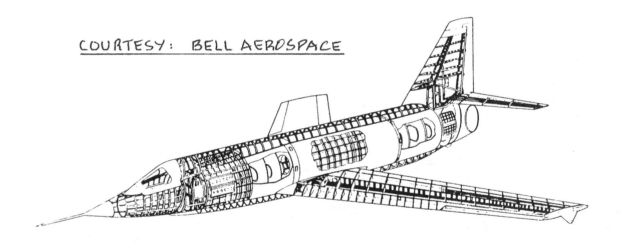

Figure 8.47 Bell X2

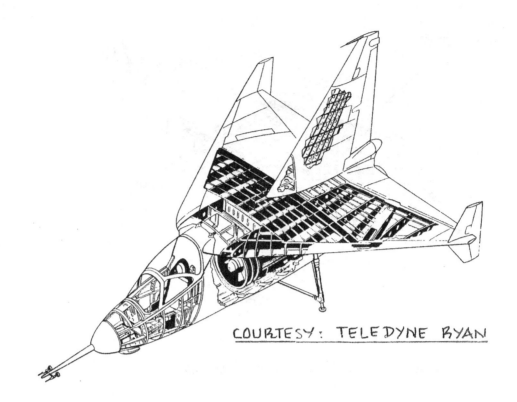

Figure 8.48 Ryan X13

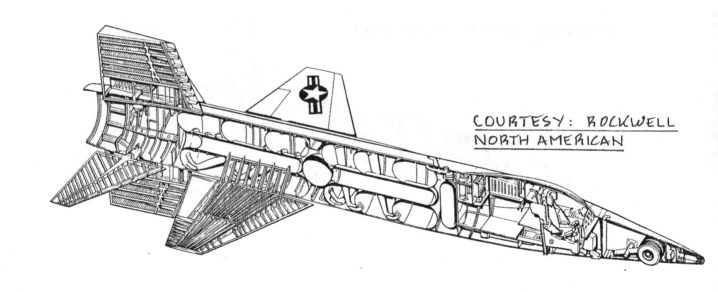

Figure 8.49 North American X15

9. REFERENCES

1. Roskam, J., Airplane Design: Part I, Preliminary Sizing of Airplanes.

2. Roskam, J., Airplane Design: Part II, Preliminary Configuration Design and Integration of the Propulsion System.

3. Roskam, J., Airplane Design: Part IV, Layout Design of Landing Gear and Systems.

4. Roskam, J., Airplane Design: Part V, Component Weight Estimation.

5. Roskam, J., Airplane Design: Part VI, Preliminary Calculation of Aerodynamic, Thrust and Power Characteristics.

6. Roskam, J., Airplane Design: Part VII, Determination of Stability, Control and Performance Characteristics: FAR and Military Requirements.

7. Roskam, J., Airplane Design: Part VIII, Airplane Cost Estimation and Optimization: Design, Development Manufacturing and Operating.

Note: These books are all published by: Design, Analysis and Research Corporation, 1440 Wakarusa Drive, Suite 500, Lawrence, KS, 66049. Tel. (785) 832-0434

8. Human Factors, Flight Safety Foundation, 468 Park Ave. South, New York, N.Y., 10016.

9. Damon, A., Stoudt, H.W. and McFarland, R.A., The Human Body in Equipment Design, Harvard University Press, Cambridge, Mass., 1966.

10. Anon., AFSC Design Handbook, Series DH2-2, Crew Station and Passenger Accomodations, AFWAL, WPAFB, Ohio, 45433.

11. Anon., Federal Aviation Regulations, Part 23 and Part 25, Department of Transportation, Federal Aviation Administration, Distribution Requirements Section, M-482.2, Washington, D.C., 20590.

12. Torenbeek, E., Synthesis of Subsonic Airplane Design, Kluwer Boston Inc., Hingham, Maine, 1982.

13. Küchemann, D., The Aerodynamic Design of Aircraft, Pergamon Press Ltd, Oxford, England, 1978.

14. Stinton, D., The Design of the Aeroplane, Granada, London, England, 1983.

15. Hoerner, S.F., Fluid Dynamic Drag, Hoerner Fluid Dynamics, P.O. Box 342, Brick Town, N.J., 08723, 1965.

16. Roskam, J., Proceedings of the NASA/Industry/University General Aviation Drag Reduction Workshop, Space Technology Center, University of Kansas, Lawrence, Kansas, 66045, 1975.

17. Thurston, D.B., Design for Flying, McGraw Hill Book Co., N.Y., 1978.

18. Olson, R.E. and Land, N.S., Effect of Afterbody Length and Keel Angle on Minimum Depth of Step for Landing Stability and on Take-off Stability of a Flying Boat, NACA TR 923, 1949.

19. Anon., MIL-A-8861(ASG), Military Specification, Airplane Strength and Rigidity, Flight Loads, May, 1960.

20. Bruhn, E.F., Analysis and Design of Flight Vehicle Structures, Tri-State Offset Co., Cincinnati, Ohio, 45202, 1965.

21. Anon., Fatigue Design Handbook, Society of Automotive Engineers, 400 Commonwealth Drive, Warrendale, Pennsylvania, 15096, 1968.

22. Smith, H.W., Airplane Designer's Checklist for Occupant Injury Prevention, AIAA Paper 84-2520, 1984.

23. Anon., British Civil Airworthiness Regulations, Civil Aviation Authority, United Kingdom.

24. Anon., Energy Absorbing Structure Improves Aircraft Safety, Automotive Engineering, Volume 89, Number 12, 1981, Society of Automotive Engineers.

25. Carden, H.D., Impulse Analysis of Airplane Crash Data with Consideration Given to Human Tolerance, SAE Paper 830748, 1983.

26. Snyder, R.G., Impact Protection in Air Transport Passenger Seat Design, SAE Paper 821391, 1982.

27. Saczalski, K., Singley, G.T., Pilkey, W.D. and Huston, R.L., Aircraft Crashworthiness, University Press of Virginia, Charlottesville, 1975.

28. Andrews, D.G. et al, Application of Advanced Technologies to Small Short Haul Aircraft, NASA CR 152089, March 1, 1978.

29. Nicolai, L.M., Fundamentals of Aircraft Design, METS, Inc., 6520 Kingsland Court, San Jose, CA, 95120.

30. Taylor, J.W.R., Jane's All The World Aircraft, Published annually by: Jane's Publishing Company, 238 City Road, London, EC1V 2PU, England.

31. Weisshaar, T., Aeroelastic Stability and Performance Characteristics of Aircraft with Advanced Composite Swept Forward Wing Structures, AFFDL-TR-78-116, 1978.

32. Cook, E., On the Feasibility of Aluminum Forward Swept Wings, Bristol Forward Swept Wing Conference, Bristol, England, 1985.

33. Loftin, L.K., Jr., Quest for Performance, NASA SP-468, 1985.

34. Von Mises, R., Theory of Flight, Dover Publishing Co, N.Y., N.Y., 1960.

35. Wolkovitch, J., Principles of the Joined Wing, Engel Engineering Report No. 80-1, 28603 Trailriders Drive, Rancho Pales Verdes, CA, 90274.

36. Schlichting, H. and Truckenbrodt, E., Aerodynamics of the Airplane, McGraw-Hill International Book Company, N.Y., N.Y., 1979.

37. Roskam, J., Airplane Flight Dynamics and Automatic Flight Controls, Part I, Roskam Aviation and Engineering Corp., Rt 4, Box 274, Ottawa, Ks, 66067, 1984

38. Hsu, C.H. and Lan, C.E., Theory of Wing Rock, Journal of Aircaft, Vol.22, No.10, pp 920-924, October 1985.

39. Whitcomb, R.T., A Study of the Zero-Lift Drag-Rise Characteristics of Wing-Body Combinations Near the Speed of Sound, NACA TR 1273, 1956.

40. Abbott, I.H. and Von Doenhoff, A.E., Theory of Wing Sections, Dover Publications, N.Y., N.Y., 1959.

41. Hoerner, S. and Borst, H.V., Fluid Dynamic Lift, Hoerner Fluid Dynamics, P.O. Box 342, Brick Town, N.J., 08723

42. Holmes, B.J., Obara, C.J. and Yip, L.P., Natural Laminar Flow Experiments on Modern Airplanes, NASA TP 2256, 1984.

43. Gratzer, B., Results of Natural Laminar Flow Flight Tests on a Boeing 757, Oral Presentation at the AIAA Aircraft Design and Operations Meeting, Oct., 1985.

44. Williams, K.L., Vijgen, P. and Roskam, J., Natural Laminar Flow and Regional Aircraft, SAE Paper 850864, General Aviation Aircraft Meeting and Exposition, April 16-19, 1985, Wichita, Ks.

45. Holmes, B.J. et al, Manufacturing Tolerances for Natural Laminar Flow Airframe Surfaces, SAE Paper 850863, as ref. 44.

46. Croom, C.C. and Holmes, B.J., Flight Evaluation of an Insect Contamination Protection System for Laminar Flow Wings, SAE Paper 850860, as ref. 44.

47. Warwick, G., Jetstar Smooths the Way, Flight International, September 1985, pp 32-34.

48. Kendall, E.R., The Minimum Induced Drag, Longitudinal Trim and Static Longitudinal Stability of Two-Surface and Three-Surface Airplanes, AIAA Paper 84-2164, 2nd Applied Aerodynamics Conference, August, 1984.

49. Rokhsaz, K. and Selberg, B.P., Analytical Study of Three-Surface Lifting Systems, SAE Paper 850866, as ref. 44.

50. Steiner, J.E. et al, Case Study in Aircraft Design, The Boeing 727, Professional Study Series, AIAA, September 14, 1978.

51. Anon., Military Specification for Flying Qualities of Piloted Airplanes, MIL-F-8785, B and C, 1984.

52. Pelikan, R.J., Evaluation of Aircraft Departure Divergence Critaria with a Six-Degree-of-Freedom Digital Simulation Program, AIAA Paper 74-68, 1974.

53. Chambers, J.R., Overview of Stall/Spin Technology, AIAA Paper 80-1580, 1980.

54. Shortal, J.A. and Maggin, B., Effect of Sweepback and Aspect Ratio on Longitudinal Stability Characteristics of Wings at Low Speeds, NACA TN 1093, '46.

55. Larrabee, E.E., Five Years Experience with Minimum Loss Propellers-Part I: Theory, SAE Paper 840026, International Congress and Exposition, 1984.

56. Larrabee, E.E., Five Years Experience with Minimum Loss Propellers-Part II: Applications, SAE Paper 840027, as Ref.55.

57. Lan, C.E. and Roskam, J., Airplane Aerodynamics and Performance, Roskam Av. and Engrg Corp. as ref.7.

58. McKinley, J.L. and Bent, R.D., Powerplants for Aerospace Vehicles, McGraw Hill Book Co., NY, NY, 1965.

59. Taylor, C.F., The Internal Combustion Engine in Theory and Practice, Volumes 1 and 2, MIT Press, 1966.

60. Kerrebrock, J.L., Aircraft Engines and Gas Turbines, MIT Press, 1977.

61. Treager, I.E., Aircraft Gas Turbine Engine Technology, Mc Graw Hill Book Co., NY, NY, 1970.

62. Godston, J. and Reynolds, C.N., Prop-fan Powered Aircraft, Journal of Aircraft, December, 1985.

63. Whitlow, J.B. and Sievers, G.K., Fuel Savings Potential of the NASA Advanced Turboprop Program, NASA TM 83736, September, 1984.

64. Keiter, I.D., Impact of Advanced Technology on Aircraft/Mission Characteristics of Several General Aviation Aircraft, SAE Paper 810584, 1981.

65. Covert, E.E. et al, Thrust and Drag: Its Prediction and Verification, Volume 98, Progress in Aeronautics and Astronautics, AIAA, NY, NY, 1985.

66. Anon., MIL-STD-1474(MI), Military Standard, Noise Limits for Army Materiel, March 1973.

67. Wilby, J.F., The Prediction of Interior Noise of Propeller Driven Aircraft: A Review, SAE Paper 830737, 1983.

68. Klatte, R.J., General Aviation Propeller Noise Reduction: Penalties and Potential, SAE Ppr 810585, 1981.

69. Waterman, E.H., Kaptein, D. and Sarin, S.L., Fokker's Activities in Cabin Noise Control for Propeller Aircraft, SAE Paper 830736, 1983.

70. Smith, M.H., A Prediction Procedure for Propeller Aircraft Flyover Noise Based on Empirical Data, SAE Paper 810604, 1981.

71. Anon., Prediction Procedure for Near-Field and Far-Field Propeller Noise, AIR 1407, SAE Aerospace Information Report, May 1977.

72. Feiler, C.E. and Conrad, W., Noise from Turbomachinery, AIAA Paper 73-815, AIAA 5th Aircraft Design and Operations Meeting, August 1973.

73. Drell, H., Impact of Noise on Subsonic Transport Design, SAE Paper 700806.

74. Carlson, H.W., Simplified Sonic Boom Prediction, NASA TP-1122, 1978.

75. Roskam, J. et al, Proceedings of the NASA/Industry/University General Aviation Drag Reduction Workshop, Space Technology Center, The University of Kansas, Lawrence, Kansas, 1975.

76. Bingelis, T., Firewall Forward, Engine Installation Methods, Tony Bingelis, 8509 Greenflint Lane, Austin, Texas, 78759, 1983.

77. Anon., Lord Kinematics Catalog, Technical Reference LB-571, Lord Kinematics, 1635 West 12th Street, Erie, Pennsylvania, 16512, 1974.

78. Stinton, D., The Anatomy of the Aeroplane, Granada Publishing, England, 1980.

79. Warwick, G., Aluminum Fights Back, Flight International, March 2, 1985.

80. Brahney, J., Polyamide-imide Resins Examined, Aerospace Engineering, December, 1985.

81. Langston, P.R., Design and use of Kevlar in Aircraft Structures, SAE Paper 850893, General Aviation Aircraft Meeting and Exposition, Wichita, April, 1985.

82. Heinemann, E., Rausa, R. and Van Every, K., Aircraft Design, The Nautical and Aviation Publ. Co., 1985.

10. INDEX

Access diagram	84
Accessibility	82
Aerodynamic coupling	207,206
Aisle width	52
Area rule	204,41,39
Aspect ratio	185
Baggage hold	76
Baggage density	76
Base drag	38,36
Bi-plane	184
Bluntness	40
Braced wing	184
Cabin design	107-91,85
Cabin window	147,146,143
Canard configurations	268
Cargo arrangement	80
Cargo container	79,77,76
Cargo hold	122-108,77,76,53
Cargo pallet	79,76
Cargo volume	76
Cockpit accessibility	7
Cockpit, center-stick controlled	18,16,14
Cockpit, civil layouts	16,15,14,12
Cockpit controls	12
Cockpit, ejection seat	22,21,20
Cockpit layout design	3
Cockpit layout examples	34,33,32,31,30,29
Cockpit, military layouts	19,18,13
Cockpit seating	12
Cockpit, side-stick control	17
Cockpit visibility	23,3
Cockpit, wheel controlled	19
Cockpit window	143
Compressibility drag	273,216,175,36
Container, see cargo	
Crew member dimensions	8,6,4
Crew member, scaled views	11,10,9
Cruise performance	167
Cutaway drawings	444-399
Deep stall	269
Directional control mechanizations	287
Door(s)	85,68
Door design	139-137
Double-bubble cross section	78
Drainage	151

Ejection seats	22,21,20
Emergency exits	71,70,69,68
Empennage configuration	250
Empennage cross section design	287
Empennage layout design	272,249
Empennage structural design	275
Empennage structural arrangements	287-279,277
Engine data	327-303,291
Engine disposition (lateral)	337
Engine installations	375-357,356
Engine integration	335
Engine mounts	335
Engine shutdown rates	349
Escape chute deployment	72
Extension shafts	337
Eye vectors	25
Fire wall	344
Flap (high lift) mechanizations	232
Flaps	206
Flightdeck	3
Float	42
Floor design	152,151,150,149,148
Flutter	339
Flying boat hull	44,43,42
Foreign object damage	346
Frame depth	124
Frame spacing	385,124
Freight floor	77
Friction drag	273,214,36
Fuselage, aerodynamic design	36
Fuselage bluntness	40,39
Fuselage bulge	39
Fuselage cross section	90,89,88,87,86,85,45,41
Fuselage fineness ratio	37,36
Fuselage interior layout	45
Fuselage layout design	122-86,85,35
Fuselage layout design procedure	35
Fuselage shell and skin layout	136-132
Fuselage structural arrangement	131-126,125
Fuselage upsweep	40,39
Galley dimensions	75,73
Galley layout	75,74
Inboard profiles	162-153
Induced drag	216 40,36
Inspection considerations	232,82
Interference drag	273,217

Joined wing	184
Lateral controls	262,261,213,211,208
Lateral control mechanizations	232
Longitudinal control mechanizations	287
Lavatory dimensions	73
Leading edge strakes	199
Lift-to-drag ratio	167,166
Loading and unloading	81,80,77
Longeron spacing	124
Longitudinal control considerations	260
Longitudinal stability considerations	260
Maintenance and access requirements	343-340,288,232,82
Manufacturing breakdown(s)	398-393
Material selection	386-381
Military design considerations	239
Noise, interior and exterior	350
Nozzle	376
Oblique wing	181,180
Overhead storage dimensions	78
Pallet, see cargo	
Parachute dimensions	58
Particle separator	348,346
Passenger cabin	46
Passenger dimensions and views	51,50,49,48,47
Piston engines	300
Pitch stability (pitch-up)	269,266,265
Preliminary structural arrangement	385-381
Pressure bulkhead	139
Primary flight controls	22
Profile drag	273,217,38,36
Propeller(s)	292
Propeller blade excitation	337
Propeller data	299-296,291
Propfan engines	302
Propulsion system	291
Rib spacings	384,277,275,220,218
Ride through turbulence	168
Restraint system	67,57
Safety considerations	344
Seat, dimensions	65,64,63,62,57,52
Seating layout	61,60,58,57
Seat limit load factors	66
Seat pitch	59,58
Seat weights	66

Servicing considerations	83,82
Shoulder harness	67
Slewed wing	182,180
Soldier, see troop	
Spar locations	382,275,275,218
Spin characteristics	271,263
Stair design	142-140
Stall behavior	269,263,175
Stiffener spacings	277,275,218
Structural arrangements	248-244,239
Structural design considerations	123
Structural layout design	123
Take-off and landing field length	166,165
Thrust reverser(s)	380-376
Transmission of thrust	335
Troop dimensions	56,55,54
T-tail	254
Turbojet/Turbofan engines	301
Turbopropeller engines	301
Variable camber	199
Variable sweep	178
Visibility	23,3
Visibility pattern	28,27,24
Visibility requirements	26
Vortex separation	41
V-tail	254
Wardrobe dimensions	73
Windows	147,146,145,143,85,68
Windshield	145,144
Wing configuration	170,164
Wing cross section design	226
Wing dihedral	194
Wing folding	242-240,239
Wing: high, mid or low	170
Wing incidence	195
Wing layout design	163
Winglets	186,185
Wing loading	169,166,165
Wing pivot	243,178
Wing skin gages	232
Wing span	204
Wing structural arrangements	225-221
Wing structural design	218
Wing sweep	175
Wing taper ratio	191,189
Wing thickness ratio	187
Wing twist	193,192

Airplane Design & Analysis Book Descriptions

All books can be ordered from our on-line store at www.darcorp.com.

Airplane Aerodynamics & Performance
C.T. Lan & Jan Roskam

The atmosphere • basic aerodynamic principles and applications • airfoil theory • wing theory • airplane drag • airplane propulsion systems • propeller theory • fundamentals of flight mechanics for steady symmetrical flight • climb performance and speed • take-off and landing performance • range and endurance • maneuvers and flight

Airplane Flight Dynamics & Automatic Flight Controls Part I
Jan Roskam

General steady and perturbed state equations of motion for a rigid airplane • concepts and use of stability & control derivatives • physical and mathematical explanations of stability & control derivatives • solutions and applications of the steady state equations of motion from a viewpoint of airplane analysis and design • emphasis on airplane trim, take-off rotation and engine-out control • open loop transfer functions • analysis of fundamental dynamic modes: phugoid, short period, roll, spiral and dutch roll • equivalent stability derivatives and the relation to automatic control of unstable airplanes • flying qualities and the Cooper-Harper scale: civil and military regulations • extensive numerical data on stability, control and hingemoment derivatives

Airplane Flight Dynamics & Automatic Flight Controls Part II
Jan Roskam

Elastic airplane stability and control coefficients and derivatives • method for determining the equilibrium and manufacturing shape of an elastic airplane • subsonic and supersonic numerical examples of aeroelasticity effects on stability & control derivatives • bode and root-locus plots with open and closed loop airplane applications, and coverage of inverse applications • stability augmentation systems: pitch dampers, yaw dampers and roll dampers • synthesis concepts of automatic flight control modes: control-stick steering, auto-pilot hold, speed control, navigation and automatic landing • digital control systems using classical control theory applications with Z-transforms • applications of classical control theory • human pilot transfer functions

Airplane Design Part I
Preliminary Sizing of Airplanes
Jan Roskam

Estimating take-off gross weight, empty weight and mission fuel weight • sensitivity studies and growth factors • estimating wing area • take-off thrust and maximum clean, take-off and landing lift • sizing to stall speed, take-off distance, landing distance, climb, maneuvering and cruise speed requirements • matching of all performance requirements via performance matching diagrams

Airplane Design Part II
Preliminary Configuration Design and Integration of the Propulsion System
Jan Roskam

Selection of the overall configuration • design of cockpit and fuselage layouts • selection and integration of the propulsion system • Class I method for wing planform design • Class I method for verifying clean airplane maximum lift coefficient and for sizing high lift devices • Class I method for empennage sizing and disposition, control surface sizing and disposition, landing gear sizing and disposition, weight and balance analysis, stability and control analysis and drag polar determination

Design • Analysis • Research
1440 Wakarusa Drive, Suite 500, Lawrence, Kansas 66049, USA - Tel: (785) 832-0434
info@darcorp.com – www.darcorp.com

Airplane Design & Analysis Book Descriptions

All books can be ordered from our on-line store at www.darcorp.com.

Airplane Design Part III
Layout Design of Cockpit, Fuselage, Wing and Empennage: Cutaways and Inboard Profiles
Jan Roskam

Cockpit (or flight deck) layout design • aerodynamic design considerations for the fuselage layout • interior layout design of the fuselage • fuselage structural design considerations • wing aerodynamic and operational design considerations • wing structural design considerations • empennage aerodynamic and operational design considerations • empennage structural and integration design consideration • integration of propulsion system • preliminary structural arrangement, material selection and manufacturing breakdown

Airplane Design Part IV
Layout Design of Landing Gear and Systems
Jan Roskam

Landing gear layout design • weapons integration and weapons data • flight control system layout data • fuel system layout design • hydraulic system design • electrical system layout design • environmental control system layout design • cockpit instrumentation, flight management and avionics system layout design • de-icing and anti-icing system layout design • escape system layout design • water and waste systems layout design • safety and survivability considerations

Airplane Design Part V
Component Weight Estimation
Jan Roskam

Class I methods for estimating airplane component weights and airplane inertias • Class II methods for estimating airplane component weights, structure weight, powerplant weight, fixed equipment weight and airplane inertias • methods for constructing v-n diagrams • Class II weight and balance analysis • locating component centers of gravity

Airplane Design Part VI
Preliminary Calculation of Aerodynamic, Thrust, and Power Characteristics
Jan Roskam

Summary of drag causes and drag modeling • Class II drag polar prediction methods • airplane drag data • installed power and thrust prediction methods • installed power and thrust data • lift and pitching moment prediction methods • airplane high lift data • methods for estimating stability, control and hingemoment derivatives • stability and control derivative data

Airplane Design Part VII
Determination of Stability, Control, and Performance Characteristics: FAR and Military Requirements
Jan Roskam

Controllability, maneuverability and trim • static and dynamic stability • ride and comfort characteristics • performance prediction methods • civil and military airworthiness regulations for airplane performance and stability and control • the airworthiness code and the relationship between failure states, levels of performance and levels of flying qualities

Airplane Design Part VIII
Airplane Cost Estimation: Design, Development, Manufacturing, and Operating
Jan Roskam

Cost definitions and concepts • method for estimating research, development, test and evaluation cost • method for estimating prototyping cost • method for estimating manufacturing and acquisition cost • method for estimating operating cost • example of life cycle cost calculation for a military airplane • airplane design optimization and design-to-cost considerations • factors in airplane program decision making

Design • Analysis • Research
1440 Wakarusa Drive, Suite 500, Lawrence, Kansas 66049, USA - Tel: (785) 832-0434
info@darcorp.com – www.darcorp.com

Made in the USA
Monee, IL
04 April 2025